Frank Cowan

Curious facts in the history of insects: Including spiders and scorpions

A complete collection of the legends, superstitions, beliefs, and ominous signs

Frank Cowan

Curious facts in the history of insects: Including spiders and scorpions
A complete collection of the legends, superstitions, beliefs, and ominous signs

ISBN/EAN: 9783337153335

Printed in Europe, USA, Canada, Australia, Japan

Cover: Foto ©berggeist007 / pixelio.de

More available books at **www.hansebooks.com**

IN THE

HISTORY OF INSECTS;

INCLUDING

SPIDERS AND SCORPIONS.

A

COMPLETE COLLECTION OF THE LEGENDS, SUPERSTITIONS, BELIEFS,
AND OMINOUS SIGNS CONNECTED WITH INSECTS: TOGETHER
WITH THEIR USES IN MEDICINE, ART, AND AS FOOD;
AND A SUMMARY OF THEIR REMARKABLE
INJURIES AND APPEARANCES.

BY

FRANK COWAN.

———

PHILADELPHIA:
J. B. LIPPINCOTT & CO.
1865.

TO

MISS CATHARINE STOY

THE FOLLOWING PAGES

ARE RESPECTFULLY INSCRIBED

BY HER FRIEND,

THE AUTHOR.

PREFACE.

In the early part of the winter of 1863–4, having the free use of
the Congressional Library at Washington, I began the compilation of the present work. It was my prime intent, and one which
I have endeavored to follow most carefully, to attach some fact,
whatever might be its nature, to as many Insects as possible, to
increase the interest, in a commonplace way, of the science of
Entomology. I noticed the pleasurable satisfaction I invariably
felt when I came accidentally upon any extra-scientific fact, and
how the association fixed the particular Insect, to which it related,
ineffaceably upon my memory. To collect and group, then, all
these facts together, to remember many Insects as easily as one,
—was a natural thought; and as this had never been done, but to
a very limited extent, I undertook it myself.

The facts contained in this volume are supposed to be purely
historical, or rather not to belong to the natural history of Insects,
namely, their anatomy, habits, classification, etc. They have been
collected mostly from Chronicles, Histories, Books of Travels, and
such like works, which, at first view, seem to be totally foreign to
Insects: and were only discovered by examination of the indexes
and tables of contents.

But are my facts *facts?*—it may be asked. They are; but I do
not vouch for each one's containing more than one truth. It is a
fact, or truth if you will, that Pliny, Nat. Hist. xi. 34, says,
"Folke use to hang Beetles about the neck of young babes, as
present remedies against many maladies;" but that this statement
is entitled to credit, and that these Insects, hung about the necks
of young babes, *are* a present remedy against many maladies, are
two things which may be very true or far otherwise. I confine
myself to the fact that Pliny says so, and only wish to be understood in that sense, unless when otherwise stated.

(v)

The classification of Mr. Westwood, in the arrangement of the orders and families, I have followed as closely as was possible, except in one or two instances: and where Insects have common and familiar names, they have been given together with their scientific ones.

To Dr. J. M. Toner, of Washington, for his suggestions and assistance in collecting material, I tender my thanks; the same also to N. Bushnell, Esq., and Hon. O. H. Browning, of Quincy, Ill., for the use of their several libraries.

I am much indebted, too, to Mrs. A. L. Ruter Dufour, of Washington, for many superstitions and two pieces of poetry contained in this volume. I beg her to accept my thanks.

GREENSBURG, PENNA.,
July 10th, 1865.

CONTENTS.

HYMENOPTERA.

LEPIDOPTERA.

HOMOPTERA.

HETEROPTERA.

DIPTERA.

APHANIPTERA.

ANOPLEURA.

ARACHNIDÆ.

AUTHORS QUOTED.

ALEXANDER, SIR JAS. EDW. Exped. of Disc. into Interior of Africa. 2 v. 12mo., London, 1838.

ANDERSON, CHAS. ROSS. Lake Ngami; or, Explor. and Disc. during four years wanderings in S. W. Africa. 8vo., New York, 1856.

ANDREWS, JAMES PETTIT. Anecdotes, etc., Ancient and Modern. New edit. 8vo., London, 1790.

ASIATICK MISCELLANY. 2 v. 4to., Calcutta, 1785, 1786.

ASTLEY, THOMAS. New Gen. Collection of Voyages and Travels in Europe, Asia, Africa, and America. 4 v. 4to., London, 1745–1747.

AUBREY, JOHN. Miscellanies upon various subjects. 16mo. 4th edit., London, 1857.

BACKHOUSE, JAMES. Narrat. of Visit to Mauritius and S. Africa. 8vo., London, 1844.

BAIRD, WILLIAM. Cyclopædia of Natural Sciences. 8vo., London and Glasgow, 1858.

BANCROFT, EDWARD. Essay on the Nat. Hist. of Guiana, in S. America. 8vo., London, 1769.

BANCROFT, EDWARD. On Permanent Colours. 2 v. 8vo., London, 1813.

BARTER, CHARLES. The Dorp and the Veld. 16mo., London, 1852.

BARTH, HENRY. Travels and Discov. in North and Central Africa, from 1849 to 1855. 5 v. 8vo., London, 1857–1858.

BIOGRAPHIE UNIVERSELLE, ANCIENNE ET MODERNE. 84 v. 8vo., Paris, 1811–1857.

BJÖRNSTJERNA, COUNT M. Theogony of the Hindoos. 8vo., London, 1844.

BOSMAN, WILLIAM. New and Accurate Desc. of Coast of Guinea. 8vo., London, 1705.

BOYLE, ROBERT. Works. New edit. 6 v. royal 4to., London, 1772.

BRANDE, JOHN. Observations on the Popular Antiquities of Great Britain. 3 v. 12mo., London, 1853–5.

DARWIN, CHARLES. Journ. of Research. into Nat. Hist. and Geol. of Countries visited during Voy. of H. M. S. Beagle, round the world. New edit. 12mo., London, 1852.

DIAZ DEL CASTILLO, BERNAL. Memoirs and Disc. and Conq. of Mexico and New Spain. Tr. by John J. Lockhart. 2 v. 8vo., London, 1844.

DIODORUS THE SICILIAN, Historical Library of, in fifteen books; Fragments, etc. Tr. by G. Booth. 2 v. 8vo., London, 1814.

DONOVAN, EDWARD. Nat. Hist. of Insects of China. 4to., London, 1842.

DRAYSON, ALFRED W. Sporting Scenes in S. Africa. 8vo., London, 1858.

DU HALDE, J. B. General Hist. of China, etc. 4 v. 8vo., London, 1836.

FABYAN, ROBERT. New Chronicles of England and France. 4to., London, 1811.

FLEMING, FRANCIS. Kaffraria. 12mo., London, 1853.

FORBES, JAMES. Oriental Memoirs. 4 v. 4to., London, 1813.

FOSBROKE, THOS. DUDLEY. Encyclopædia of Antiquities. 2 v. 4to., London, 1825.

GASSENDUS, PETRUS. Mirrour of true Nobility and Gentility. Life of Peiresc. Tr. by W. Rand. 8vo., London, 1657.

GENTLEMAN'S MAGAZINE. 202 v. 8vo., London, 1731–1859.

GOLDSMITH, OLIVER. Hist. of the Earth, and Animated Nature. 4 v. 8vo., London, 1826.

GOOD, JOHN MASON. Study of Medicine. 4th edit. 4 v. 8vo., London, 1840.

GOSSE, PHILIP HENRY. Naturalist's Sojourn in Jamaica. 12mo., London, 1851.

GROSIER, ABBE J. B. G. A. Genl. Desc. of China. 2d edit. 2 v. 8vo., London, 1795.

HARLEIAN MISCELLANY. 12 v. 8vo., London, 1808–1811.

HARRIS, JOHN. Navigantium atque Itinerantium Bibliotheca; or, a Complete Col. of Voy. and Travels. 2 v. folio, London, 1744, 1748.

HAWKINS, SIR JOHN. General Hist. of the Science and Practice of Music. 5 v. 4to., London, 1776.

HAWKS, FRANCIS L. Monuments of Egypt. 8vo., New York, 1850.

HOLINSHED, RAPHAEL. Chronicles of England, Scotland, and Ireland. 6 v. 4to., London, 1807–8.

HOLMAN, JAMES. Travels in Brazil, Cape Colony, etc. 2d edit. 8vo., London, 1840.

Hone, William. Every-Day Book and Table Book. 3 v. royal
8vo., London, 1838.

Horne, Thomas Hartwell. Introd. to the Study of Bibliography.
2 v. in 1, 8vo., London, 1814.

Houdin, Robert. Autobiograpical Memoirs. 12mo., Philad., 1859,

Huber, Pierre. Nat. Hist. of Ants. Tr. by J. R. Johnson. 12mo.,
London, 1820.

Hughes, Griffith. Nat. Hist. of Barbados. Folio, London, 1750.

Insectorum sive Minimorum Animalium Theatrum. Thos. Mou-
feti operâ perfectum. Folio, Loudoni, 1634.

Jackson, James Grey. Acct. of Empire of Marocco, and Districts
of Suse and Tafilelt. 2d edit. 4to., London, 1811.

Jenkins, John S. Voy. of U. S. Exploring Squadron, commanded
by Capt. Chas. Wilkes; from 1838 to 1842. 8vo., Auburn, 1852.

Jones, John Matthew. Naturalist in Bermuda. 12mo., London,
1859.

Josephus, Flavius. Genuine Works. Tr. by William Whiston.
Folio, London, 1737.

Josselyn, John. Acct. of Two Voyages to New England. 16mo.,
London, 1674.

Kalm, Peter. Travels into North America. Tr. by John R. Foster.
2 v. 8vo., London, 1859.

Kidder, Danl. P., and J. C. Fletcher. Brazil and the Brazilians.
Royal 8vo., Philad., 1857.

Kirby, R. S. Wonderful and Eccentric Museum; or, Mag. of Remarkable Characters. 6 v. 8vo., London, 1820.

Kirby, William, and William Spence. Introduction to Entomology.
5th edit. 4 v. 8vo., London, 1829.

Knox, Robert. Hist. Relation of the Island of Ceylon. 4to., London, 1817.

Kolben, Peter. Pres. State of Cape of Good Hope. Tr. by Mr.
Medley. 2d edit. 2 v. 8vo., London, 1731, 1738.

Koran, The: commonly called the Alcoran of Mohammed. Tr. by
Geo. Sale. 8vo., Philad., 1850.

Latrobe, Chas. Jos. Journ. of Visit to S. Africa, in 1815 and
1816. 8vo., New York, 1818.

Langstroth, L. L. Prac. Treatise on the Hive and Honey-Bee.
3d edit. 12mo., New York, 1860.

Layard, Austen H. Disc. among the Ruins of Nineveh and Babylon; with Travels in Armenia, etc. 8vo., New York, 1853.

Lepsius, Richard. Desc. in Egypt, Ethiopia, and Penins. of Sinai,
in 1842–1845. 2d edit. 8vo., London, 1853.

LINNÆUS, CAROLUS. Lachesis Lapponica; or, a Tour in Lapland. Tr. by J. E. Smith. 2 v. 8vo., London, 1811.

LIVINGSTONE, DAVID. Missionary Travels and Researches in S. Africa. 8vo., New York, 1858.

LIVIUS, TITUS. History of Rome. Tr. by George Barker. 2d edit. 6 v. 8vo., London, 1814.

MAGAZINE OF NATURAL HISTORY. Cond. by J. C. Loudon. 9 v. 8vo., London, 1829–1836.

MARTYR, PETER. De Nouo Orbe; or, The Hist. of the West Indies. Tr. by R. Eden and M. Lok. 4to., London, 1612.

MAYHEW, HENRY. London Labor and the London Poor. 4 v. 8vo., London, 1861, 1862.

MIRROR OF LITERATURE, AMUSEMENT, AND INSTRUCTION. 40 v. 8vo., London, 1823–1842.

MOFFAT, ROBT. Missionary Labors and Scenes in S. Africa. 8vo., London, 1842.

MONTFAUCON, BERNARD DE. L'Antiquité Expliquée et Représentée en Figures. 2e édition, revue et corrigée. Lat. et Fr. 5 v. en 10, folio, Paris, 1722.

MONTAIGNE, MICHAEL DE. Works. By William Hazlitt. 8vo., Philad., 1850.

MOUFET, THOMAS. Insectorum sive Minimorum Animalium Theatrum. Londoni, 1634.

————— ————. The same, translated. See Topsel's Hist. of Beasts, etc.

NATURAL HISTORY OF INSECTS. Vols. 64 and 65 of John Murray's Fam. Library. 18mo., London, 1830–1842.

NEWELL, ROBT. HASELL. Zoology of the English Poets. 16mo., London, 1845.

OCKLEY, SIMON. History of the Saracens. 3d ed. 2 v. 8vo., Cambridge, 1757.

OGILBY, JOHN. America. Folio, London, 1671.

OLIN, STEPHEN. Travels in Egypt, Arabia Petræa, and the Holy Land. 8th edit. 2 v. 8vo., New York, 1846.

OLIPHANT, LAURENCE. Narrat. of Earl of Elgin's Mission to China and Japan, in 1857–9. 8vo., New York, 1860.

OWEN, REV. T. Geoponika; or, Agricultural Pursuits. 2 v. 8vo., London, 1805.

PERCY SOCIETY PUBLICATIONS. 30 v. 12mo., London, 1840–'52.

PETTIGREW, THOS. JOS. History of Egyptian Mummies. 4to., London, 1834.

PHILOSOPHICAL TRANSACTIONS. Royal Society of London. 1665 to 1858. 147 v. 4to., London, 1665–1858.

PHILOSOPHICAL TRANSACTIONS. Royal Society of London, abridged. 1665 to 1750. 11 v. 4to., London, 1749-1756.

PIERIUS VALERIANUS, IOANNIS. Hicroglyphica. Folio, Lugduni, 1626.

PINKERTON, JOHN. General Collection of Voyages and Travels in all parts of the World. 17 v. 4to., London, 1808-1814.

PLINY, by J. Bostock and H. T. Riley. 6 v. Bohn's Classical Library.

PLINIUS SECUNDUS, CAIUS. Historie of the World: commonly called the Nat. Hist. of C. Plinius Secundus. Tr. by Philemon Holland. 2 v. in 1, folio, London, 1657.

PRICHARD, JAMES COWLES. Analysis of the Egyptian Mythology. 8vo., London, 1819.

PRINGLE, THOMAS. Narrat. of Resid. in S. Africa. New edit. 8vo., London, 1851.

PURCHAS, SAMUEL. Hakluytus Posthumus; or, Purchas his Pilgrimes. 5 v. folio, London, 1625, 1626.

RHIND, A. HENRY. Thebes; its Tombs and their Tenants, anct. and modern. 8vo., London, 1862.

RICHARDSON, JAMES. Travels in Great Desert of Sahara, in 1845-6. 2 v. 8vo., London, 1848.

RILEY, JAMES. Authen. Narrat. of Loss of Amer. Brig Commerce, wrecked on western coast of Africa, in 1815. 8vo., Hartford, 1850.

RIVERO, MARIANO EDWARD, and JNO. JAS. VON TSCHUDI. Peruvian Antiquities. Tr. by Francis L. Hawks. 8vo., New York, 1853.

ROBBINS, ARCHIBALD. Journ. of Advent. in Africa, in 1815-'17. 12mo., Hartford, 1851.

SAMOUELLE, GEORGE. Entomological Cabinet. 2d edit. 16mo., London, 1841.

SATURDAY MAGAZINE. Folio. From 1833 to 1844, London.

SCHOMBURGK, SIR ROBERT H. Hist. of Barbados. 8vo., London, 1847.

SHAW, GEORGE. General Zoology; or, Syst. Nat. Hist. 14 v. 8vo., London, 1800-1826.

SILLIMAN, BENJAMIN. Amer. Journ. of Sci. and Art. 78 v. 8vo., New York and New Haven, 1819-1859.

SIMMONDS, PETER LUND. Curiosities of Food; or, the Dainties and Delicacies of different nations obtained from the Animal Kingdom. 12mo., London, 1859.

SLOANE, HANS. Voy. to Islands of Madeira, Barbados, Nieves, S. Christophers, and Jamaica; with the Nat. Hist. of Jamaica. 2 v. folio, London, 1707-1725.

SMITH, THOMAS. Wonders of Nature and Art; or. a Concise Acct. of whatever is most curious and remarkable in the world. 12 v. 16mo., Philad., 1806-1807.

SPARRMAN, ANDERS. Voy. to C. of G. Hope, towards Antarc. Circle, and Round the World. From 1772 to 1776. 2 v. 12mo., Perth, 1789.

SOUTHEY, ROBT. Common-Place Book. 4th series. In 4v. 8vo., London, 1849-1851.

———————————— Hist. of Brazil. 3 v. 4to., London, 1817-1822.

STANLEY, THOMAS. History of Philosophy. 3d edit. Folio, London, 1701.

STEDMAN, J. G. Narrat. of five years' Exped. against revolted Negroes of Surinam, in Guiana, in 1772-1777. 2 v. 4to., London, 1796.

STEEDMAN, ANDREW. Wanderings and Advent. in Interior of S. Africa 2 v. 8vo., London, 1835.

ST. JOHN, JOHN AUG. Hist. of Manners and Customs of Ancient Greece. 3 v. 8vo., London, 1842.

STRABO, by H. C. Hamilton and W. Falconer. 3 v. Bohn's Classical Library.

STRONG, A. B. Illustr. Nat. Hist. of the Three Kingdoms. New ser. 2 v. 8vo., New York, 1853.

STUART, J. View of Past and Present State of Island of Jamaica. 8vo., Edinburgh, 1823.

SWAMMERDAM, JAN. Book of Nature; or, the Hist. of Insects. Tr. by Thos. Floyd. Folio, London, 1758.

TAYLOR, FITCH W. Voy. Round the World, and Visits to foreign countries, in the U. S. Frigate Columbia. 9th edit. 8vo., 2 v. in 1, New Haven, 1848.

TENNENT, SIR J. EMERSON. Sketches of the Nat. Hist. of Ceylon. 12mo., London, 1861.

THEODORET and EVAGRIUS. Hist. of the Church, from A.D. 322 to A.D. 594. 12mo., London, 1854.

THEVENOT, MONSIEUR DE. Travels into the Levant. Folio, London, 1687.

THORPE, BENJ. Northern Mythology. 3 v. post 8vo., London, 1851, 1852.

THUNBERG, KARL PETER. Travels in Europe, Africa, and Asia, bet. 1770-9. 4 v. 8vo., London, 1795, 1796.

TOPSEL, EDWARD. The Hist. of Four-footed Beasts and Serpents. Whereunto is added The Theater of Insects: by T. Moufet. Folio, London, 1658.

TREASVRIE OF AVNCIENT AND MODERNE TIMES. Tr. from Pedro Mexia, M. Francesco Sansovino, Anthony du Verdier, etc., by Thomas Milles. Folio, London, 1613.

——————— ——— ———————— ————————————— Containing Ten following Bookes to the former. Folio, London, 1619.

TWELVE YEARS IN CHINA. The People, the Rebels, and the Mandarins. By a British Resident. 12mo., Edinburgh, 1860.

UNIVERSAL HISTORY. Ancient Part. 21 v. 8vo., London, 1747–1754.

VOLNEY, COMTE C. F. CHASSEBŒUF DE. Travels through Syria and
    Egypt, in 1783–'85. 2 v. 8vo., London, 1787.

WALTON, WILLIAM, JR. Pres. State of the Spanish Colonies. 2 v.
    8vo., London, 1810.

WANLEY, NATHANIEL. Wonders of the Little World; or, a General
    Hist. of Man. 2 v. 8vo., London, 1806.

WELD, ISAAC. Travels through States of N. America, and Canadas,
    in 1795–'97. 3d edit. 2 v. 8vo., London, 1800.

WESTWOOD, JOHN OBAD. Introd. to Mod. Classif. of Insects. 2 v.
    8vo., London, 1840.

WHITE, GILBERT. Nat. Hist. of Selborne. 8vo., London, 1854.

WILKINSON, SIR J. G. Manners and Customs of the Anct. Egyp-
    tians. 6 v. 8vo., London, 1837–1841.

WILLIAMS, S. WELLS. The Middle Kingdom; or, Survey of Chinese
    Empire. 3d edit. 2 v. 8vo., New York, 1853.

WOOD, WILLIAM. Zoography. 3 v. 8vo., London, 1807.

CURIOUS HISTORY OF INSECTS.

ORDER I.

COLEOPTERA—BEETLES.

Coccinellidæ—Lady-birds.

THE Lady-bird, *Coccinella septempunctata*, in Scandinavia was dedicated to the Virgin Mary, and is there to this day called *Nyckelpiga*—Our Lady's Key-maid,[1] and (in Sweden, more particularly) *Jung-fru Marias Gullhona*—the Virgin Mary's Golden-hen.[2] A like reverence was paid to this beautiful insect in other countries: in Germany they have been called *Frauen* or *Marien-käfer*—Lady-beetles of the Virgin Mary; and in France are now known by the names of *Vaches de Dieu*—Cows of the Lord, and *Bêtes de la Vierge*—Animals of the Virgin.[3] The names we know them by, *Lady-bird*, *Lady-bug*, *Lady-fly*, *Lady-cow*,[4] *Lady-clock*, *Lady-couch* (a Scottish name),[5] etc., have reference also to this same dedication, or, at least, respect.

The Lady-bird in Europe, and particularly in Germany, where it probably is the greatest favorite, and whence most of the superstitions connected with it are supposed to have originated, is always connected with fine weather. At Vienna, the children throw it into the air, crying,—

---

[1] Thorpe's *Northern Mythol*, ii. 104.

[2] Jamieson's *Scot. Dict.* Another designation, in Sweden, is not so honorable, for it is that of *Laettfaerdig kona*, the Wanton Quean.—*Ibid.* The term Lady-bird, in England, has been also applied to a prostitute.—Wright's *Provinc. Dict.*

[3] Jaeger, *Life of Amer. Ins.*, p. 22.

[4] It is curious to notice the association of this insect with the cow in the English and French names.

[5] Jamieson's *Scot. Dict.*

> Käferl', käferl', käferl',
> Flieg nach Mariabrunn,
> Und bring uns ü schone sun.

Or,—

> Little birdie, birdie,
> Fly to Marybrunn,
> And bring us a fine sun.

Marybrun being a place about twelve English miles from the Austrian capital, with a miracle-working image of the Virgin (still connected with the Virgin), who often sends good weather to the merry Viennese.[1]

And, from the marsh of the Elbe, to this little insect the following words are addressed:

> Maikatt,
> Flug weg,
> Stuff weg,
> Bring me morgen goet wedder med.

Or,—

> May-cat,
> Fly away,
> Hasten away,
> Bring me good weather with you to-morrow.[2]

In England, the children are wont to be afraid of injuring the Lady-bird lest it should rain.

With the Northmen the Lady-bird—Our Lady's Key-maid—is believed to foretell to the husbandman whether the year shall be a plentiful one or the contrary: if its spots exceed seven, bread-corn will be dear; if they are fewer than seven, there will be an abundant harvest, and low prices.[3]   And, in the following rhyme from Ploen, this insect is invoked to bring food:

> Marspäert (Markpäert) fleeg in Himmel!
> Bring my'n Sack voll Kringeln, my een, dy een,
> Alle lütten Engeln een.

Or,—

> Marspäert, fly to heaven!
> Bring me a sack full of biscuits, one for me, one for thee,
> For all the little angels one.[4]

In the north of Europe it is thought lucky when a young girl in the country sees the Lady-bird in the spring ; she

---

[1] Chambers' *Pop. Rhymes*, 1841, p. 170–1.
[2] Thorpe's *North. Mythol.*, iii. 182.
[3] *Ibid.*, ii. 104.
[4] *Ibid.*, iii. 182.

then lets it creep about her hand, and says : "She measures me for wedding gloves." And when it spreads its little wings and flies away, she is particular to notice the direction it takes, for thence her sweetheart shall one day come.[1] The latter part of this notion obtains in England; and it has been embodied by Gay in one of his Pastorals, as follows :

> This Lady-fly I take from off the grass,
> Whose spotted back might scarlet red surpass.
> Fly, Lady-bird, north, south, or east or west,
> Fly where the man is found that I love best.
> He leaves my hand, see to the west he's flown,
> To call my true-love from the faithless town.[2]

In Norfolk, too, where this insect is called the Bishop Barnabee, the young girls have the following rhyme, which they continue to recite to it placed upon the palm of the hand, till it takes wing and flies away :[3]

> Bishop, Bishop Barnabee,
> Tell me when my wedding be :
> If it be to-morrow day,
> Take your wings and fly away!
> Fly to the east, fly to the west,
> Fly to him that I love best.[4]

Why the Lady-bird is called Bishop Barnabee, or Burnabee, there is great difference of opinion. Some take it to be from St. Barnabas, whose festival falls in the month of June, when this insect first appears; and others deem it but a corruption of the Bishop-that-burneth, in allusion to its fiery color.[5]

The following metrical jargon is repeated by the children in Scotland to this insect under the name of Lady Lanners, or Landers :[6]

> Lady, Lady Lanners,
> Lady, Lady Lanners,
> Tak' up your clowk about your head,
> An' flee awa' to Flanners (Flanders).

---

[1] Thorpe's *North. Mythol.*, ii. 104.

[2] 4th Pastoral, ll. 83–8.

[3] It probably is induced to fly away by the warmth of the hand.

[4] *Notes and Queries*, i. 132.

[5] *Ibid.*, i. 28, 55, 78.

[6] Jamieson supposes this word to be derived from the Teutonic *Land-heer*, a petty prince.—*Scot. Dict.*

> Flee ower firth, and flee ower fell,
> Flee ower pule and rinnan' well,
> Flee ower muir, and flee ower mead,
> Flee ower livan, flee ower dead,
> Flee ower corn, and flee ower lea,
> Flee ower river, flee ower sea,
> Flee ye east, or flee ye west,
> Flee till him that lo'es me best.

So it seems that also in Scotland, the Lady-bird, which is still a great favorite with the Scottish peasantry, has been used for divining one's future helpmate. This likewise appears from a rhyme from the north of Scotland, which dignifies the insect with the title of Dr. Ellison:

> Dr. Dr. Ellison, where will I be married?
> East, or west, or south, or north?
> Take ye flight and fly away.

It is sometimes also termed Lady Ellison, or knighted Sir Ellison; while other Scottish names of it are Mearns, Aberd, The King, and King Galowa, or Calowa. Under this last title of dignity there is another Scottish rhyme, which evinces also the general use of this insect for the purpose of divination:

> King, King Calowa,
> Up your wings and flee awa'
> Over land, and over sea;
> Tell me where my love can be.[1]

There is a Netherlandish tradition that to see Lady-birds forebodes good luck;[2] and in England it is held extremely unlucky to destroy these insects. Persons killing them, it is thought, will infallibly, within the course of the year, break a bone, or meet with some other dreadful misfortune.[3]

In England, the children are accustomed to throw the Lady-bird into the air, singing at the same time,—

> Lady-bird, lady-bird, fly away home;
> Your house is on fire, your children's at home,
> All but one that ligs under the stone,—
> Ply thee home, lady-bird, ere it be gone.[4]

---

[1] Jamieson's *Scot. Dict.* Cf. Chambers' *Pop. Rhymes*, 1841, p. 170-1.

[2] Thorpe's *North. Mythol.*, iii. 328.

[3] Grose, *Antiq.* (*Prov. Gloss.*) p. 121.

[4] Chambers' *Pop. Rhymes*, 1841, p. 170.

Or, as in Yorkshire and Lancashire,—

> Lady-bird, lady-bird, eigh thy way home;
> Thy house is on fire, thy children all roam,
> Except little Nan, who sits in her pan,
> Weaving gold laces as fast as she can.[1]

Or, as most commonly with us in America,—

> Lady-bird, lady-bird, fly away home,
> Your house is on fire, and your children all burn.

The meaning of this familiar, though very curious couplet, seems to be this: the larvæ, or young, of the Lady-bird feed principally upon the aphides, or plant-lice, of the vines of the hop; and fire is the usual means employed in destroying the aphides; so that in killing the latter, the former, which had come for the same purpose, are likewise destroyed.

Immense swarms of Lady-birds are sometimes observed in England, especially on the southeastern coast. They have been described as extending in dense masses for miles, and consisting of several species intermixed.[2] In 1807, these flights in Kent and Sussex caused no small alarm to the superstitious, who thought them the forerunners of some direful evil. They were, however, but emigrants from the neighboring hop-grounds, where, in their larva state, they had been feasting upon the aphides.[3]

The Lady-bird was formerly considered an efficacious remedy for the colic and measles;[4] and it has been recommended often as a cure for the toothache: being said, when one or two are mashed and put into the hollow tooth, to immediately relieve the pain. Jaeger says he has tried this application in two instances with success.[5]   .

In the northern part of South America—the Spanish Main—a species of Lady-bug, Captain Stuart tells me, is extensively worn as jewels and ornaments. He may, however, refer to some species of the Gold-beetles—*Chrysomelidæ*, next mentioned.

Hurdis, who has frequently, in his Poems, availed himself of the modern discoveries in Natural History, has

---

[1] *Notes and Queries*, iv. 53.
[2] Baird's *Cyclop. of Nat. Sci.*
[3] Kirby and Spence, *Introd.*, ii. 9.
[4] Newell's *Zool. of the Poets*, p. 48.
[5] *Life of Amer. Ins.*, p. 21.

drawn the following accurate and beautiful picture of the Lady-bird in his tragedy of Sir Thomas More:

SIR JOHN.

What d'ye look at?

CECILIA.

A little animal, that round my glove,
And up and down to every finger's tip,
Has traveled merrily, and travels still,
Tho' it has wings to fly: what its name is
With learned men I know not; simple folk
Call it the Lady-bird.

SIR JOHN.

Poor harmless thing!
Save it.

CECILIA.

I would not hurt it for the world;
Its prettiness says, Spare me; and it bears
Armor so beautiful upon its back,
I could not injure it to be a queen:
Look, sir, its coat is scarlet dropp'd with jet,
Its eyes pure ivory.

SIR JOHN.

Child, I'm not blind
To objects so minute: I know it well;
'Tis the companion of the waning year,
And lives among the blossoms of the hop;
It has fine silken wings enfolded close
Under that coat of mail.

CECILIA.

I see them, sir,
For it unfurls them now—'tis up and gone.[1]

Southey, also, in his lines addressed to this insect under the name of the Burnie-Bee, has thus elegantly described it:

---

[1] A. 1, sc. iii.

Back o'er thy shoulders throw thy ruby shards,
　With many a tiny coal-black freckle deck'd ;
My watchful eye thy loitering saunter guards,
　My ready hand thy footsteps shall protect.

So shall the fairy train, by glow-worm light,
　With rainbow tints thy folding pennons fret,
Thy scaly breast in deeper azure dight,
　Thy burnish'd armor deck'd with glossier jet.[1]

Chrysomelidæ—Gold-beetles.

In Chili and Brazil, the ladies form necklaces of the golden *Chrysomelidæ* and brilliant Diamond-beetles, with which their countries abound, which are said to be very beautiful.[2]　The wing-cases of our common Gilded-Dandy, *Eumolpus auratus*, the metallic colors of which are pre-eminently brilliant and showy, have been recommended as ornaments for fancy boxes, and such like articles.[3]　A closely allied species, I have seen upon the finest Parisian artificial flowers.

Carabidæ.

In some parts of Africa, a rather curious benefit is derived from a large beetle belonging to this family, the *Chlænius saponarius*, for it is manufactured by the natives into a soap.[4]

Pausidæ.

The etymology of the word *Pausus*, Dr. Afzelius imagines to be from the Greek παυσις, signifying a pause, cessation, or rest; for Linnæus, now (in 1796) old and infirm, and sinking under the weight of age and labor, saw

---

[1] Quot. with preceding in Newell's *Zool. of the Poets*, p. 50–2.
[2] Kirb. and Sp. *Introd.*, i. 317.
[3] Jaeger, *Life of Amer. Ins.*, p. 61.
[4] Kirb. and Sp. *Introd.*, i. 316.

no probability of continuing any longer his career of glory. He might therefore be supposed to say *hic meta laborum*, as it in reality proved, at least with regard to insects, for Pausus was the last he ever described.[1]

Dermestidæ—Leather-beetles.

In one of the stone coffins exhumed from the tumuli in the links of Skail, were found several small bags, which seemed to have been made of rushes. They all contained bones, with the exception of one, which is said to have been full of beetles belonging to the genus *Dermestes*. Both the bag and beetles were black and rotten.[2]

Four species of *Dermestes* were found in the head of one of the mummies brought by Sir J. Gardner Wilkinson from Thebes—the *D. vulpinus* of Fabricius, and the *pollinctus*, *roei*, and *elongatus* of Hope.[3]

It is a remarkable coincidence that two peoples should bury beetles of the same genus with their dead, and much the more so, when they differ so widely, as did the ancient Britons and Egyptians. Was it for the same reason—the result of any communication?

At one time the ravages of the *Dermestes vulpinus* were so great in the skin-warehouses of London, that a reward of £20,000 was offered for an available remedy.[4]

Lucanidæ—Stag-beetles.

The etymology of the word Lucanus, as well as its application to a species of insect, it is interesting to notice. The ancients gave the name of *Lucas*, *Lucana*, to the ox and elephant. It is said that Pyrrhus had thus named the

---

[1] Shaw's *Zool.*, vi. 42.

[2] Gough's *Sepul. Mon.*, vol. i. p. xii.—These sepulchral tumuli, or burrows, are of the remotest antiquity, and continued in use till the twelfth century.—*Ibid.*

[3] Wilkin. *Anct. Egypt.* ii. (2d S.) 261; and Pettig. *Hist. of Mummies*, p. 53–5.

[4] Baird's *Cyclop. of Nat. Sci.*

elephant the first time that he saw it, because this word signified *ox* in his own language, and that he thus gave it the name of the largest animal which he had ever before seen. According to Pliny, who employed the word *Lucani*, in speaking of the Horn-beetles, Nigridius was the first who gave the name to these insects; and this he did, most probably, from their large size, and the resemblance of their mandibles to horns. Dalechamp, however, thinks that the name *Lucanus* was given to the Horn-beetle only because this insect was very common among the Lucanians, a people of Italy. But it is probable, after what has been above said, that the Lucanians themselves were thus named, in consequence of the great numbers of oxen which they reared. The common name, *Flying-bull*, given to this insect in different languages, corresponds very well with that given by Nigridius.[1]

A popular belief in Germany is, that the Stag-beetle, *Lucanus cervus*, carries burning coals into houses by means of its jaws, and that it has thus occasioned many fearful fires.[2]

In the New Forest of England, the Stag-beetle by the rustics is called the *Devil's Imp*, and is believed to be sent to do some evil to the corn; and woe be to this unfortunate insect when met by these superstitious foresters, for it is immediately stoned to death. A writer, in the Notes and Queries,[3] states that he saw one of these insects actually thus destroyed.

Professor Bradley, of Cambridge, mentions the following remarkable instance of insect strength in a Stag-beetle. He asserts that he saw the beetle carry a wand a foot and a half long, and half an inch thick, and even fly with it to the distance of several yards.[4] Linnæus observes, that if the elephant was as strong in proportion as the Stag-beetle, it would be able to tear up rocks and level mountains.[5]

Bingley has the following marvelous story of the supposed rapacity of the Stag-beetle, which, it has been remarked, if not gravely stated by the reverend editor of the

---

[1] Cuvier's *Animal Kingd.—Ins.*, i. 530.
[2] *The Mirror*, xix. 180; and *Saturday Mag.*, xvi. 144.
[3] N. & Q., 2d S., ii. 83.
[4] Bradley, *Phil. Account*, p. 184.
[5] *N. Dict. d'Hist. Nat.*, xxii. 81.

Animal Biography, as related to him by one of his own intimate and intelligent friends, might have been supposed by the general reader to have been borrowed from the Travels of the veracious Munchausen. "An intimate and intelligent friend of the editor informed him that he had often found several heads of these insects together, all perfectly alive, while the abdomens were gone, and the trunks and heads were left together. How this circumstance took place he never could discover with any certainty. He supposes, however, that it must have been in consequence of the severe battles that sometimes take place among the fiercest of the insect tribes; but their mouths not seeming formed for animal food, he is at a loss to guess what becomes of their abdomens. They do not fly till most of the birds have retired to rest, and indeed if we were to suppose that any of them devoured them, it would be difficult to say why the heads or trunks should be rejected."[1]

Moufet says: "When the head (of the Stag-beetle) is cut off, the other parts of the body live long, but the head (contrary to the usual custom of insects) lives longer. This is said to be dedicated to the moon, and the head and horns of it wax with the moon, and do wane with the moon, but it is the opinion of vain astrologers."[2]

The mandibles of the Stag-beetle were formerly employed in medicine, under the name of Horns of Scarabæi. This remedy was administered as an absorbent, in case of pains or convulsions supposed to be produced by acidity in the *primæ viæ*.[3] This is the insect most probably alluded to by Pliny, when he says, "Folke use to hang Beetles about the neck of young babes, as present remedies against many maladies."[4] The *Scarabæus cornutus* of Schröder (v. 345) is also, perhaps, the *Lucanus cervus*. We learn from this gentleman that it has been recommended to be worn as an amulet for an ague, or pains and contractions of the tendons, if applied to the part affected. He tells us also, that if tied about the necks of children, it enables them to retain their urine. An oil, prepared by infusion of these insects,

<hr>

[1] *Nat. Hist. of Ins.*, *Lond.*, 1838, ii. 156.
[2] *Theatr. Ins.*, p. 149. Topsel's *Hist. of Beasts*, p. 1006.
[3] Cuvier, *An. King.—Ins.*, i. 533.
[4] *Nat. Hist.*, xi. 34. Holl. *Trans.*, p. 326. K.

is recommended by the same author, in pains of the ears, if dropped into them.[1]

The *Cossus* of the Greeks and Romans, which, at the time of the greatest luxury among the latter, was introduced at the tables of the rich, was the larva, or grub, of a large beetle that lives in the stems of trees, particularly the oak; and was, most probably, the larva of the Stag-beetle, *Lucanus cervus*. On this subject, however, entomologists differ very widely, for it has been supposed the larva of the *Calandra palmarum* by Geoffroy and Keferotein; of the *Prionus damicornis* by Drury; but of the *Lucanus cervus* by Roesel, Scopoli, and most others. The first two, being neither natives of Italy nor inhabiting the oak, are out of the question. But the larva of the *Lucanus cervus*, and perhaps also the *Prionus coriarius*, which are found in the oak as well as in other trees, may each have been eaten under this name, as their difference could not be discernible either to collectors or cooks. Linnæus, following the opinion of Ray, supposed the caterpillar of the great Goat-moth to be the cossus.[2]

Pliny tells us that the epicures, who looked upon these *cossi* as delicacies, even fed them with meal, in order to fatten them.[3]

Our children, who call the Stag-beetles and the *Passalus cornutus*, oxen, are wont to hitch them with threads to chips and small sticks, and, for their amusement, make them drag the wood along as if they were oxen.

Scarabæidæ—Dung-beetles.

The *Coprion*, *Cantharus*, and *Heliocantharus* of the ancients were evidently the *Scarabæus* (*Ateuchus*) *pilurarius*, or, as it is commonly called, the Tumble-dung, or one nearly related to it, for it is described as rolling backward large

---

[1] James' *Med. Dict.* Cf. Brookes' *Nat. Hist. of Ins.*, p. 321.

[2] *Amoreux*, p. 154. Burmeister's *Manl. of Entomol.*, p. 561. Keferot. *Uber den unmittelbaren Nutzen der Insekten*, Erfurt, 1829, 4to, p. 8–10. Kirb. and Sp. *Introd.*, i. 303, note. Shaw's *Zool.*, vi. 28, note.

[3] *Nat. Hist.*, xvii. 37.

masses of dung; and in doing this it attracted such general
attention as to give rise to the proverb *Cantharus pipulam.*
From the name, derived from a word signifying an ass, it
should seem the Grecian beetle made, or was supposed to
make, its pills of *asses'* dung; and this is confirmed by a
passage in one of the plays of Aristophanes, the Irene,
where a beetle of this kind is introduced, on which one of
the characters rides to heaven to petition Jupiter for peace.
The play begins with one domestic desiring another to feed
the Cantharus with some bread, and afterward orders his
companion to give him another kind of bread made of *asses'*
dung.[1]

Illustrative of the great strength of the Tumble-bug, the
following anecdote may be related: Dr. Brichell was sup-
ping one evening in a planter's house of North Carolina,
when two of these beetles were placed, without his knowledge, under the candlestick. A few blows were struck on
the table, when, to his great surprise, the candlestick began
to move about, apparently without any agency, except that
of a spiritual nature; and his surprise was not lessened
when, on taking one of them up, he discovered that it was
only a chafer that moved.[2]

In Denmark, the common Dung-beetle, *Geotrupes stercorarius,* is called *Skarnbosse* or *Tor(Thor)bist,* and an
augury as to the harvest is drawn by the peasants from
the mites which infest it. The notion is, that if there are
many of these mites between the fore feet, there will be an
early harvest, but a late one if they abound between the
hind feet.[3]

In Gothland, where Thor was worshiped above and more
than the other gods, the *Scarabæus (Geotrupes) stercorarius* was considered sacred to him, and bore the name of
Thorbagge—Thor's-bug. "Relative to this beetle," says
Thorpe, "a superstition still exists, which has been transmitted from father to son, that if any one finds in his path
a Thorbagge lying helpless upon its back, and turns it on
its feet, he expiates seven sins; because Thor in the time
of heathenism was regarded as a mediator with a higher

---

[1] Kirb. and Sp. *Introd.,* i. 255, note.
[2] *Ins. Archit.,* p. 252.
[3] Detharding *de Ins. Coleop. Danicis,* 9. Quot. by Kirb. and Sp.
Introd., i. 33.

power, or All-father. On the introduction of Christianity, the priests strove to terrify the people from the worship of their old divinities, pronouncing both them and their adherents to be evil spirits, and belonging to hell. On the poor Thorbagge the name was now bestowed of Thordjefvul or Thordyfvel—Thor-devil, by which it is still known in Sweden Proper. No one now thinks of Thor, when he finds the helpless creature lying on its back, but the good-natured countryman seldom passes it without setting it on its feet, and thinking of his sin's atonement."[1]

A common symbol of the Creator among the Hindoos (from whom it passed into Egypt, and thence into Scandinavia, says Bjornstjerna) was the *Scarabæus (Ateuchus) sacer*, commonly called the Sacred-beetle of the Egyptians.[2] Of this insect we next treat at length.

Of the many animals worshiped by the ancient Egyptians, one of the most celebrated, perhaps, is the insect commonly known as the Sacred-scarab—*Scarabæus sacer*. This name was given it by Linnæus, but later writers know it as the *Ateuchus sacer*.[3] The insect is found throughout all Egypt, in the southern part of Europe,[4] in China, the East Indies, in Barbary, and at the Cape of Good Hope.[5]

The *Ateuchus sacer*, however, is not the only insect that was regarded as an object of veneration by the Egyptians; but another species of the same genus, lately discovered in the Sennâri by M. Caillaud de Nantes, appears to have first fixed the attention of this people, in consequence of its more brilliant colors, and of the country in which it was found, which, it is supposed, was their first sojourn.[6] This species, which Cuvier has named *Ateuchus Ægyptorum*, is green, with a golden tint, while the first is black.[7] The *Buprestis* and *Cantharus*, or *Copris*, were also held in high repute by the Egyptians, and used as synonymous emblems of the same deities as the Scarabæus. This is further confirmed by the fact of S. Passalacqua having found a species of

---

[1] *Northern Mythol.*, ii. 53.

[2] Bjornstj. *Theog. of Hindoos*, p. 108.

[3] Oliv. Col. I. 3, viii. 59. Cuvier, *An. King.—Ins.*, i. 452.

[4] Cuvier, *qua supra*.

[5] Donovan's *Ins. of China*, p. 4.

[6] Cuvier, *qua supra*.

[7] De Pauw's Sacred-beetle of the Egyptians was "the great golden Scarabee, called by some the Cantharides."—ii. 104.

Buprestis embalmed in a tomb at Thebes.[1] But the *Scarabæus*, or *Ateuchus sacer*, is the beetle most commonly represented, and the type of the whole class; and the one referred to in this article under the general name of *Scarabæus*, unless when otherwise particularly mentioned.

The Scarabæus, according to the beliefs of the ancient Egyptians, was sacred to the Sun and to Pthah, the personification of the creative power of the Deity; and it was adopted as an emblem or symbol of—

1. The World.—According to P. Valerianus, the Scarab was symbolical of the world, on account of the globular form of its pellets of dung, and from an odd notion that they were rolled from sunrise to sunset.[2]

2. The Sun.—P. Valerianus supposes this insect to have been a symbol of the sun, because of the angular projection from its head resembling rays, and from the thirty joints of the six tarsi of its feet answering to the days of an (ordinary) solar month.[3] According to Plutarch, it was because these insects cast the seed of generation into round balls of dung, as a genial nidus, and roll them backward with their feet, while they themselves look directly forward. And as the sun appears to proceed in the heavens in a course contrary to the signs, thus the Scarabæi turn their balls toward the west, while they themselves continue creeping toward the east; by the first of these motions exhibiting the diurnal, and by the second the annual, motion of the earth and the planets.[4] Porphyry gives the same reason as Plutarch why the beetle was considered, as he calls it, "a living image of the sun."[5] Horapollo assigns two reasons for the

---

[1] Wilkinson, *Anct. Egypt.*, ii. (2d S.) 259.

[2] Val. *Hieroglyphica*, p. 93–5.

[3] *Ibid.*

[4] Plut. *Of Isis and Osiris*, p. 220. The translation of this passage as given by Philemon Holland is as follows: "The Fly called the Beetill they (the Egyptians) reverence, because they observe in them I wot not what little slender Images (like as in drops of water we see the resemblance of the Sun) of the Divine power. . . . . . As for the Beetills, they hold, that throughout all their kinds there is no female, but all the males do blow or cast their seed into a certain globus or round matter in the form of balls, which they drive from them and roll to and fro contrariwise, like as the Sun, when he moveth himself from the West to the East, seemeth to turn about the Heaven clean contrary."—p. 1071, ed. of 1657.

[5] Quot. by Montfaucon, *Antiq.*, vol. ii., Part 2, p. 322.

Scarab being taken as an emblem of the sun. He tells us there are three species of beetles: one of which has the form of a cat, and is radiated;[1] and this one from a supposed analogy the Egyptians have dedicated to the Sun, because, first, the statue of the Deity of Heliopolis (City of the Sun) has the form of a cat![2] In this, however, Wilkinson asserts, that Horapollo is wrong; for the Deity of Heliopolis, under the form of a cat, was the emblem of Bubastis, and not of Rê, a type of the sun; and the presence of her statue is explained by the custom of each city assigning to the Divinities of neighboring places a conspicuous post in its own temples; and Bubastis was one of the principal contemplar Deities of Heliopolis.[3] The second reason of Horapollo is, that this insect has thirty fingers, which correspond to the thirty days of a solar month.[4]

3. The Moon.—The second of the three species of beetles, described by Horapollo, has, according to this writer, two horns, and the character of a bull; and it was consecrated to the moon; whence the Egyptians say, that the bull in the heavens is the elevation of this Goddess. This statement of beetle "with two horns" (the *Copris Isidis*) consecrated to the moon, Wilkinson says is not confirmed by the sculptures where it is never introduced.[5]

It is said the Egyptians believed that the pellet of the Scarabæus remained in the ground for a period of twenty-eight days. May not this have some connection with their choosing the insect as a symbol of the moon which divides the year into months of twenty-eight days each; or, of the month itself (of which we shall notice it was also a symbol) for the same reason? I have seen, too, a Scarabæus engraved upon a seal, the joints of whose tarsi numbered but twenty-eight.

Conformable to this supposition, the following quotation may be given from that chapter of the Treasvrie of Auncient and Modern Times devoted to the "Many mervailous (marvelous) properties in sundrie things; and to what

---

[1] De Pauw tells us that the description of the Scarabæus as given by Orus Apollo (Horapollo) is, that "it resembles the sparkling luster of the eye of a cat in the dark."(!)—ii. 104.

[2] Horap., i. 10.

[3] *Anct. Egypt.*, i. (1st S.) 296.

[4] Horap., *Hierogl.*, i. 10.

[5] *Anct. Egypt.*, ii. (2d S.) 258.

Stars and Planets they are subjected naturally," where we find mention of the Scarab as being subject to the moon: " The *Scarabe*, which is otherwise commonly called the Beetle-flye, a little old Creature, is maruelously subject to the Moon, and thereof is found both written, and by experience: That she gathereth or little pellets, or little round bals, and therein encloseth her young Egges, keeping the Pellets hid in the ground eight and twenty daies; during which time the Moone maketh her course, and the nine and twentieth day shee taketh them forth, and then hideth them againe vnder the Earth.   Then, at such time as the Moone is conioyned with the Sunne, which wee vsually tearme the New Moone: they all issue forth aliue, and flye about."[1]

4. Mercury.—The third of the three species of beetles, described by Horapollo, has one horn, and a peculiar form ; and it is supposed, like the Ibis, to refer to Mercury.[2]

5. A Courageous Warrior.—As such they forced all the soldiers to wear rings, upon each of which a beetle was engraved, *i.e.* an animal perpetually in armor, who went his rounds in the night.[3]   Plutarch thus alludes to this custom :   " In the signet or seal-ring of their martial and military men, there was engraven the portraeture of the great Fly called the Beettil ;" and assigns this curious and ridiculous reason, " because in that kinde there is no female, but they be all males."[4]   The custom is also mentioned by Ælian ;[5] and some Scarabs have been found perfect, set in gold, with the ring attached.[6]   The Romans adopted this emblem and made it a part of some legionary standards.

6. Pthah, the Creative Power.—Plutarch says, that in consequence of there being no females of this species, but all males, they were considered fit types of the creative power, self-acting and self-sufficient.[7]   Some, too, have supposed that its position upon the female figure of the heavens, which encircles the zodiacs, refers to the same singular idea of its generative influence.[8]

---

1 *Treasvrie*, B. 7. c. 14, p. 662.   Printed 1613.
2 Horap. *Hierog.*, i. 10.
3 Fosbroke, *Encycl. of Antiq.*, i. 208.
4 *Of Isis, &c.*   Holl. *Transl.*, p. 1051.
5 Ælian, x. 15.
6 Wilkinson, *Anct. Egypt.*, ii. (2d S.) 257.
7 *Of Isis, &c., qua supra.*
8 Wilkin. *Anct. Egypt.*, ii. (2d S.) 256.

7. Pthah Tore, another character of the creative power.[1]

8. Pthah-Sokari-Osiris.—Of this pigmy Deity of Memphis, it was adopted as a distinctive mark, being placed on his head.[2]

9. Regeneration, or reproduction, from the fact of its being the first living animal observed upon the subsidence of the waters of the Nile.[3]

10. Spring.[4]

11. The Egyptian month anterior to the rising of the Nile, as it appears first in that month.[5] It also may have been a symbol of a lunar month from an above-mentioned belief, namely, that its pellets remain twenty-eight days in the ground. It is sometimes found with the joints of its tarsi numbering but twenty-eight instead of thirty, hence the supposition is that it was held as a symbol of a lunar, as well as a solar, month.

12. Fecundity.—Dr. Clarke informs us that these beetles are even yet eaten by the women to render them prolific.[6]

13. With the eyes pierced by a needle, of a man who died from fever.[7]

14. Surrounded by roses, of a voluptuary, because they thought that the smell of that flower enervated, made lethargic, and killed the beetle.[8]

15. An only son; because, says Fosbroke, they believed that every beetle was "both male and female."[9] Was it not because they imagined these insects were all males, as above stated upon the authority of Plutarch, and hence the analogy in a family of an only son since it could be but of the masculine gender?

The Scarabæus was also connected with astronomical subjects, occurring in some zodiacs in the place of Cancer; and with funereal rites.[10]

To no place in particular, as the dog at Cynopolis, the

---

[1] Wilkin. *Anct. Egypt.*, ii. (2d S.) 256.　　[2] *Ibid.*

[3] Pettigrew, *Hist. of Mum.*, p. 220.　[4] *Ibid.*　[5] *Ibid.*

[6] Travels. ii. 306 (?).

[7] Fosbroke, *Encycl. of Antiq.*, i. 208.

[8] *Ibid.* Vide Pierius' *Hieroglyph.*, p. 76–80. Solis operum similitudo; Mundus; Generatio; Vnigenitus; Deus in humano corpore; Vir, paterve; Bellator strenuus; Sol; Luna; Mercurius; Febris lethalis a sole; Virtus enervata deliciis.

[9] Fosbroke, *Encycl. of Antiq.*, i. 208.

[10] Wilkin. *Anct. Egypt.*, ii. (2d S.) 257.

ichneumon at Heracleopolis, was the worship of the beetle confined; but traces of it are found throughout the whole of Egypt. It is probable, however, it received the greatest honors at Memphis and Heliopolis, of which cities Pthah and the Sun were the chief Deities.[1] The worship is also of great antiquity, for in many of the above-mentioned characters, the beetle occurs upon the royal sepulchers of Biban-el-Moluc, which are said to be more ancient than the Pyramids.[2] Scarabæi are, in fact, to be retraced in all their monuments and sculptures, and under divers positions, and often depicted of gigantic dimensions. Mr. Hamilton tells us that in the most conspicuous part of the magnificent temple which marks the site of the ancient Ombite nome, priests are represented paying divine honors to this beetle, placed upon an altar; and, that it might have a character of more mysterious sanctity, it was generally figured with two mitered heads—that of the common hawk, and that of the ram with the horn of Ammon.[3] It may be remarked here, that the Scarabæus, when represented with the head of a hawk, or of a ram, is meant to be an emblem of the sun; and as such emblem it is most commonly found. It often occurs in a boat with extended wings, holding the globe of the sun in its claws, or elevated in the firmament as a type of that luminary in the meridian. Figures too of other Deities are often seen praying to it when in this character.[4]

In the cabinet of Montfaucon, there is a Scarabæus in the middle of a large stone, with outspread feet; and two men, or women, who are perhaps priests, or priestesses, stand before it with clasped hands as if in adoration.[5] This gentleman also has remarked that on the Isiac table, there is the figure of a man in a sitting posture, who holds his hands toward a beetle which has the head of a man with a crescent upon it.[6] On this table there is another Scarab with the head of Isis.[7] Besides these Scarabæi with the heads of hawks, rams, men, and the goddess Isis, Mr.

---

[1] Wilkin. *Anct. Egypt.*, ii. (2d S.) 257.
[2] De Pauw, ii. 104.
[3] Pettig. *Hist. of Mum.*, p. 220.
[4] Wilkin. *Anct. Egypt.*, ii. (2d S.) 256.
[5] Montf. *Antiq.*, ii. (Pt. II.) 322.
[6] *Ibid.*, ii. (Pt. II.) 339.
[7] Wilkin. *Anct. Egypt.*, ii. (2d S.) 259, note.

Hertz has in his possession a small Scarabæus in stone with the head of a cow.[1]

The mode of representing the Scarabæi on the monuments was frequently very arbitrary. Some are figured with, and some without the scutellum ; and others are sometimes introduced with two scutella, one on either clypeus. An instance of this mode of representation, of which no example is to be found in nature, occurs in a large Scarabæus in the British museum.[2]

Among the ideographics of the hieroglyphic writing, the Scarabæus is found under several forms : seated with closed and spread wings upon the head of a god, it signifies the name of a god—a Creator ;[3] and with the head and legs of a man, it is emblematic of the same creative power, or of Pthah. Another emblem of Pthah is supported by the arms of a man kneeling on the heavens, and surmounted by a winged Scarab supporting a globe or sun.[4]

The Scarabæus likewise belongs to the hieroglyphic signs as a syllabic phonetic ; and with complement a mouth, signifies type, form, and transformation : flying, to mount—a phonetic of the later alphabet, with sound of H in the name of Pthah. Another phonetic of the later alphabet, belonging to the XXVI. dynasty, of the time of Domitianus and Trajanus, was a Scarabæus in repose.[5]

The Scarabæus entered also into the royal scutcheons. It first appeared in the XI. dynasty, and is found afterward in the XII., XIII., XIV., XVIII., XIX., XX., XXI., XXII. XXIII., and XXX.[6]

The most important monuments of the great edifice of Amenophis—the so-called Palace of Luxon,—in an historical sense, are said to be four great Scarabæi. They contain statements as to the frontier of the Egyptian empire under Amenophis at the time of his marriage with Taja. Rosellini has given copies and explanations of two of them. A third, now in the Louvre, states that the King, conqueror of the Lybian Shepherds, husband of Taja, made the foreign

---

[1] Wilkin. *Anct. Egypt.*, ii. (2d S.) 259, note.

[2] *Ibid.*

[3] Bunsen, *Egypt's Place*, i. 504, fig. 116; i. 508, fig. 169.

[4] Wilkin. *Anct. Egypt.*, i. (2d S.) 258, fig.

[5] Bunsen, *Ibid.*, i. 572, fig. 12; i. 576, fig. 9; i. 582, fig. 3.

[6] Bunsen, *Ibid.*, i. 617–632.

country of the Karai his southern frontier, the foreign land of Nharina (Mesopotamia) his northern. The inscription of the other Scarabæus, now in the Vatican, states that in the eleventh year and third month of his reign, King Amenhept made a great tank or lake to celebrate the festival of the waters; on which occasion he entered it in a barge of "the most gracious Disc of the Sun." This substitution, by the King, of the barge of the Disc of the Sun for the usual barge of Amun-Ra, is the *first* indication of an heretical sun-worship.[1]

Such historical Scarabæi, Champollion and Rosellini have happily compared to commemorative coins; and, in fact, those which record the names of the kings might perhaps be considered as small Egyptian coins.[2]

Besides being ensculped upon monuments and tablets, Scarabæi, as images in baked earth, are found in great numbers with the mummies of Egypt. These little figures also present an intermingling of several animal forms; for some are found with the heads of men, others with those of dogs, lions, and cats, and others are figures entirely fantastical. Father Kirker says, they were interred with the dead to drive away evil spirits; and there is much probability, he continues, that these were put here for no other purpose than to protect their relatives.[3] The largest of these rude images of Scarabæi, thus used for funereal purposes, frequently had a prayer, or legend connected with the dead, engraved upon them; and a winged Scarabæus was generally placed on those bodies which were embalmed according to the most extensive process.[4] These latter are found in various positions, but generally upon the eye and breast of the body.[5] Placed over the stomach, it was deemed a never-failing talisman to shield the "soul" of its wearer against the terrific genii of Amenthi.[6]

A small, closely cut, glazed limestone Scarabæus has been found tied like a ring by a twist of plain cord on the fourth finger of the left hand. This has occurred twice. Another has been found fastened around the left wrist.[7]

[1] Bunsen's *Egypt's Place*, iii. 142.   [2] *Ibid.*
[3] Quot. by Montf. *Antiq.*, ii. (Pt. II.) 323.
[4] Wilkin. *Anct. Egypt.*, ii. (2d S.) 257.
[5] Pettig. *Hist. of Mum.*, p. 220.
[6] Maury's *Indig. Races*, p. 156.
[7] Phind's *Thebes*, p. 130.

It has been remarked before that the Scarabæus was connected with astronomical subjects.   Donovan tells us that
"when sculptured on astronomical tables, or on columns, it
expressed the divine wisdom which regulated the universe
and enlightened man."[1]

From another point of view we will look now upon the
worship of the Scarabæus.   When the hieroglyphics of the
ancient Egyptians, by reason of their antiquity, became unintelligible, and, in consequence, to the superstitious people,
sacred, they were formed into circles and borders, after the
manner of cordons, and engraved upon precious stones and
gems, by way of amulets and trinkets.   It is thought this
fashion was coeval with the introduction of the worship of
Serapis by the Ptolemies.[2]   In the second century, that sect
of the Egyptians called the Basilidians, intermingling the
new-born Christianity with their heathenism, introduced
that particular kind of mysterious hieroglyphics and figures
called Abraxas, which were supposed to have the singular
property of curing diseases.[3]   These abraxas are generally
oval, and made of black Egyptian basalt.   They are sometimes covered with letters and characters, fac-similes of the
ancient hieroglyphics, but more commonly with the inscriptions in the more modern letters.   Besides these inscriptions,
figures of animals and scenes were also frequently represented; and among the animals, one of frequent occurrence
was the Scarabæus.   For this insect the Basilidians had the
same great veneration as their forefathers; and they paid
to it almost the same divine honors.   This appears in many
abraxas, and particularly in one in the cabinet of Montfaucon, where two women are seen standing before a beetle,
with uplifted hands, as if supplicating it to grant them some
favor.   Above is a large star, or, more probably, the sun,
of which the beetle was the well-known symbol.[4]   On another abraxas, figured by Montfaucon, there are two birds
with human heads, which stand before a Scarab.   These
figures are surrounded by a snake the ends of which meet.
Upon the other side is written in Greek characters the word
φρή (Phre or Phri), which in the Coptic or Egyptian language signifies the sun.[5]   Chifflet has figured an abraxas

---

[1] Donovan, *Ins. of China*, p. 3.
[2] Fosbroke, *Encyclop. of Antiq.*, i. 208.   [3] *Ibid.*
[4] Montf. *Antiq.*, ii. (Pt. II.) 339.   [5] *Ibid.*

which contains a Scarabæus having the sun for its head, and the arms of a man for legs.[1] Another, in the cabinet of M. Capello, is remarkable for having a woman on its reverse, who holds two infants in her arms.[2] Montfaucon has also figured two others, given by Fabreti; and Count Caylus has engraved one, which represents a woman's head upon the body of a Scarab. The head is that of Isis.[3] As these beetles differ much in form, it may be there are several species. To the abraxas succeeded the talismans, which were of the highest estimation in the East.

Carved Scarabæi of all sizes and qualities are quite common in the cabinets of Europe. They were principally used for sets in rings, necklaces, and other ornamental trinkets, and are now called Scarabæi gems,[4] though some suppose them to have been money. All of these gems, Winkleman says, which have a beetle on the convex side, and an Egyptian

---

[1] Montf. *Antiq.*, ii. (P.t II.) 339.   [2] *Ibid.*

[3] Fosbroke, *Encycl. of Antiq.*, i. 208.

[4] There is now at Thebes an arch-forger of Scarabæi—a certain Ali Gamooni, whose endeavors, in the manufacture of these much-sought-after relics, have been crowned with the greatest success. For the coarser description of these, he has, as well as chance European purchasers, an outlet in a native market: for they are bought from him to be carried up the river into Nubia, where they are favorite amulets and ornaments, as mothers greatly delight to patch one or two to the girdles by short thongs, which constitute the only article of dress of their children. Through this very medium, too, it sometimes happens that these spurious Scarabæi come into the possession of unsophisticated travelers, who are not likely to suspect their origin in that remote country, and under such circumstances.
Scarabæi also of the more elegant and well-finished descriptions are not beyond the range of this curious counterfeiter. These he makes of the same material as the ancients themselves used,—a close-grained, easily-cut limestone, which, after it is graven into shape and lettered, receives a greenish glaze by being baked on a shovel with brass filings.
Ali, not content with closely imitating, has even aspired to the creative; so antiquarians must be on their guard lest they waste their time and learning on antiquities of a very modern date.— *Vide* Rhind's Thebes, p. 253–5. Mr. Gliddon, in an incidental note, *Indig. Races*, p. 192, takes credit for having furnished this same Ali, some twenty-four years ago (as it would appear), with broken penknives and other appliances to aid his already-manifested talent, in the somewhat fantastic hope of flooding the local market with such curiosities, and so saving the monuments from being laid under contribution!

deity on the concave, are of a date posterior to the Ptolemies; and, moreover, all the ordinary gems, which represent the figures or heads of Serapis, or Anubis, are of the Roman era.[1] According to C. Caylus, the Egyptians used these gems for amulets, and made them of all substances except metal. They preferred, however, those of pottery, covered with green and black enamel. Cylinders, squares, and pyramids were first used; then came the Scarabæi, which were the last forms. They now began to have the appearance of seals or stamps, and many believe them to have been such. The body of the beetle being a convenient hold for the hand, and the base a place of safety and facility to engrave whatsoever was wished to be stamped or printed. Many of these characters are as yet unintelligible. These seals are made of the most durable stones, and their convex part commonly worked without much art.

The Egyptian form of the Scarabæus, which somewhat resembled a half-walnut, the Etruscans adopted in the manufacture of their gems. These scarcely exceed the natural size of the Scarabæus which they have on the convex side. They have also a hole drilled through them lengthwise, for suspension from the neck, or annexation to some other part of the person. They are generally cornelians. Some are of a style very ancient, and of extremely precious work, although in the Etruscan manner, which is correctness of design in the figures, and hardness in the turn of the muscles.

The Greeks also made use of the Scarabæus in their gems; but in the end they suppressed the insect, and preserved alone the oval form which the base presented, for the body of the sculpture. They also mounted them in their rings.[2]

Several Egyptian Scarabæi were among the relics discovered by Layard at Arban on the banks of the Khabour; and similar objects have been brought from Nimroud, and various other ruins in Assyria.[3]

---

[1] Winkleman, *Art.* 2, c. 1.

[2] Paraph. from Fosbroke's *Encycl. of Antiq.*, i. 208.

[3] Of those deposited in the British Museum, Mr. Birch has made the following report:

1. A Scarabæus having on the base *Ra-men-Chepr*, a prenomen of

Layard has figured a bronze cup, and two bronze cubes, found among the ruins of Nimroud, on which occur as

Thothmes III. Beneath is a Scarab between two feathers, placed on the basket *sub*.

2. A Scarabæus in dark steaschist, with the figure of the sphinx (the sun), and an emblem between the fore paws of the monster. The sphinx constantly appears on the Scarabæi of Thothmes III., and it is probably to this monarch that the one here described belongs. (On many Scarabæi in the British Museum, and on those figured by Klaproth from the Palin Collection, in Leeman's Monuments, and in the "Description de l'Egypt," Thothmes is represented as a sphinx treading foreign prisoners under him.—*Layard*.) After the Sphinx on this Scarab are the titles of the king, "The sun-placer of creation," of Thothmes III.

3. Small Scarabæus of white steaschist, with a brownish hue; reads *Neter nefer nebta Ra-neb-ma*, "The good God, the Lord of the Earth, the Sun the Lord of truth, rising in all lands." This is Amenophis III., one of the last kings of the XVIII. dynasty, who flourished about the fifteenth century B. C.

4. Scarabæus in white steaschist, with an abridged form of the prenomen of Thothmes III., *Ra-men-cheper at en Amen*, "The sun-placer of creation, the type of Ammon." This monarch was the greatest monarch of the XVIII. dynasty, and conquered Naharaina and the Saenkar, besides receiving tribute from Babel or Babylon and Assyria.

5. Scarabæus in pale white steaschist, with three emblems that cannot well be explained. They are the sun's disk, the ostrich feather, the uræus, and the guitar nablium. They may mean "Truth the good goddess," or "lady," or *ma-nefer*, "good and true."

6. Scarabæus in the same substance, with a motto of doubtful meaning.

7. Scarabæus, with a hawk, and God holding the emblem of life, and the words *ma nefer*, "good and true." The meaning very doubtful.

8. A Scarabæus with a hawk-headed gryphon, emblem of *Menta-Ra*, or Mars. Behind the monster is the goddess Sati, or Nuben. The hawk-headed lion is one of the shapes into which the sun turns himself in the hours of the day. It is a common emblem of the Aramæan religion.

9. Scarabæus with hawk-headed gryphon, having before in the uræus and the *nabla* or guitar, hieroglyphic of good. Above it are the hieroglyphics "Lord of the earth."

10. Small Scarabæus in dark steaschist, with a man in adoration to a king or deity, wearing a crown of the upper country, and holding in the left hand a lotus flower. Between this is the emblem of life.

11. Scarabæus, with the hawk-headed Scarabæus, emblem of *Ra-cheper*, "the creator Sun," flying with expanded wings, four in number, which do not appear in Egyptian mythology till after the

ornaments the figures of Scarabs. Those on the cubes are with outstretched wings, inlaid with gold. The cubes have much the appearance of weights.[1]

The Scarabæus was not only venerated when alive, but embalmed after death. In that state they are found at Thebes. It, however, was not the only insect thus honored, for in one of the heads brought by Mr. Wilkinson from Thebes, several others were discovered. These were submitted to Mr. Hope for examination; and the species ascertained by this gentleman, Mr. Pettigrew has enumerated as follows:

1. Corynetes violaceous, *Fab.*
2. Necrobia mumiarum, *Hope.*
3. Dermestes vulpinus, *Fab.*
4. ——————— pollinctus, *Hope.*
5. ——————— roci, *Hope.*
6. ——————— elongatus, *Hope.*
7. Pimelia spinulosa, *Klug?*
8. Copris sabæus? "found by Passalacqua; so named on the testimony of Latrielle."
9. Midas, *Fab.*
10. Pithecius, *Fab.*
11. A species of Cantharis in Passalacqua's Collection, No. 442.[2] The House-fly has also been found embalmed at Thebes.[3]

Concerning the worship in general of the Scarabæus, many curious observations have been made besides the ones above recorded.

Pliny, in the words of his ancient translator, Philemon Holland, tells us "The greater part of Ægypt honour all beetles, and adore them as gods, or at leastwise having

---

time of the Persians, when the gods assume a more Pantheistic form. Such a representation of the sun, for instance, is found in the Torso Borghese.

It will be observed, adds Layard, that most of the Egyptian relics discovered in the Assyrian ruins are of the time of the XVIII. Egyptian dynasty, or of the fifteenth century before Christ; a period when, as we learn from Egyptian monuments, there was a close connection between Assyria and Egypt.—Layard's *Babylon and Nineveh*, p. 239-240.

[1] Layard's *Babylon and Nineveh*, p. 157, 166.
[2] *Hist. of Mum.*, 53-5; Wilkin. *Anct. Egypt.*, ii. (2d S.) 261, note.
[3] Wilkin. *Anct. Egypt.*, ii. (2d S.) 156.

some divine power in them: which ceremoniall devotion
of theirs, Appion giveth a subtile and curious reason of;
for he doth collect, that there is some resemblance between
the operations and works of the Sun, and this flie; and this
he setteth abroad, for to colour and excuse his countrymen.”[1]

Dr. Molyneux, in the conclusion of his article on the
swarms of beetles that appeared in Ireland in 1688, makes
the following allusion to the worship of the Scarabæus by
the Egyptians: “It is also more than probable that this
same destructive Beetle (Hedge-chafer—*Melontha vulgaris*)
we are speaking of, was that very kind of *Scarabæus* the
idolatrous *Ægyptians* of old had in such high veneration,
as to pay divine worship to it. For nothing can be supposed more natural, than to imagine a Nation addicted to
Polytheism, as the *Ægyptians* were, in a Country frequently suffering great Mischief and Scarcity from Swarms
of devouring Insects, should from a strong Sense and Fear
of Evil to come (the common Principle of Superstition and
Idolatry) give sacred worship to the visible Authors of
these their Sufferings, in hopes to render them more pro-
pitious for the future. Thus ’tis allowed on all hands, that
the same People adored as a God the ravenous Crocodile of
the River Nile; and thus the *Romans*, though more polite
and civilized in their Idolatry, *Febrem ad minas nocendam
venerabantur, eamque variis* Templis extructis *colebant*,
says Valerius Maximus, L. 2, c. 5.”[2]

It is curious to observe how the reason is affected by circumstances. The mind of Dr. Molyneux being long engaged
upon the destruction caused by insects, worked itself insen-
sibly into certain grooves, out of which it was afterward impossible to act. The same may be remarked of Mr. Henry
Baker, as appears from his article, “On a *Beetle* that lived
three years without Food.” In conclusion, this gentleman
says, “As the *Egyptians* were a wise and learned people,
we cannot imagine they would show so much regard to a
creature of such a mean appearance (as the Beetle) without some extraordinary reason for so doing. And is it not
possible they might have discovered its being able to subsist

---

[1] Pliny, *Nat. Hist.*, xxx 11; Holland, ii. 395. K.
[2] *Phil. Trans. Abridg.*, ii. 785; *Gent. Mag.*, xix. 264–5.

a very long time without any visible sustenance, and therefore made it a symbol of the Deity?"[1]

---

[1] *Phil. Trans. Abridg.*, ix. 11. Concerning the worship of animals
in general by the Egyptians, the following remarks in a note may not
be inappropriate, as they embrace the worship of the Scarabœus.

1. A class of animals, to which may be referred the cow, dog,
sheep, and ibis, were *at first* naturally protected and respected out
of gratitude for the benefits derived from them. But in time, it is
supposed, this respect, by unthoughtful descendants believing too
implicitly the teachings of their fathers, was gradually enlarged to
so great extent that it became reverence, and at last, perhaps after
centuries, worship. For example, at A time, the ibis is respected
on account of its destroying noxious serpents; at B, reverenced;
and at C, worshiped

2. When at C time, the ibis is worshiped, suppose the masses have
lost the reason (which in the case of the Egyptians is an allowable
supposition, since it is an historical fact that but the initiated knew
the reasons for their manner of worship), and serpents are its food, is
it plain then that if the food be taken away the sacred bird cannot
live? Hence at C time are serpents preserved and protected as food
for the ibis; and as this protecting care increases as above, till at D
they are reverenced, and at E worshiped. To this second class may
be referred the crocodile, which was preserved, etc. as food for the
ichneumon, a sacred animal of the first class.

3. Analogies between animals, and even plants, and certain sources
of goodness, or objects of wonder, as the sun, and motion of the
stars, were at A time, noticed; at B, respected or reverenced; and
at C, worshiped. Thus, among plants, became the onion sacred, from
the resemblance of the laminæ which compose it, in a transverse
section, to circles—to the orbits of the planets. And thus the Scarabœus from the analogies between its movements and shape and the
motions of the sun, traced, as we have before remarked on the authority of several ancient writers, became also an object of adoration.

4. A fourth reason may also be given, which follows as a consequence of the latter. If such analogy, as. for example, that between
the beetle and the sun, had been observed in the time of picture and
hieroglyphic writing, to represent the sun, the beetle would have
been taken. Now, it is a well-authenticated fact, that these hieroglyphics in time became sacred, and, if the beetle was found among
them, it for this, if for no other reason, would have been looked upon
with the same veneration.

5. Good men, too, to preserve the lives of animals oftentimes wan-
tonly taken, introduce them into fables and poetry, and connect
pleasing tales with them. The "Babes in the Wood" have so fixed
the respect for the tameness of the robin, that it is even now deemed
a sacrilege with our boys to stone this bird. And may there not
have been such good men, and such tender stories, among the Egyptians, and the remembrance of whom and which long lost by the
lapse of time?

In parts of Europe the ladies string together for necklaces the burnished violet-colored thighs of the *Geotrupes stercorarius* and such like brilliant species of insects.[1]

Under *Copris molossus*, in Donovan's Insects of China, it is mentioned that the larvæ of the larger kinds of coleopterous insects, abounding in unctuous moisture, are much esteemed as food by the Chinese "Under the roots of the canes is found a large, white grub, which, being fried in oil, is eaten as a dainty by the Chinese." Donovan suggests that perhaps this is the larvæ of the *Scarabæus* (*copris*) *molossus*, the general description and abundance of which insect in China favors such an opinion.[2]

Insects belonging to the family Scarabæidæ have been used also in medicine. Pliny says the green Scarabæus has the property of rendering the sight more piercing of those who gaze upon it, and that hence, engravers of precious stones use these insects to steady their sight.[3]

Again, he says: "And many there be, who, by the directions of magicians, carrie about them in like manner," *i.e.* tied up in a linen cloth with a red string, and attached to the body, "for the quartan ague, one of these flies or beetles that use to roll up little balls of earth."[4] We learn from Schroder (v. 345) that the powder of the *Scarabæus pilurarius* "sprinkled upon a protuberating eye or prolapsed anus, is said to afford singular relief;" and that "an oil prepared of these insects by boiling in oil till they are consumed, and applied to the blind hæmorrhoids, by means of a piece of cotton, is said to mitigate the pains thereof."[5] Fabricius states that the *Scarabæus* (*copris*) *molossus* is medicinally employed in China.[6]

We quote the following from Moufet: "The Beetle engraven on an emerald yeelds a present remedy against all witchcrafts, and no less effectual than that moly which Mercury once gave Ulysses. Nor is it good only against these, but it is also very useful, if any one be about to go before the king upon any occasion, so that such a ring ought especially to be worn by them that intend to beg of

---

[1] Kirb. and Sp. *Introd.*, i. 33.

[2] *Ins. of China*, p. 6.

[3] *Nat. Hist.*, xxix. 6 (38).

[4] *Nat. Hist.*, xxx. 11 (80). Holland, *Trans.*, ii. 390.

[5] James' *Med. Dict.*

[6] Donovan's *Ins. of China*, p. 6.

noblemen some jolly preferment or some rich province.　It keeps away likewise the head-ach, which, truly, is no small mischief, especially to great drinkers. . . ..

"The magicians will scarce finde credit, when foolishly rather than truly, they report and imagine that the precious stone Chelonitis, that is adorned with golden spots, put into hot water with a Beetle, raiseth tempests. *Pliny, l.* 37, *c.* 10.

"The eagle, the Beetle's proud and cruel enemy, does no less make havock of and devour this creature of so mean a rank, yet as soon as it gets an opportunity, it returneth like for like, and sufficiently punisheth that spoiler.　For it flyeth up nimbly into her nest with its fellow-soldiers, the Scara-beetles, and in the absence of the old she eagle bringeth out of the nest the eagle's eggs one after another, till there be none left; which falling, and being broken, the young ones, while they are yet unshapen, being dashed miserably against the stones, are deprived of life, before they can have any sense of it.　Neither do I see indeed how she should more torment the eagle than in her young ones.　For some who slight the greatest torments of their own body, cannot endure the least torments of their sons."[1]

Pliny says that in Thrace, near Olynthus, there is a small locality, the only one in which the beetle[2] cannot exist; from which circumstance it has received the name of "Cantharolethus—Fatal-to-the-Beetle."[3]

Dynastidæ—Hercules-beetle, etc.

The Hercules-beetle, *Dynastes Hercules*, is four, five, or even sometimes six inches long, and a native of South America.　It is said great numbers of these immense insects are sometimes seen on the Mammæa-tree, rasping off the rind of the slender branches by working nimbly round them with their horns, till they cause the juice to flow, which they drink to intoxication, and thus fall senseless to

---

[1] *Theatr. Ins.*, p. 160.　Topsel's *Hist. of Beasts*, p. 1012.

[2] Cuvier suggests that the *Scarabæus nasicornis* of Linnæus, which haunts dead bark, or the *S. auratus*, may be the insect here referred to.

[3] *Nat. Hist.*, xi. 28 (34).

the ground! These stories, however, as the learned Fabricius has well observed, seem not very probable; since the thoracic horn, being bearded on its lower surface, would undoubtedly be made bare by this operation.[1]

Col. St. Clair, though he confesses he never could take one of these insects in the act of sawing off the limbs of trees, or ascertain what they worked for, gravely repeats the above old story, and says that during the operation they make a noise exactly like that of a knife-grinder holding steel against the stone of his wheel; but a thousand knife-grinders at work at the same moment, he continues, could not equal their noise! He calls this beetle hence the knife-grinder.[2]

The Goliath-beetle, *Dynastes Goliathus*, is said to be roasted and eaten by the natives of South America and Africa.[3]

The enormous prices of £30, £40, and even £50 used to be asked for these latter beetles a piece; fine specimens for cabinets even now bring from five to six pounds.[4]

The large pulpy larva of a species of Dynastidæ—the *Oryctes rhinoceros*, called by the Singhalese *Gascoo-roo-miniya*—is, notwithstanding its repulsive aspect, esteemed a luxury by the Malabar coolies.[5]

Immediately after mentioning the above fact, Tennent records the following interesting superstition respecting a beetle when found in a house after sunset:

"Among the superstitions of the Singhalese arising out of their belief in demonology, one remarkable one is connected with the appearance of a beetle when observed on the floor of a dwelling-house after nightfall. The popular belief is that in obedience to a certain form of incantation (called *cooroominiya-pilli*) a demon in shape of a beetle is sent to the house of some person or family whose destruction it is intended to compass, and who presently falls sick and dies. The only means of averting this catastrophy is, that some one, himself an adept in necromancy, should perform a counter-charm, the effect of which is to send back the disguised beetle to destroy his original employer; for in such a conjuncture the death of one or the other

<hr>

[1] Shaw's *Zool.*, vi. 20.   Baird's *Cyclop. of Nat. Sci.*
[2] St. Clair, *West Indies, etc.*, i. 152.
[3] Simmond. *Curiosities of Food,* p 295.   [4] *Ibid.*
[5] Tennent. *Nat. Hist. of Ceylon.* p. 407.

is essential to appease the demon whose intervention has
been invoked.   Hence the discomfort of a Singhalese on
finding a beetle in his house after sunset, and his anxiety
to expel but not kill it."[1]

The *Dynastes Goliathus*, Moufet says, "like to beetles
(*Ateuchus sacer*), hath no female, but it shapes its own
form itself.   It produceth its young one from the ground
by itself, which Joach. Camerarius did elegantly express,
when he sent to Pennius the shape of this insect out of the
storehouse of natural things of the Duke of Saxony ; with
these verses :

> A bee begat me not, nor yet did I proceed
> From any female, but myself I breed.

For it dies once in a year," continues Moufet, "and from its
own corruption, like a Phœnix, it lives again (as Moninus
witnesseth) by heat of the sun.

> A thousand summers' heat and winters' cold
> When she hath felt, and that she doth grow old,
> Her life that seems a burden, in a tomb
> O' spices laid, comes younger in her room."[2]

Melolonthidæ—Cock-chafers.

The family of insects, commonly called *Cock-chafers*,
Hedge-chafers, *May-bugs*, and *Dorrs* (from the Irish
dord, humming, buzzing, or from the Anglo-Saxon *dora*,
a locust or drone) have been included by Fabricius in the
genus *Melolontha*,—a word which retains an odd notion of
the Greeks respecting them, viz., that they were produced
from or with the flowers of apple-trees.   It is a name
also by which the Greeks themselves used to distinguish
the same kind of insects.

In Sweden the peasants look upon the grub of the Cock-
chafer, *Melolontha vulgaris*, as furnishing an unfailing
prognostic whether the ensuing winter will be mild or
severe; if the animal have a bluish hue (a circumstance

---

[1] Tennent, *Nat. Hist. of Ceylon*, p. 407.
[2] *Theatr. Ins.*, p. 152.   Topsel's *Hist. of Beasts*, p. 1009.

which arises from its being replete with food), they affirm it will be mild, but on the contrary if it be white, the weather will be severe: and they carry this so far as to foretell, that if the anterior be white and the posterior blue, the cold will be most severe at the beginning of the winter. Hence they call this grub *Bemärkelse-mask*—prognostic worm.[1]

An absurd notion obtains in England that the larvæ of the May-bugs are changed into briers.[2]

The following quotation is from the Chronicle of Hollingshed: "The 24 day of Februarie (1575), being the feast of Saint Matthie, on which dai the faire was kept at Tewkesburie, a strange thing happened there. For after a floud which was not great, but such as therby the medows neere adioning were covered with water, and in the after noone there came downe the river of Seuerne great numbers of flies and beetles (*Melolontha vulgaris?*), such as in summer evenings use to strike men in the face, in great heapes, a foot thicke above the water, so that to credible mens judgement there were seene within a paire of buts length of those flies above a hundred quarters. The mils there abouts were dammed up with them for the space of foure daies after, and then were clensed by digging them out with shovels: from whence they came is yet unknowne but the daie was cold and a hard frost."[3]

Such another remarkable phenomenon is recorded to have occurred in Ireland, in the summer of 1688. The Cock-chafers, in this instance, were in such immense numbers, "that when," as the chronicler, Dr. Molyneux, relates, "towards evening or sunset, they would arise, disperse, and fly about, with a strange humming noise, much like the beating of drums at some distance; and in such vast incredible numbers, that they darkened the air for the space of two or three miles square. The grinding of leaves," he continues, "in the mouths of this vast multitude altogether, made a sound very much resembling the sawing of timber."[4]

In a short time after the appearance of these beetles in

---

[1] De Geer, iv. 275-6. Kirb and Sp. *Introd.*, i. 33.
[2] *Hist. of Ins.* (Murray, 1830) ii. 296.
[3] *Chronicles*, iv. 326.—The water overflowing the low grounds brought the beetles for air to the surface, whence they were swept away by the current.
[4] *Phil. Trans. Abridg.*, ii. 781-3.

these immense numbers, they had so entirely eaten up and destroyed the leaves of the trees, that the whole country, for miles around, though in the middle of summer, was left as bare as in the depth of winter.

During the unfavorable seasons of the weather, which followed this plague, the swine and poultry would watch under the trees for the falling of the beetles, and feed and fatten upon them; and even the poorer sort of the country people, the country then laboring under a scarcity of provision, had a way of dressing them, and *lived upon them as food.* In 1695, Ireland was again visited with a plague of this same kind.[1]

In Normandy, according to Mouffet, the Cock-chafers make their appearance every third year.[2] In 1785, many provinces of France were so ravaged by them, that a premium was offered by the government for the best mode of destroying them.[3] During this year, a farmer, near Blois, employed a number of children and the poorer people to destroy the Cock-chafers at the rate of two liards a hundred, and in a few days they collected fourteen thousand.[4]

The county of Norfolk in England seems occasionally to have suffered much from the ravages of these insects; and Bingley tells us that "about sixty years ago, a farm near Norwich was so infested with them, that the farmer and his servants affirmed they had gathered eighty bushels of them; and the grubs had done so much injury, that the court of the city, in compassion to the poor fellow's misfortune, allowed him twenty-five pounds."[5]

The seeming blunders and stupidity of these insects have long been proverbial, as in the expressions, "blind as a beetle," and "beetle-headed."

Cetoniidæ—Rose-chafers.

A very pretty species of the *Cetoniidæ*, the *Agestrata luconica*, is of a fine brilliant metallic green, and found in

---

[1] *Phil. Trans. Abridg.*, ii. 782.
[2] Shaw, *Zool.*, vi. 25.
[3] Kirb. and Sp. *Introd.*, i. 179.
[4] Anderson's *Recr. in Agric.*, iii. 420.
[5] *Anim. Biog.*, iii. 233.

the Philippine Islands. These the ladies of Manilla keep as pets in small bamboo cages, and carry them about with them wheresoever they may go.[1]

Buprestidæ—Burn-cows.

Many species of the *Buprestidæ* are decorated with highly brilliant metallic tints, like polished gold upon an emerald ground, or azure upon a ground of gold; and their elytra, or wing-coverings, are employed by the ladies of China, and also of England, for the purpose of embroidering their dresses.[2] The Chinese have also attempted imitations of these insects in bronze, in which they succeed so well that the copy may be sometimes mistaken for the reality.[3] In Ceylon[4] and throughout India,[5] the golden wing-cases of two of this tribe, the *Sternocera chrysis* and *S. sternicornis*, are used to enrich the embroidery of the Indian zenana, while the lustrous joints of the legs are strung on silken threads, and form necklaces and bracelets of singular brilliancy. The *Buprestis attenuata, ocellata* and *vittata* are also wrought into various devices and trinkets by the Indians. The *B. vittata* is much admired among them. This insect is found in great abundance in China, and thence exported into India, where it is distributed at a low price.[6]

Mr. Osbeck saw in China a *Buprestis maxima*, which had been dried, and to which were fastened leaden wings so painted as to make them look like the wings of butterflies. This artificial monster, he adds, was to be sold in the vaults among other trifles.[7] The *B. maxima* is set up along with Butterflies in small boxes, and vended in the streets of Chinese cities.[8]

So many species of the *Buprestidæ* are clothed with with such brilliant colors, that Geoffroy has thought proper

---

[1] Baird's *Cyclop. of Nat. Sci.*   [2] *Ibid.*
[3] Shaw's *Zool.*, vi. 88.
[4] Tennent, *Nat. Hist. of Ceylon*, p. 405.
[5] Donovan, *Ins. of India*, p. 5.
[6] Donovan, *Ins. of China*, p. 13.
[7] Travels, i. 384.   [8] *Ibid.*, i. 331.

to designate them all under the generic appellation of *Rich-ard*. The origin of this name is as singular as its applica-tion is fantastical. It was originally given to the Jay, in consequence of the facility with which that bird was taught to pronounce the word.[1]

Modern writers have been much divided in their opinion as to what genus the celebrated *Buprestis* of the ancients belongs. All indeed have regarded it as of the order Cole-optera, but here their agreement ceases. Linnæus seems to have looked upon it as a species of the genus to which he has given its name. Geoffroy thinks it to be a *Carabus* or *Cicindela;* M. Latrielle, to the genus *Melöe;* and Kirby and Spence to *Mylabris*.[2]

Of this Buprestis, Pliny says: "Incorporat with goat sewer, it taketh away the tettars called lichenes that be in the face."[3] And Dr. James says that insects of this family "are all in common, inseptic, exulcerating, and (possess) a heating quality; for which reason, they are mixed up with medicines adapted to the cure of a Carcinoma, Lepra, and the malignant Lichen. Mixed in emollient pessaries, they provoke the Catamenial discharges."[4]

The Greeks, it is said, commended the Buprestis in food.[5]

Elateridæ—Fire-flies, Spring-beetles, etc.

In an historical sense, the most interesting species of the family *Elateridæ* is the *Elater noctilucus*, a native of the West Indies, and called by the inhabitants, *Cucujus*. From an ancient translation of Peter Martyr's History of the West Indies, we make the following quotation, which contains many curious facts relative to this insect:

"Whoso wanteth *Cucuji*, goeth out of the house in the first twilight of the night, carrying a burning fier-brande in his hande, and ascendeth the next hillocke, that the *Cucuji* may see it, and swingeth the fier-brande about calling

---

[1] Cuvier, An. King.—*Ins.*, i. 356.
[2] *Introd* , i. 156.
[3] Pliny, xxx. 4; Holland, ii. 377. E.
[4] *Med. Dict.*   [5] *Ibid.*

Cucuji aloud, and beating the ayre with often calling and crying out *Cucuji, Cucuji.* . . . Beholde the desired number of *Cucuji,* at what time, the hunter casteth the fier-brande out of his hande. Some *Cucuji* sometimes followeth the fier-brande, and lighteth on the grounde, then is he easily taken. . . . The hunter havinge the hunting *Cucuius,* returneth home, and shutting the doore of the house, letteth the praye goe. The *Cucuius* loosed, swiftly flyeth about the whole house seeking gnatts, under their hanging bedds, and about the faces of them that sleepe, whiche the gnattes used to assayle, they seem to execute the office of watchmen, that such as are shut in, may quietly rest. Another pleasant and profitable commodity proceedeth from the *Cucuji.* As many eyes as every *Cucuius* openeth, the host enjoyeth the light of so many candles : so that the Inhabitants spinne, sewe, weave, and daunce by the light of the flying *Cucuji.* The Inhabitants think that the *Cucuius* is delighted with the harmony and melodie of their singing, and that he also exerciseth his motion in the ayre according to the action of their dancing. . . . Our men also read and write by that light, which always continueth untill hee have gotten enough gnatts whereby he may be well fedd. . . . There is also another wonderfull commodity proceeding from the *Cucuius :* the Islanders, appoynted by our menn, goe with their good will by night with 2 *Cucuji* tyed to the great tooes of their feete : (for the travailer[1] goeth better by direction of the lights of the *Cucuji,* then if hee brought so many candels with him, as the *Cucuji* open eyes) he also carryeth another *Cucuius* in his hande to seeke the Utiae by night (Utiae are a certayne kind of Cony, a little exceeding a mouse in bignesse.) . . . . They also go a fishing by the lights of the *Cucuji.* . . . In sport, and merriment, or to the intent to terrifie such as are affrayed of every shaddow, they say that many wanton wild fellowes sometimes rubbed their faces by night with the fleshe of a *Cucuius* being killed, with purpose to meete their neighbors with a flaming countenance . . . . for the face being annointed with the lumpe or fleshy parte of the *Cucuius,* shineth like a flame of fire."[2]

---

[1] Peruvians travel by the light of the *Cucujus Peruvianus.*—See Kirby's *Wond. Museum,* ii. 151.

[2] *Hist. of West Indies,* p. 274.

At Cumana, the use of the Cucujus is forbidden, as the young Spanish ladies used to carry on a correspondence at night with their lovers by means of the light derived from them.[1]

Captain Stedman tells us, that one of his sentinels, one night, called out that he saw a negro, with a lighted tobacco-pipe, cross a creek near by in a canoe. At which alarm they lost no time in leaping out of their hammocks, and were not a little mortified when they found the pipe was nothing more than a Fire-fly on the wing.[2]

An individual of this species, brought to Paris in some wood, in the larva or nymph state, there underwent its metamorphosis, and by the light which it emitted, excited the greatest surprise among many of the inhabitants of the Faubourg St. Antoine, to whom such a phenomenon had hitherto been unknown.[3]

When Cortes and Narvaez were at war with one another in Mexico, Bernal Diaz relates "that one night in the midst of darkness numbers of shining Beetles (*Elater noctilucus*) kept continually flying about, which Narvaez's men mistook for the lighted matches of our fire-arms, and this gave them a vast idea of the number of our matchlocks."[4] Thomas Campanius tells us that one night the Cucuji frightened all the soldiers at Fort Christina, in New Sweden (Pennsylvania ?) : they thought they were enemies advancing toward them with lighted torches.[5] Another such like story, which is not incredible by any means, is told us by Mouffet. He says that when Sir Thomas Cavendish and Sir Robert Dudley first landed in the West Indies, and saw an infinite number of moving lights in the woods, which were merely these Elaters, they supposed that the Spaniards were advancing upon them with lighted matches, and immediately betook themselves to their ships.[6]

The Indians of the Carribbee Islands, Ogleby remarks, "anoint their bodies all over (at certain solemnities wherein candles are forbidden) with the juice squeezed out of them

---

[1] Baird's *Cyclop. of Nat. Sci.*
[2] Stedm. *Surinam*, i. 140.
[3] Cuvier, *An. King.—Ins.*, i. 321.
[4] *Conq. of Mex.*, i. 327.
[5] *Hist. of New Swed.*, p. 162.
[6] *Theatr. Insect.*, p. 112.

(Cucuji), which causes them to shine like a flame of fire."[1] And in the Spanish Colonies, on certain festival days in the month of June, these insects are collected in great numbers, and tied as decorations all over the garments of the young people, who gallop through the streets on horses similarly ornamented, producing on a dark evening the effect of a large moving body of light. On such occasions the lover displays his gallantry by decking his mistress with these living gems.[2]

At the present day, the poorer classes of Cuba and the other West India Islands, make use of these luminous insects for lights in their houses. Twenty or thirty of them put into a small wicker-work cage, and dampened a little with water, will produce quite a brilliant light. Throughout these islands, the Cucujus is worn by the ladies as a most fashionable ornament. As many as fifty or a hundred are sometimes worn on a single ball-room dress. Capt. Stuart tells me he once saw one of these insects upon a lady's white collar, which at a little distance rivaled the Kohinoor in splendor and beauty. The insect is fastened to the dress by a pin through its body, and only worn so long as it lives, for it loses its light when dead.

The statement of Humboldt is, that at the present day in the habitations of the poorer classes of Cuba, a dozen of Cucuji placed in a perforated gourd suffice for a light during the night. By shaking the gourd quickly, the insect is roused, and lights up its luminous disks. The inhabitants employ a truthful and simple expression, in saying that a gourd filled with Cucuji is an ever-lighted torch; and in fact it is only extinguished by the death of the insects, which are easily kept alive with a little sugar cane. A lady in Trinidad told this great traveler, that during a long and painful passage from Costa Firme, she had availed herself of these phosphorescent insects whenever she wished to give the breast to her child at night. The captain of the ship would not permit any other light on board at night, for fear of the privateers.[3]

Southy has happily introduced the Cucujus in his

---

[1] *Hist. of Amer.*, p. 378.
[2] Walton, *Pres. St. of Span. Col.*, i. 128.
[3] Humboldt's *Cuba*, p. 395.

"Madoc" as furnishing the lamp by which Coatel rescued the British hero from the hands of the Mexican priests :

> She beckon'd and descended, and drew out
> From underneath her vest a cage, or net
> It rather might be called, so fine the twigs
> Which knit it, where, confined, two fire-flies gave
> Their lustre. By that light did Madoc first
> Behold the features of his lovely guide.

Darwin says : " In Jamaica, at some seasons of the year, the Fire-flies are seen in the evening in great abundance. When they settle on the ground, the bull-frog greedily devours them, which seems to have given origin to a curious, though very cruel, method of destroying these animals : if red-hot pieces of charcoal be thrown toward them in the dusk of the evening, they leap at them, and hastily swallow them, mistaking them for Fire-flies, and are burnt to death." (!)[1]

Beetles belonging to the family *Elateridæ* have been so called from a peculiar power they have of leaping up like a tumbler when placed on their backs, and for this reason they have received the English appellations of *Spring-beetles* and *Skip-jacks*, and from the noise which the operation makes when they leap, they are also called *Snap*, *Watch*, or *Click-beetle*, and likewise *Blacksmiths*.

If a Blacksmith beetle enters your house, a quarrel will ensue which may end in blows.

This superstition obtains in Maryland.

Lampyridæ.—Glow-worms.

Antonius Thylesius Bonsentinus, following his elegant description of the Glow-worm, gives a pretty fable of its origin. As translated in Moufet's Theater of Insects, his words are these :

> This little fly shines in the air alone,
> Like sparks of fire, which when it was unknown
> To me a boy, I stood then in great fear,
> Durst not attempt to touch it, or come near.

---

[1] *Saturday Mag.*, ix. 229.

> May be this worm from shining in the night,
> Borrow'd its name, shining like candle bright.
> The cause is one, but divers are the names,
> It shines or not, according as she frames
> Herself to fly or stand; when she doth fly,
> You would believe 'twere sparkles in the skie,
> At a great distance you shall ever finde
> Prepar'd with light and lanthorn all this kinde.
> Darkness cannot conceal her, round about
> Her candle shines, no winds can blow it out.
> Sometimes she flies as though she did desire
> Those that pass by to observe her fire;
> Which being nearer, seem to be as great,
> As sparks that fly when smiths hot iron beat.
> When Pluto ravish'd Proserpine, that rape,
> For she was waiting on her, chang'd her shape,
> And since that time, she flyeth in the night
> Seeking her out with torch and candle light.[1]

The following anecdote is related by Sir J. E. Smith, of the effect of the first sight of the Italian Glow-worms upon some Moorish ladies ignorant of such appearances. These females had been taken prisoners at sea, and, until they could be ransomed, lived in a house in the outskirts of Genoa, where they were frequently visited by the respectable inhabitants of the city; a party of whom, on going one evening, were surprised to find the house closely shut up, and their Moorish friends in the greatest consternation. On inquiring into the cause, they found that some Glow-worms—*Pygolampis Italica*—had found their way into the building, and that the ladies within had taken it into their heads that these brilliant guests were no other than the troubled spirits of their relations; of which curious idea it was some time before they could be divested.—The common people of Italy have a superstition respecting these insects somewhat similar, believing that they are of a spiritual nature, and proceed out of the graves, and hence carefully avoid them.[2]

Cardan, Albertus, Gaudentinus, Mizalduo, and many others have asserted that perpetual lights can be produced from the Glow-worm; and that waters distilled from this insect afford a lustre in the night. It is needless to say these assertions are without foundation.[3]

---

[1] *Theatr. Ins.*, p. 111. Topsel's *Hist. of Beasts*, p. 977.
[2] *Tour on the Continent*, 2d. Edit., iii. 85.
[3] Browne's *Vulg. Err.*, B. iii. c. 17. *Works*, ii. 531.

In India, the ladies have recourse to Fire-flies for ornaments for their hair, when they take their evening walks. They inclose them in nets of gauze.[1] And the beaux of Italy, Sir J. E. Smith tells us, are accustomed in the summer evenings to adorn the heads of the ladies with Glow-worms, by sticking them also in their hair.[2]

Never kill a Glow-worm, if you do, the country people say, you will put "the light out of your house,"—*i.e.* happiness, prosperity, or whatever blessing you may be enjoying.

A Glow-worm, in your path, denotes brilliant success in all your undertakings. If one enters a house, one of the heads of the family will shortly die. These superstitions obtain in Maryland.

Of the Glow-worm—*Noctiluca terrestris*, Col. Ecphr., i. 38—Dr. James says: "The whole insect is used in medicine, and recommended by some against the Stone. Cardan ascribes an anodyne virtue to it."[3]

Mr. Ray, in his travels through the State of Venice, says: "A discovery made by a certain gentleman, and communicated to me by Francis Jessop, Esq., is, that those reputed meteors, called in Latin *Ignis fatui*, and known in England by the conceited names of *Jack with a Lanthorn*, and *Will with a Wisp*, are nothing else but swarms of these flying Glow-worms. Which, if true, we may give an easy account of those phenomena of these supposed fires, *viz.*, their sudden motion from place to place, and leading travelers that follow them into bogs and precipices."[4] It has been suggested[5] also that the mole-cricket, *Gryllotalpa vulgaris*,[6] which in its nocturnal peregrinations was supposed to be luminous, is this notorious "Will-o'-the-wisp."

Pliny says: "When Glow-worms appear, it is a common

---

[1] Kirb. and Sp. *Introd.*, i. 317.
[2] *Tour on Continent*, iii. 85. 2d Edit.
[3] *Med. Dict.*
[4] Harris' *Col. of Voy. and Trav.*, ii. 688.
[5] Harris, *Farm Insects*, p. 372.
[6] This insect has received its English names, of *Mole-cricket* and *Earth-crab*, from its burrowing like a mole, and some species of W. Indian crabs; and, from its supposed jarring song at night, it is also called *Eve-churr*, *Churr-worm*, and *Jarr-worm.*—*Ibid.*

sign of the ripenesse of barley, and of sowing millet and pannick. . . . .   And Mantuan sung to the same tune:

Then is the time your barley for to mow,
When Glow-worms with bright wings themselves do show."[1]

____

Ptinidæ—Death-watch, etc.

The common name of *Death-watch*, given to the *Anobium tesselatum*, sufficiently announces the popular prejudice against this insect; and so great is this prejudice, that, as says an editor of Cuvier's works, the fate of many a nervous and superstitious patient has been accelerated by listening, in the silence and solitude of night, to this imagined knell of his approaching dissolution.[2]   The learned Sir Thomas Browne considered the superstition connected with the Death-watch of great importance, and remarks that "the man who could eradicate this error from the minds of the people would save from many a cold sweat the meticulous heads of nurses and grandmothers,"[3] for such persons are firm in the belief, that

The solemn Death-watch clicks the hour of death.

The witty Dean of St. Patrick endeavored to perform this useful task by means of ridicule.   And his description, suggested, it would appear, by the old song of "A cobbler there was, and he lived in a stall," runs thus:

————A wood worm
That lies in old wood, like a hare in her form,
With teeth or with claws, it will bite, it will scratch;
And chambermaids christen this worm a Death-watch;
Because, like a watch, it always cries click.
Then woe be to those in the house that are sick!
For, sure as a gun, they will give up the ghost,
If the maggot cries click when it scratches the post.
But a kettle of scalding hot water injected,
Infallibly cures the timber affected:
The omen is broken, the danger is over,
The maggot will die, and the sick will recover.

____

[1] Moufet, *Theatr. Ins.*, p. 110.   Topsel's *Hist. of Beasts*, p. 977.
[2] Cuvier, *An. King.—Ins.*, i. 382.
[3] Cf. *Works*, ii. 375.

Grose, in his Antiquities, thus expresses this superstition: "The clicking of a Death-watch is an omen of the death of some one in the house wherein it is heard." Watts says: "We learn to presage approaching death in a family by ravens and little worms, which we therefore call a Death-watch."[1] Gay, in one of his Pastorals, thus alludes to it:

> When Blonzelind expired, ...
> The solemn Death-watch click'd the hour she died.[2]

And Train,—

> An' when she heard the Dead-watch tick,
>   She raving wild did say,
> "I am thy murderer, my child;
>   I see thee, come away."

And Pope,—

> Misers are muck-worms, silkworms beaux,
> And Death watches physicians.[3]

"It will take," says Mrs. Taylor, a writer in Harper's New Monthly Magazine, "a force unknown at the present time to physiological science to eradicate the feeling of terror and apprehension felt by almost every one on hearing this small insect." She herself, an entomologist, confesses to have been very much annoyed at times by coming in contact with this "strange nuisance;" but she was cured by an overapplication. "I went to pay a visit," says she, "to a friend in the country. The first night I fancied I should have gone mad before morning. The walls of the bed-room were papered, and from them beat, as it were, a thousand watches—tick, tick, tick! Turn which way I would, cover my head under the bedclothes to suffocation, every pulse in my body had an answering tick, tick, tick! But at last the welcome morning dawned, and early I was down in the library; even here every book, on shelf above shelf, was riotous with tick, tick, tick! At the breakfast table, beneath the plates, cups, and dishes, beat the hateful sound. In the parlor, the withdrawing-room, the kitchen, nothing

---

[1] Johnson's *Eng. Dict.*

[2] 4th Past., l. 101.

[3] In Kirby's *Wonderful Museum*, ii. 309, there is an article on the Death-watch, headed "A curious Description and Explanation of the Death-watch, so commonly listened to with such dread."

but tick, tick ! The house was a huge clock, with thousands of pendulums ticking from morning till night. I was careful not to allow my great discomfort to annoy others. I argued what they could tolerate, surely I could; and in a few days habit had rendered the fearful, dreaded ticking a positive necessity."[1]

The Death-watch commences its clicking, which is nothing more than the call or signal by which the male and female are led to each other, chiefly when spring is far advanced. The sound is thus produced: Raising itself upon its hind legs, with the body somewhat inclined, it beats its head with great force and agility upon the plane of position. The prevailing number of distinct strokes which it beats in succession is from seven to nine or eleven; which circumstance, thinks Mr. Shaw, may perhaps still add, in some degree, to the ominous character which it bears. These strokes follow each other quickly, and are repeated at uncertain intervals. In old houses, where these insects abound, they may be heard in warm weather during the whole day.[2]

Baxter, in his World of Spirits, p. 203, most sensibly observes, that "there are many things that ignorance causeth multitudes to take for prodigies. I have had many discreet friends that have been affrighted with the noise called a Death-watch, whereas I have since, near three years ago, oft found by trial that it is a noise made upon paper by a little, nimble, running worm, just like a louse, but whiter and quicker; and it is most usually behind a paper pasted to a wall, especially to wainscot; and it is rarely, if ever, heard but in the heat of summer." Our author, however, relapses immediately into his honest credulity, adding: "But he who can deny it to be a prodigy, which is recorded by Melchior Adamus, of a great and good man, who had a clock-watch that had layen in a chest many years unused; and when he lay dying, at eleven o'clock, of itself, in that chest, it struck eleven in the hearing of many."

In the British Apollo, 1710, ii. No. 86, is the following query: "Why Death-watches, crickets, and weasels do come more common against death than at any other time? A. We look upon all such things as idle superstitions, for were

---

[1] Harper's *Mag.*, xxiii. 775.
[2] Shaw, *Zool.*, vi. 34. *Nat. Misc.*, iii. 104.

any thing in them, bakers, brewers, inhabitants of old houses, &c., were in a melancholy condition."

To an inquiry, ibid. vol. ii. No. 70, concerning a Death-watch, whether you suppose it to be *a living creature*, answer is given : " It is nothing but a little worm in the wood."

" How many people have I seen in the most terrible palpitations, for months together, expecting every hour the approach of some calamity, only by a little worm, which breeds in old wainscot, and, endeavoring to eat its way out, makes a noise like the movement of a watch !" Secret Memoirs of the late Mr. Duncan Campbell, 8vo. Lond. 1732, p. 61.[1]

Authors were formerly not agreed concerning the insect from which this sound of terror proceeded, some attributing it to a kind of wood-louse, others to a spider.

M. Peiguot mentions an instance where, in a public library that was but little frequented, *twenty-seven folio* volumes were perforated in a straight line by one and the same larva of a small insect (*Anobium pertinax* or *A. striatum?*) in such a manner that, on passing a cord through the perfectly round hole made by the insect, these twenty-seven volumes could be raised at once.[2]

Bostrichidæ—Typographer-beetles.

The Typographer-beetle, *Bostrichus typographus*, is so called on account of a fancied resemblance between the paths it erodes and letters. This insect bores into the fir, and feeds upon the soft inner bark; and in such vast numbers that 80,000 are sometimes found in a single tree. The ravages of this insect have long been known in Germany under the name of *Wurm trökniss*—decay caused by worms; and in the old liturgies of that country the animal itself is formally mentioned under its common appellation, *The Turk*. About the year 1665, this pest was particularly prevalent and caused incalculable mischief. In the beginning of the last century it again showed itself in the Hartz forests; it reappeared in 1757, redoubled its injuries in

---

[1] Brand's *Pop. Antiq.*, iii. 226-7.
[2] Horne's *Introd. to Bibliog.*, i. 311.

1769, and arrived at its height in 1783, when the number of trees destroyed by it in the above-mentioned forests alone was calculated at a million and a half, and the whole number of insects at work at once one hundred and twenty thousand millions. The inhabitants were threatened with a total suspension of the working of their mines, for want of fuel. At this period these *Bostrichi*, when arrived at their perfect state, migrated in swarms like bees into Suabia and Franconia. At length a succession of cold and moist seasons, between the years 1784 and 1789, very sensibly diminished the numbers of this scourge. In 1790 it again appeared, however, and so late as 1796 there was great reason to fear for the few fir-trees that were left.[1]

Cantharidæ—Blister-flies.

Many species of this family of insect possess strong vesicating powers, and are employed externally in medicine to produce blisters, and internally as a powerful stimulant. Taken internally, Pliny considered them a poison, and mentions the following instance of their causing death : Cossinus, a Roman of the Equestrian order, well known for his intimate friendship with the Emperor Nero, being attacked with lichen, that prince sent to Egypt for a physician to cure him; who recommended a potion prepared from Cantharides, and the patient was killed in consequence.[2] But there is no doubt, however, Pliny adds,

---

[1] Wilhelm's *Recr. from Nat. Hist.*, quot. by Latrielle, *Hist. Nat.*, ix. 194. Quot. by Kirb. and Sp. *Introd.*, i. 213. Carpenter, *Zool.*, ii. 133.

[2] Brookes informs us that Dr. Greenfield, a practitioner in London, was sent to Newgate, by the college, for having given Cantharides inwardly. This happened in the year 1698; but he was soon after released, by a superior authority, when he published a work upon the good effects of these insects taken inwardly for strangury, and other disorders of the kidneys and bladder. We are also told by Ambrose Parry, that a courtezan, having invited a young man to supper, had seasoned some of the dishes with the powder of Cantharides, which the very next day produced such an effect, that he died with an evacuation of blood, which the physicians were not able to stop. Many other instances might be brought, continues Brookes, of persons that have been either

that applied externally they are useful, in combination with juice of Taminian grapes, and the suet of a sheep or she-goat. They are extremely efficacious, too, continues Pliny, for the cure of leprosy and lichens; and act as an emmena-gogue and diuretic, for which last reason Hippocrates used to prescribe them for dropsy.[1]

The vesicatory principle of the Blister-fly is called *Can-tharidine*, and has been ascertained by experiment to reside more particularly in the wings than in other parts of the body. Our officinal insect is the *Cantharis vesicatoria;* and since the principal supply is from Spain, we call them commonly *Spanish-flies.* In Italy, the *Mylabris cichorii*, a native of the south of Europe, is used; and the *M. pus-tulata*, a native of China, is used by the Chinese, who also export it to Brazil, where it is the only species employed. In India also a species of *Meloe* is used,[2] possessing all the properties of the Spanish-fly.

At one time in Germany, the genus Meloe—Oil-beetles (so called from their emitting from the joints of the legs an oily yellowish liquor, when alarmed)—were extolled as a specific against hydrophobia; and the oil which is expressed from them is used in Sweden, with great success, in the cure of rheumatism, by anointing the affected part.[3] Dr. James thus enumerates the medicinal virtues of these in-sects: " The Oil-beetle (*Scarabæus unctuosus* of Schroder) is much of the nature of Cantharides, forces urine and blood, and is of extraordinary efficacy against the bite of a mad dog. Taken in powder, it cures the vari, or wandering gout, as we are assured by Wierus. The liquor is, by some, esteemed of efficacy in wounds; it is an ingredient also in plaisters for the pestilential bubo and carbuncle, and in antidotes; an oil is prepared by infusion of the living

---

killed, or brought to death's door, by a wanton use of these Flies, which had been given them privately, with a design to cause love. Some go so far as to affirm, that people have been thrown into a fever, only by sleeping under trees on which were a great number of Cantharides; and Mr. Boyle informs us, after authors worthy of credit, that some persons have felt considerable pains about the neck of the bladder, only by holding Cantharides in their hands.— *Nat. Hist. of Ins.*, p. 50–1.

[1] Pliny, *Nat. Hist.*, xxix. 30.
[2] *Asiatic Res.*, v. 213.
[3] Baird's *Cyclop. of Nat. Sci.*

animals in common oil, which some use instead of oil of Scorpions."[1] In some parts of Spain, they are mingled with the Cantharides, for the same purposes as these latter insects. Farriers also employed, in some cases, oil in which these insects had been macerated.[2]

Pliny tells us that Cato of Utica was one time reproached for selling poison, because when disposing of a royal property by auction, he sold a quantity of Cantharides, at the price of sixty thousand sesterces.[3]

The natives of Guiana and Jamaica make ear-rings and other ornaments of the elytra, or wing-coverings, of the *Cantharis maxima;* the brilliant metallic colors of which beetles, says Sloane, sparkle with an extraordinary lustre, when worn by the Indians dancing in the sun.[4]

Zoroaster says, that "Cantharides" will not hurt the vines, if you macerate some in oil, and apply it to the whet-stone on which you are going to set your pruning-knives.[5]

Cantharides are comparatively rare in Germany; yet we are told in the German Ephemerides, says Brookes, that in June, 1667, there were found about the town of Heldeshiem, such a great number of them, that they covered all the willow-trees. Likewise that in May, 1685, when the sky was serene and the weather mild, a great number of Cantharides were seen to settle upon a privet-tree, and devour all the leaves; but they did not meddle with the flowers. We are also told that the country people expect the return of these insects every seven years. It is very certain, adds Brookes, that such a number of these insects have been together in the air, that they appeared like swarms of bees; and that they have so disagreeable smell, that it may be perceived a great way off, especially about sunset, though they are not seen at that time. This bad smell is a guide for those who make it their business to catch them.[6]

---

[1] *Med. Dict.*
[2] Cuvier, *An. King.—Ins.*, i. 569.
[3] Pliny, *Nat. Hist.*, xxix. 30.
[4] Sloane, *Hist. of Jamaica*, ii. 206.
[5] Owen's *Geoponika*, ii. 156.
[6] *Nat. Hist. of Ins.*, p. 49.

Tenebrionidæ—Meal-worms.

The larvæ of the *Tenebrio molitor*, commonly called
Meal-worms, which are found in carious wood, are bred by
bird-fanciers, to feed nightingales, and constitute the only
bait by which these shy birds can be taken : a fact the more
curious when it is considered that the nightingale, in a state
of nature, can seldom or never see these larvæ.   They are
also used to feed cameleons which are exhibited.[1]

Blapsidæ—Church-yard beetle, etc.

We learn from Linnæus that in Sweden the appearance
of the Church-yard beetle, *Blaps mortisaga*, produces the
most violent alarm and trepidation among the people, who,
on account of its black hue and strange aspect, regard it
as the messenger of pestilence and death.   Hence is this
insect called *mortisaga*—the prophesier of death.[2]

A common species in Egypt, the *Blaps sulcata*, is made
into a preparation which the Egyptian women eat with the
view of acquiring what they esteem a proper degree of
plumpness ! The beetle they broil and mash up in clarified
butter; then add honey, oil of sesame, and a variety of
aromatics and spices pounded together.[3] Fabricius reports
that the Turkish women also eat this insect, cooked with
butter, to make them fat.   He also tells us that they use it
in Egypt and the Levant, as a remedy for pains and maladies in the ears, and against the bite of scorpions.[4] Carsten
Niebuhr also mentions this curious practice of the women
of Turkey, and adds, the women of Arabia likewise make
use of these insects for the same purpose, taking three of
them, every morning and evening, fried in butter.[5]

The Blatta mentioned by Pliny is evidently, from his description, the Church-yard beetle, *Blaps mortisaga*, instead

---

[1] Cuvier, *An. Kingd.—Ins.*, i. 569.
[2] Linn. *Faun. Suec.*, p. 822.
[3] Lane's *Mod. Egypt.*, i. 237, ii. 275.
[4] Cuvier, *An. King.—Ins.*, i. 568.
[5] Pinkerton's *Voy. and Trav.*, x. 190.

of the insect we now call by that name—the Cockroach:
and may very properly be here introduced. "There is
kind of fattinesse," says this author in the words of his
translator, Philemon Holland, " to bee found in the Flie or
insect called Blatta, when the head is plucked off, which,
if it be punned and mixed with Oile of Roses, is (as they
say) wonderful good for the ears: but the wooll wherein
this medicine is enwrapped, and which is put into the ears,
must not long tarrie there, but within a little while drawne
forth againe; for the said fat will very soone get life and
prove a grub or little worme. Some writers there be who
affirme, that two or three of these flies called Blattæ sodden
in oile, make a soveraigne medicine to cure the eares, and if
they be stamped and spread upon a linen rag and so applied, they will heale the eares, if they be hurt by any bruise
or contusion: Certes this is but a nastie and ill-favoured
vermine, howbeit in regard of the manifold and admirable
properties which naturally it hath, as also of the industrie
of our auncestours in searching out the nature of it, I am
moved to write thereof at large and to the full in this
place. For they have described many kinds of them. In
the first place, some of them be soft and tender, which
being sodden in oile, they have proved by experience to be
of great efficacie in fetching off werts, if they be annointed
therewith. A second sort there is, which they call Myloe-
con, because ordinarily it haunteth about mils and bake-
houses, and there breedeth: these by the report of *Musa*
and *Picton*, two famous Physicians, being bruised (after
their heads were gone) and applied to a bodie infected with
the leprosie, cured the same persitely. They of a third
kind, besides that they bee otherwise ill-favoured ynough,
carrie a loathsome and odious smell with them: they are
sharp rumped and pin buttockt also; howbeit, being incorporat with the oile of pitch called Pisselæon, they have
healed those ulcers which were thought *nunquam sana*, and
incurable. Also within one and twenty daies after this plas-
tre laid too, it hath been knowne to cure the swelling wens
called the King's evil: the botches or biles named Pani,
wounds, contusions, bruises, morimals, scabs, and fellons:
but then their feet and wings were plucked off and cast
away. I make no doubt or question, but that some of us
are so daintie and fine-eared, that our stomacke riseth at
the hearing onely of such medicines: and yet I assure you,

Diodorus, a renowned Physician, reporteth, that he has given these foure flies inwardly with rozin and honey, for the jaundise, and to those that were so streight-winded that they could not draw their breath but sitting upright. See what libertie and power over us have these Physicians, who to practise and trie conclusions upon our bodies, may exhibit unto their patients, what they list, be it never so homely, so it goe under the name of a medicine."[1]

The following extraordinary case of insects introduced into the human stomach, which is of rare occurrence, has been completely authenticated, both by medical men and competent naturalists. It was first published by Dr. Pickells, of Cork, in the Dublin Transactions.[2]

Mary Riordan, aged 28, had been much affected by the death of her mother, and at one of her many visits to the grave seems to have partially lost her senses, having been found lying there on the morning of a winter's day, and having been exposed to heavy rain during the night. It appears that when she was about fifteen, two popular Catholic priests had died, and she was told by some old woman, that if she would drink daily, for a certain time, a quantity of water, mixed with clay taken from their graves, she would be forever secure from disease and sin. So following this absurd and disgusting prescription, she took from time to time large quantities of the draught; and, some time afterward, being affected with a burning pain in the stomach (*cardialgia*), she began to eat large pieces of chalk, which she sometimes also mixed with water and drank. In all these draughts, it is most probable, she swallowed the eggs of the enormous progenies of apterous, dipterous, and coleopterous insects, which she for several years continued to throw up alive and moving. Dr. Pickells asserts that altogether he himself saw nearly 2000 of these larvæ, and that there were many he did not see, for, to avoid publicity, she herself destroyed a great number, and many, too, escaped immediately by running into holes in the floor. Of this incredible number, the greatest proportion were larvæ of the Church-yard beetle, *Blaps mortisaga*, and of a dipterous insect, an *Ascarides;* and two were specimens

---

[1] Pliny, *Nat. Hist.*, xxix. 6. Holl., p. 370.
[2] *Trans. of Assoc. Phys. in Ireland*, iv., vii., and v., p. 177, 8vo., Dublin, 1824-8.

of the Meal-worm—the larvæ of the Darkling—*Tenebrio molitor.* It may be interesting to learn that, by means of turpentine in large doses, this unfortunate woman was at length entirely rid of her pests.[1]

Curculionidæ—Weevils.

At Rio Janeiro, the brilliant Diamond-beetle, *Eutimis nobilis*, is in great request for brooches for gentlemen, and ten piasters are often paid for a single specimen. In this city many owners send their slaves out to catch insects, so that now the rarest and most brilliant species are to be had at a comparatively trifling sum. Each of these slaves, when he has attained to some adroitness in this operation, may, on a fine day, catch in the vicinity of the city as many as five or six hundred beetles. So this trade is considered there very lucrative, since six milresis (four rix dollars, or about fourteen shillings) are paid for the hundred. For these splendid insects there is a general demand; and their wing-cases are now sought for the purpose of adorning the ladies of Europe—a fashion, it is said, which threatens the entire extinction of this beautiful tribe.[2]

Messrs. Kidder and Fletcher tell us that in Brazil "a commerce is carried on in artificial flowers made from beetles' wings, fish-scales, sea-shells, and feathers, which attract the attention of every visitor. These are made," they continue, "by the *mulheres* (women) of almost every class, and thus they obtain not only pin-money, but some amass wealth in the traffic."[3] Among the beetles referred to by these gentlemen may be placed no doubt the *Eutimis nobilis.*

Among the largest of the species of this family is the Palm-weevil, *Calandra palmarum*, which is of an uniform black color, and measures more than two inches in

---

[1] In Kirby's *Wonderful Museum*, iv. 360, there are several instances of living insects being found in the human stomach, quite as extraordinary as the above.

[2] *The Mirror*, xxviii. 304.

[3] *Hist. of Brazil*, p. 346.

length. Its larva, called the *Grou-grou*,[1] or Cabbage-tree worm, which is very large, white, of an oval shape, resides in the tenderest part of the smaller palm-trees, and is considered, fried or broiled, as one of the greatest dainties in the West Indies. "The tree," says Madame Merian, "grows to the height of a man, and is cut off when it begins to be tender, is cooked like a cauliflower, and tastes better than an artichoke. In the middle of these trees live innumerable quantities of worms, which at first are as small as a maggot in a nut, but afterward grow to a very large size, and feed on the marrow of the tree. These worms are laid on the coals to roast, and are considered as a highly agreeable food."[2] Capt. Stedman tells us these larvæ are a delicious treat to many people, and that they are regularly sold at Paramaribo. He mentions, too, the manner of dressing them, which is by frying them in a pan with a very little butter and salt, or spitting them on a wooden skewer; and, that thus prepared, in taste they partake of all the spices of India—mace, cinnamon, cloves, nutmegs, etc.[3] This gentleman also says he once found concealed near the trunk of an old tree a "case-bottle filled with excellent butter," which the rangers told him the natives made by melting and clarifying the fat of this larva.[4] Dr. Winterbottom states this grub is served up at all the luxurious tables of West Indian epicures, particularly of the French, as the greatest dainty of the western world.[5]

Dobrizhoffer doubtless refers to the larva of the *Calandra palmarum*, when he says: "The Spaniards of Santiago in Tucuman, when they go seeking honey in the woods, cleave certain palm-trees upon their way, and on their return find large grubs in the wounded trees, which they fry as a delicious food."[6] The same is said of the Guaraunos of the Orinoco—"that they find these grubs in great numbers in the palms, which they cut down for the sake of their juice. After all has been drawn out that will flow, these grubs

---

[1] Jamieson gives Grou-grou as a Scottish name for the Corn-grub. —*Scot. Dict.*, iii. 516.

[2] Shaw, *Zool.*, vi. 62   Cuvier, *An. Kingd.—Ins.*, ii. 80.

[3] Stedm. *Surinam*, ii. 23.

[4] *Ibid.*, ii. 115.

[5] *Acct. of the Sierra Leone Africans*, i. 314, note.

[6] Travels, i. 410.

breed in the incisions, and the trunk produces, as it were, a second crop."[1]

The Creoles of the Island of Barbados, says Schomburgk, consider the Grou-grou worm a great delicacy when roasted, and say it resembles in taste the marrow of beef-bones.[2]

Antonio de Ulloa, in his *Noticias Americanas*, says this grub has the singular property of producing milk in women.[3] The Argentina, the historic poem of Brazil, adds an assertion which is more certainly fabulous, viz., that they first become butterflies, and then mice.[4]

They have a similar dainty in Java in the larva of some large beetle, which the natives call *Moutouke*.—"A thick, white maggot which lives in wood, and so eats it away, that the backs of chairs, and feet of drawers, although apparently sound, are frequently rotten within, and fall into dust when it is least expected. This creature may sometimes be heard at work. It is as big as a silk-worm, and very white, . . . a mere lump of fat. Thirty are roasted together threaded on a little stick, and are delicate eating."[5]

Ælian speaks of an Indian king, who, for a dessert, instead of fruit set before his Grecian guests a roasted worm taken from a plant, probably the larva of the *Calandra palmarum*, a native of Persia and Mesopotamia as well as of the West Indies, which he says the Indians esteemed very delicious—a character that was confirmed by some of the Greeks who tasted it.[6]

The trunk of the grass-tree, or black-boy, *Xanthorea arborea*, when beginning to decay, furnishes large quantities of marrow-like grubs, which are considered a delicacy by the aborigines of Western Australia. They have a fragrant, aromatic flavor, and form a favorite food among the natives, either raw or roasted. They call them *Bardi*. They are also found in the wattle-tree, or mimosa. The presence of these grubs in the *Xanthorea* is thus ascertained: if the top of one of these trees is observed to be dead, and it contain

<hr>

1 *Gummila*, i. 9.   See also Southey's *Hist. of Brazil*, i. 110.
2 *Hist. of Barbados*, p. 646.
3 *Entretenimiento*, vi. § 11.
4 Canto iii.
5 *Sketches of Java*, 310.
6 Ælian, *Hist. L.* xiv. c. 13.

any bardi, a few sharp kicks given to it with the foot will
cause it to crack and shake, when it is pushed over and the
grubs taken out, by breaking the tree to pieces with a ham-
mer. The bardi of the Xanthorea are small, and found
together in great numbers; those of the wattle are cream-
colored, as long and thick as a man's finger, and are found
singly.[1]

Dr. Livingstone states that in the valley of Quango, S.
Africa, the natives dig large white larvæ out of the damp
soil adjacent to their streams, and use them as a relish to
their vegetable diet.[2]

In the latter part of the eighteenth century, there was
published at Florence, by Prof. Gergi, the history of a remarkable insect which he names *Curculio anti-odontalgicus.*
This insect, as he assures us, not only in the name he has
given it, but also in an account of the many cures effected
by it, is endowed with the singular property of curing the
toothache. He tells us, that if fourteen or fifteen of the
larvæ be rubbed between the thumb and fore-finger, till the
fluid is absorbed, and if a carious aching tooth be but
touched with the thumb or finger thus prepared, the pain
will be removed; a finger thus prepared, he says in conclusion, will, unless it be used for tooth-touching, retain its
virtue for a year! This remarkable insect is only found on
a nondescript plant, the *Carduus spinosis-simus.*[3]

It is said, also by Prof. Gergi, that the Tuscan peasants
have long been acquainted with several insects which furnish a charm for the toothache, as the *Curculio jæcac, C.
Bacchus,* and *Carabus chrysocephaluo.*

The curious facts contained in the following quotation,
from Chambers' Book of Days, were among the first that
led me to attempt the present compilation. The scientific
name of the insect here mentioned is, in the opinion of
Prof. Gill and other scientists, a misprint for *Rhynchitus
auratus,* and, following this decision, I have here placed it
under the *Curculionidæ.*—"A lawsuit between the inhabitants of the Commune of St. Julien and a coleopterous insect, now known to naturalists as the *Eynchitus aureus,*

---

[1] Simmond's *Curiosities of Food,* p. 313.
[2] *Travels and Researches in S. Africa,* p. 389.
[3] *Monthly Mag.* ii. (Pt. II.) 792, for 1796.

lasted for more than forty-two years.   At length the inhabitants proposed to compromise the matter by giving up, in perpetuity, to the insects, a fertile part of the district for their sole use and benefit.   Of course the advocate of the animals demurred to the proposition, but the court, over-ruling the demurrer, appointed assessors to survey the land, and, it proving to be well wooded and watered, and every way suitable for the insects, ordered the conveyance to be engrossed in due form and executed.   The unfortunate people then thought they had got rid of a trouble imposed upon them by their litigious fathers and grandfathers; but they were sadly mistaken.   It was discovered that there had formerly been a mine or quarry of an ochreous earth, used as a pigment, in the land conveyed to the insects, and though the quarry had long since been worked out and exhausted, some one possessed an ancient right of way to it, which if exercised would be greatly to the annoyance of the new pro-prietors.   Consequently the contract was vitiated, and the whole process commenced *de novo*.   How or when it ended, the mutilation of the recording documents prevents us from knowing; but it is certain that the proceedings commenced in the year 1445, and that they had not concluded in 1487.   So what with the insects, the lawyers, and the church, the poor inhabitants must have been pretty well fleeced.   During the whole period of a process, religious processions and other expensive ceremonies that had to be well paid for, were strictly enjoined.   Besides, no district could commence a process of this kind unless all its arrears of tithes were paid up ; and this circumstance gave rise to the well-known French legal maxim—'The first step toward getting rid of locusts is the payment of tithes?' an adage that in all probability was susceptible of more meanings than one."[1]

Cerambycidæ—Musk-beetles.

Moufet says: "The Cerambyx, knowing that his legs are weak, twists his horns about the branch of a tree, and so he hangs at ease. . . . . . They thrust upon us some

---

[1] *Book of Days,* i.

German fables, as many as say it flies only, and when it is weary it falls to the earth and presently dies. Those that are slaves to tales, render this reason for it: Terambus, a satyrist, did not abstain from quipping of the Muses, whereupon they transformed him into a beetle called Cerambyx, and that deservedly, to endure a double punishment, for he hath legs weak that he goes lame, and like a thief he hangs on a tree. Antonius Libealis, lib. i. of his Metamorphosis, relates the matter in these words: The Muses in anger transformed Terambus because he reproached them, and he was made a Cerambyx that feeds on wood," etc.[1]

A large species of longicorn beetles, the *Acanthocinus ædilis*, is the well-known *Timerman* of Sweden and Lapland; an insect which the natives of these countries regard with a kind of superstitious veneration. Its presence is thought to be the presage of good fortune, and it is as carefully protected and cherished as storks are by the peasantry of the Low Countries.[2]

It has been found that the common cinnamon-colored Musk-beetle, *Cerambyx moschatus*, when dried and reduced to powder, and made use of as a vesicatory, in the manner of the officinal Cantharides, produces a similar effect, and in as short a space of time.[3]

The *Prionus damicornis* is a native of many parts of America and the West Indies, where its larva, a grub about three and a half inches in length, and of the thickness of the little finger, is in great request as an article of food, being considered by epicures as one of the greatest delicacies of the New World. We are informed by authors of the highest respectability, that some people of fortune in the West Indies keep negroes for the sole purpose of going into the woods in quest of these admired larvæ, who scoop them out of the trees in which they reside. Dr. Browne, in his History of Jamaica, informs us that they are chiefly found in the plum and silk-cotton trees (*Bombax*). They are commonly called by the name of *Macauco*, or *Macokkos*. The mode of dressing them is first to open and wash them, and then carefully broil them over a charcoal fire.[4]

---

[1] *Theatr. Ins.*, p. 151. Topsel's *Hist. of Beasts*, p. 1007.
[2] *The Mirror*, xxxiii. 202, note.
[3] Drury, *Ins.*, i. 9 (Pref.). Shaw's *Zool.*, vi. 73.
[4] Shaw's *Zool.*, vi. 71-2. Merian, *Ins. Sur.*, 24.

Sir Hans Sloane tells us the Indians of Jamaica boil them in their soups, pottages, olios, and pepper-pots, and account them of delicious flavor, much like, but preferable to, marrow; and the negroes of this island roast them slightly at the fire, and eat them with bread.[1]

A similar larva is dressed at Mauritius under the name of *Moutac*, which the whites as well as the negroes eat greedily.[2] According to Linnæus, the larva of the *Prionus cervicornis* is held in equal estimation; and that of the *Acanthocinus tribulus* when roasted forms an article of food in Africa.[3]

The *Cossus* of Pliny belonged most probably to this tribe, or to the *Lucanidæ.*

Wanley knew a nun in the monastery of St. Clare, who at the sight of a beetle was affected in the following strange manner. It happened that some young girls, knowing her disposition, threw a beetle into her bosom, which when she perceived, she immediately fell into a swoon, deprived of all sense, and remained four hours in cold sweats. She did not regain her strength for many days after, but continued trembling and pale.[4]

Galerucidæ—Turnip-fly, etc.

The striped Turnip-beetle, *Haltica nemorum*, commonly called the *Turnip-fly*, *Turnip-flea*, *Earth-flea-beetle*, *Black-jack*, etc., is a well known species from the ravages the perfect insect commits upon the turnip. In Devonshire, England, in the year 1786, the loss caused by these insects alone was valued at £100,000 sterling. And in the spring of 1837, the vines in the neighborhood of Montpellier were attacked to so great an extent by another species, *Haltica oleracea*, in the perfect state, that fears were entertained for the plants, and religious processions were instituted for the purpose of exorcising the insects.[5]

---

[1] *Hist. of Jamaica*, ii. 193–4.
[2] *St. Pierre, Voy*, 72.
[3] Smeatham, 32.   Kirb. and Sp. *Introd.*, i. 303.
[4] *Wonders*, i. 18.
[5] Curtis, *Farm Ins.*, p. 22.   Baird's *Cyclop. of Nat. Sci.*

Anatolius says that if the seeds of radishes, turnips, and other esculents be sown in the hide of a tortoise, the plants when grown will not be eaten by the fly, nor hurt by noxious animals or birds.[1] Paladius has also related the method of drying the seeds in the hide of this animal,[2] and of sowing them.[3]

---

[1] Owen's *Geoponika*, ii. 98.
[2] Probably the coriaceous tortoise, which is covered with a strong hide.
[3] Paladius, B. i. c. 35.

EUPLEXOPTERA.

Forficulidæ—Ear-wigs.

THE vulgar opinion that the Ear-wig, *Forficula auricularia*, seeks to introduce itself into the ear of human beings, and causes much injury to that organ, is very ancient, but not founded on fact, for they are perfectly harmless. To this opinion the names of this insect in almost all European languages point: as in English, *Ear-wig* (from Anglo-Saxon *eare*, the ear, and *wigga*, a worm; hence, also, our word *wiggle*), in French, *Perce-oreille*, and in the German, *Ohrwurm*. But, according to some writers, these names arose from the shape of the wing when expanded, which then resembles the human ear; and *ear-wig* might easily be a corruption of ear-*wing*.

Swift, in the following lines, introduces an "Ear-wig (probably a *Curculio*) in a plum," as though in allusion to some superstition:

> Doll never flies to cut her lace,
> Or throw cold water in her face,
> Because she heard a sudden drum,
> Or found an ear-wig in a plum.

"Oil of Ear-wigs," says Dr. James, "is good to strengthen the nerves under convulsive motions, by rubbing it on the temples, wrists, and nostrils. These insects, being dried, pulverized, and mixed with the urine of a hare, are esteemed to be good for deafness, being introduced into the ear."[1]

In August, 1755, in the parishes adjacent to Stroud, it is said there were such quantities of Ear-wigs, that they destroyed not only the fruits and flowers, but the cabbages, though of full growth. The houses, especially the old wooden buildings, were swarming with them: the cracks and crevices

---

[1] *Med. Dict.*

surprisingly full, so that they dropped out oftentimes in such multitudes as to literally cover the floor. Linen, of which they are fond, was likewise full, as was the furniture; and it was with caution any provisions could be eaten, for the cupboards and safes flocked with these little pests.[1]

---

[1] *Gent. Mag.*, xxv. 376.—Some authors assert that Ear-wigs are not in the least injurious to vegetation.

ORTHOPTERA.

Blattidæ—Cockroaches.

SLOANE tells us the Indians of Jamaica drink the ashes of Cockroaches in physic : bruise and mix them with sugar and apply them to ulcers and cancers to suppurate ; and are said also to give them to kill worms in children.[1] Dr. James, quoting Dioscorides, Lib. II. cap. 38, remarks : "The inside of the Blatta (*B. foetida*, Monf. 138), which is found in bake-houses, bruised or boiled in oil, and dropped into the ears, eases the pains thereof."[2] It is most probable the insect now called Blatta is not at all meant by either of the above gentlemen. The Blatta of Dioscorides is quite likely the Blatta of Pliny, which has been with good reason conjectured to be the modern *Blaps mortisaga*—the common Church-yard beetle.

In England, the hedge-hog, *Erinaceus Europæus*, from its fondness for insects and its nocturnal habits, is often kept domesticated in kitchens to destroy the Cock-roaches with which they are infested ; and the housekeepers of Jamaica, as we are informed by Sir Hans Sloane, for the same reasons and purpose, keep large spiders in their houses.[3] A species of monkey, *Simia jacchus*, and a species of lemur, *L. tardigradus*, are also made use of for destroying these insects, especially on board ships.[4] Mr. Neill, in the Magazine of Natural History, in his account of the above-mentioned species of monkey, says : " By chance we observed it devouring a large Cockroach, which it had caught running along the deck of the vessel ; and, from this time to nearly the end of the voyage, a space of four or five

---

[1] *Hist. of Jam.*, ii. 204.
[2] *Med. Dict.*
[3] *Hist. of Jam.*, ii. 204.
[4] Baird's *Cyclop. of Nat. Sci.*

weeks, it fed almost exclusively on these insects, and contributed most effectually to rid the vessel of them. It frequently ate a score of the largest kind, which are from two to two and a half inches long, and a very great number of the smaller ones, three or four times in the course of the day. It was quite amusing to see it at its meal. When he had got hold of one of the largest Cockroaches, he held it in his fore-paws, and then invariably nipped the head off first; he then pulled out the viscera and cast them aside, and devoured the rest of the body, rejecting the dry elytra and wings, and also the legs of the insect, which are covered with short stiff bristles. The small Cockroaches he ate without such fastidious nicety."[1]

The common Cockroach, or Black-beetle, as it is sometimes vulgarly called, the *Blatta orientalis*, is said originally to be a native of India, and introduced here, as well as in every other part of the civilized globe, through the medium of commerce. In England, another species, said to be a native of America, *Blatta Americana*, larger than the last, is now also becoming very common, especially in seaport towns where merchandise is stored.[2]

An old Swede, Luen Laock, one of the first Swedish clergymen that came to Pennsylvania, told the traveler Kalm, that in his younger days, he had once been very much frightened by a Cockroach, which crept into his ear while he was asleep. Waking suddenly, he jumped out of bed, which caused the insect, most probably out of fear, to strive with all its strength to get deeper into his skull, producing such excruciating pain that he imagined his head was bursting, and he almost fell senseless to the floor. Hastening, however, to the well, he drew a bucket of water, and threw some in his ear. The Roach then finding itself in danger of being drowned, quickly pushed out backward, and as quickly delivered the poor Swede from his pain and fears.[3]

The proverbial expression "Sound as a Roach" is sup-

---

[1] Quot. by Samouffle, *Ent. Cab.*, 1–3.
[2] Baird's *Cyclop. of Nat. Sci.*
[3] Pinkerton's *Voy. and Trav.*, xiii. 108. A beetle, insinuating itself in the ear of Captain Speke when in Central Africa, caused him the greatest pain imaginable. It was six or seven months before all the pieces of it were extracted.—*Blackwood's Mag.*, Sept. 1859. Barth's *Central Africa*, ii. 91, note.

posed to have been derived from familiarity with the legend and attributes of the Saint Roche,—the esteemed saint of all afflicted with the plague, a disease of common occurrence in England when the streets were narrow, and without sewers, houses without boarded floors, and our ancestors without linen. They believed that the miraculous St. Roche could make them as "sound" as himself.[1]

A quite common superstitious practice, in order to rid a house of Cockroaches, is in vogue in our country at the present time. It is no other than to address these pests a written letter containing the following words, or to this effect: "O, Roaches, you have troubled me long enough, go now and trouble my neighbors." This letter must be put where they most swarm, after sealing and going through with the other customary forms of letter writing. It is well, too, to write legibly and punctuate according to rule.

Another receipt for driving away Cockroaches is as follows: Close in an envelope several of these insects, and drop it in the street unseen, and the remaining Roaches will all go to the finder of the parcel.

It is also said that if a looking-glass be held before Roaches, they will be so frightened as to leave the premises. .

A firm, which has been established in London for seven years, and which manufactures exclusively poison known to the trade as the "Phosphor Paste for the Destruction of Black-beetles, Cockroaches, rats, mice," etc., has given to Mr. Mayhew the following information:

"We have now sold this vermin poison for seven years, but we have never had an application for our composition from any street-seller. We have seen, a year or two since, a man about London who used to sell beetle-wafers; but as we knew that kind of article to be entirely useless, we were not surprised to find that he did not succeed in making a living. We have not heard of him for some time, and have no doubt he is dead, or has taken up some other line of employment.

"It is a strange fact, perhaps; but we do not know anything, or scarcely anything, as to the kind of people and tradesmen who purchase our poison—to speak the truth, we do not like to make too many inquiries of our customers.

---

[1] Hone's *Every Day Book*, i. 1121.

Sometimes, when they have used more than their customary quantity, we have asked, casually, how it was and to what kind of business people they disposed of it, and we have always met with an evasive sort of answer. You see tradesmen don't like to divulge too much ; for it must be a poor kind of profession or calling that there are no secrets in ; and, again, they fancy we want to know what description of trades use the most of our composition, so that we might supply them direct from ourselves. From this cause we have made a rule not to inquire curiously into the matters of our customers. We are quite content to dispose of the quantity we do, for we employ six travelers to call on chemists and oilmen for the town trade, and four for the country.

"The other day an elderly lady from High Street, Camden Town, called upon us : she stated that she was overrun with black beetles, and wished to buy some of our paste from ourselves, for she said she always found things better if you purchased them of the maker, as you were sure to get them stronger, and by that means avoided the adulteration of the shopkeepers. But as we have said we would not supply a single box to any one, not wishing to give our agents any cause for complaint, we were obliged to refuse to sell to the old lady.

"We don't care to say how many boxes we sell in the year; but we can tell you, sir, that we sell more for beetle poisoning in the summer than in the winter, as a matter of course. When we find that a particular district uses almost an equal quantity all the year round, we make sure that that is a rat district; for where there is not the heat of summer to breed beetles, it must follow that the people wish to get rid of rats.

"Brixton, Hackney, Ball's Pond, and Lower Road, Islington, are the places that use most of our paste, those districts lying low, and being consequently damp. Camden Town, though it is in a high situation, is very much infested with beetles; it is a clayey soil, you understand, which retains moisture, and will not allow it to filter through like gravel. This is why in some very low districts, where the houses are built on gravel, we sell scarcely any of our paste.

"As the farmers say, a good fruit year is a good fly
8*

year; so we say, a good dull, wet summer, is a good beetle summer; and this has been a very fertile year, and we only hope it will be as good next year.

"We don't believe in rat-destroyers; they profess to kill with weasels and a lot of things, and sometimes even say they can charm them away. Captains of vessels, when they arrive in the docks, will employ these people; and, as we say, they generally use our composition, but as long as their vessels are cleared of the vermin, they don't care to know how it is done. A man who drives about in a cart, and does a great business in this way, we have reason to believe uses a great quantity of our Phosphor Paste. He comes from somewhere down the East-end or Whitechapel way.

"Our prices are too high for the street-sellers. Your street-seller can only afford to sell an article made by a person in but a very little better position than himself. Even our small boxes cost at the trade price two shillings a dozen, and when sold will only produce three shillings; so you can imagine the profit is not enough for the itinerant vendor.

"Bakers don't use much of our paste, for they seem to think it no use to destroy the vermin—beetles and bakers' shops generally go together."[1]

If a black beetle enters your room, or flies against you, severe illness and perhaps death will soon follow. I have never heard this superstition but in Maryland.

Mantidæ—Soothsayers, etc.

We now come to a very extraordinary family of insects, the *Mantidæ*. "Imagination itself," as Dr. Shaw well observes, "can hardly conceive shapes more strange than those exhibited by some particular species."[2] "They are called *Mantes;* that is, fortune-tellers," says Mouffet, "either because by their coming (for they first of all appear) they do

---

[1] *London Labor and London Poor*, iii. 40–1.
[2] *Zool.*, vi. 118.

show the spring to be at hand, as Anacreon, the poet, sang ; or else they foretell death and famine, as Cælias, the scholiast of Theocritus, writes ; or, lastly, because it always holds up its fore-feet, like hands, praying, as it were, after the manner of their divines, who in that gesture did pour out their supplications to their gods. So divine a creature is this esteemed, that if a childe aske the way to such a place, she will stretch out one of her feet and show him the right way, and seldome or never misse. As she resembleth those diviners in the elevation of her hands, so also in likeness of motion, for they do not sport themselves as others do, nor leap, nor play, but walking softly she returns her modesty, and showes forth a kind of mature gravity."[1]

The name *Mantis* is of Greek origin, and signifies diviner. In one of the Idylls of Theocritus, however, it is employed to designate a thin, young girl, with slender and elongated arms. *Præmacram ac pertenuem puellam μαντιν. Corpore prælongo, pedibus etiam prælongis, locustæ genus.*

These insects, *Mantis oratoria, religiosa*, etc., in consequence of their having, as Mouffet says, their fore-feet extended as if they were praying, are called in France, *Devin*, and *Prega-diou* or *Prechê-dieu;* and with us, *Praying-insects, Soothsayers*, and *Diviners*. They are also often called from their singular shape *Camel-crickets*.

The Mantis was observed by the Greeks in soothsaying ;[2] and the Hindoos displayed the same reverential consideration of its movements and flight.[3]

But, in modern times, the superstition respecting the sanctity of the Mantis begins in Southern Europe, and is found in almost every other quarter of the globe, at least wherever a characteristic species of the insect is found.

In the southern provinces of France, where the Mantis is very abundant, both the characters of praying and pointing out the lost way, as above mentioned by Mouffet, are still ascribed to it by the peasantry, as is evidenced by the above mentioned names they know them by. And here, as wherever else this superstition obtains, it is considered a great crime to injure the Mantis, and as, at least, a very

---

[1] *Theat. Ins.*, p. 983.
[2] Harwood, *Grec. Antiq.*, p. 200.
[3] Chamb. *Journ.*, xi. 362, 2d S.

culpable neglect not to place it out of the way of any danger to which it seems exposed.

The Turks and other Moslems have been much impressed by the actions of the common Mantis, the *religiosa*,[1] which greatly resemble some of their own attitudes of prayer. They readily recognize intelligence and pious intentions in its actions, and accordingly treat it with respect and attention, not indeed as in itself an object of reverence or superstition, but as a fellow-worshiper of God, whom they believe that all creatures praise, with more or less consciousness and intelligence.[2]

But it is in Africa, and especially in Southern Africa, that the Mantis (here the *Mantis causta*)[3] receives its highest honors. The attention of the travelers and missionaries in that quarter was necessarily much drawn to the kind of religious veneration paid to an insect, and from their accounts, though very contradictory, some curious information may be collected.

The authority of Peter Kolben, an early German traveler to the Cape of Good Hope, is as follows: That the Hottentots regard as a good deity an insect of the "beetle-kind" peculiar to their country. This "beetle-god" is described by him to be "about the size of a child's little finger, the back green, the belly speckled white and red, with two wings and two horns." He also assures us that whenever the Hottentots meet this insect, they pay it the highest honor and veneration; and that if it visits a kraal they assemble about it as if a divinity had descended among them; and even kill a sheep or two as a thank-offering, and esteem it an omen of the greatest happiness and prosperity. They believe, also, its appearance expiates all their guilt; and if the insect lights upon one of them, such person is looked upon as a saint, be it man or woman, and ever after treated with uncommon respect. The kraal then kills the fattest ox for a thank-offering; and the caul, powdered with *bukhu*, and twisted like a rope, is put on, like a collar, about the neck, and there must remain till it rots off.[4]

---

[1] Carpenter's *Zool.*, ii. 142.

[2] *Penny Mag.*, 1841, 2d S. p. 436.

[3] Cuvier, *An. Kingd.—Ins.*, ii. 190.

[4] *Present St. of the C. of Good Hope*, i. 99–100.   Astley's *Collec. of Voy. and Trav.*, iii. 366.

Kolben, in another place, describes the Mantis under the name of the *Gold-beetle*, saying that its head and wings are of a gold color, the back green, etc., as above.[1]

Mr. Kolben, again speaking of this singular reverence, remarks that the Hottentots will run every hazard to secure the safety of this fortunate insect, and are cautious to the last degree of giving it the slightest annoyance, and relates the following anecdote:

"A German, who had a country-seat about six miles from the fort, having given leave to some Hottentots to turn their cattle for awhile upon his land there, they removed to the place with their *kraal*. A son of this German, a brisk young fellow, was amusing himself in the kraal, when the deified insect appeared. The Hottentots, upon sight, ran tumultuously to adore it; while the young fellow tried to catch it, in order to see the effect such capture would produce among them. But how great was the general cry and agony when they saw it in his hands! They stared with distraction in their eyes at him, and at one another. 'See, see, see,' said they. 'Ah! what is he going to do? Will he kill it? will he kill it?' Every limb of them shaking through apprehensions for its fate. 'Why,' said the young fellow, who very well understood them, 'do you make such a hideous noise? and why such agonies for this paltry animal?' 'Ah! sir,' they replied, with the utmost concern, ''tis a divinity. 'Tis come from heaven; 'tis come on a good design. Ah! do not hurt it—do not offend it. We are the most miserable wretches upon earth if you do. This ground will be under a curse, and the crime will never be forgiven.' This was not enough for the young German. He had a mind to carry the experiment a little farther. He seemed not, therefore, to be moved with their petitions and remonstrances; but made as if he intended to maim or destroy it. On this appearance of cruelty they started, and ran to and again like people frantic; asked him, where and what his conscience was? and how he durst think of perpetrating a crime, which would bring upon his head all the curses and thunders of heaven. But this not prevailing, they fell all prostrate on the ground before the young fellow, and with streaming eyes and the loudest cries, besought

---

[1] Astley's *Col. of Voy. and Trav.*, iii. 381.

him to spare the creature and give it its liberty. The young German now yielded, and, having let the insect fly, the Hottentots jumped and capered and shouted in all the transports of joy; and, running after the animal, rendered it the customary divine honors. But the creature settled upon none of them, and there was not one sainted upon this occasion."[1]

Afterward, Mr. Kolben, discoursing with these Hottentots, took occasion to ask them concerning the utmost limit they carried the belief of the sanctity and avenging spirit of this insect, when they declared to him, that if the German had killed it, all their cattle would certainly have been destroyed by wild beasts, and they themselves, every man, woman, and child of them, brought to a miserable end. That they believed the kraal to be of evil destiny where this insect is rarely seen. Mr. Kolben asserts that they would sooner give up their lives than renounce the slightest item of their belief.[2]

Dr. Sparrman, a Swedish traveler into the country of the Hottentots and Caffres between the years 1772 and 1779, in speaking of the Mantis, called in his time the "Hottentot's God," denies the above statement of Mr. Kolben, and says the Hottentots are so far from worshiping it, that they several times caught some of them, and gave them to him to put needles through them, by way of preserving them, in the same manner as he did with the other insects. But there is, he adds, a diminutive species of this insect, which some think would be a crime, as well as very dangerous, to do any harm to, but that it was only a superstitious notion, and not any kind of religious worship.[3]

Dr. Thunberg, who traveled in South Africa about the same time as Dr. Sparrman, corroborates the latter's statement, and says he could see no reason for the supposition that the Hottentots worshiped the Mantis, but, he adds, it certainly was held in some degree of esteem, so that they would not willingly hurt, and deemed that person a creature fortunate on which it settled, though without paying it any sort of adoration.[4]

Dr. Vanderkemp, in his account of Caffraria, after describing the Mantis, says that the natives call it *oumtoani-zoulou*, the *Child of Heaven*, and adds that "the Hotten-

---

[1] *Pres. St. of the C. of Good Hope*, i. 101–2.[2] *Ibid.*
[3] *Trav.*, i. 150.[4] *Ibid.*, ii. 65.

tots regard it as almost a deity, and offer their prayers to it, begging that it may not destroy them.'"[1]

Mr. Kirchener, speaking of the same people, says they reverence a little insect, known by the name of the *Creeping Leaf*, a sight of which they conceive indicates something fortunate, and to kill it they suppose will bring a curse upon the perpetrator.[2]

Mr. Evan Evans, a missionary to the Cape of Good Hope, gives an account of a conversation which he had with the Hottentot driver of his wagon, which seems to make out the claims of the Mantis to be the God of the Hottentots—as it is even yet called. The driver directed his attention to "a small insect," which he called by its above-mentioned familiar name, and alluded to the notions he had in former times connected with it. "I asked him, 'Did you ever worship this insect then?' He answered, 'Oh, yes! a thousand times; always before I came to Bethelsdorf. Whenever I saw this little creature, I would fall down on my knees before him and pray.' 'What did you pray to him for?' 'I asked him to give me a good master, and plenty of thick milk and flesh.' 'Did you pray for nothing else?' 'No, sir; I did not then know that I wanted anything else. . . . Whenever I used to see this animal (holding the insect still in his hand) I used sometimes to fall down immediately before it; but if it was in the wagon-road, or in a foot-path, I used to push it up as gently as I could, to place it behind a bush, for fear a wagon should crush it, or some men or beasts would put it to death. If a Hottentot, by some accident, killed or injured this creature, he was sure to be unlucky all his lifetime, and could never shoot an elephant or a buffalo afterward.'"[3]

Niuhoff, in his account of his travels in Java in 1643, tells us "the Javanese set two of these little creatures (Mantes) a fighting together, and lay money on both sides, as we do at a cock-match."[4] Among the Chinese also this quarrelsome property in the genus Mantis is turned into an entertainment. They are so fond of gaming and witnessing fights between animals that, as says Mr. Barrow in his

---

[1] Quot. by *Penny Mag.*, 1841, 2d S. p. 486. [2] *Ibid.* [3] *Ibid.*
[4] Churchill's *Coll. of Voy. and Trav.*, ii. 23, and Pinkerton's *Voy. and Trav.*, xiv. 720.

Travels, "they have even extended their inquiries after fighting animals into the insect tribe, and have discovered a species of Gryllus or Locust that will attack each other with such ferocity as seldom to quit their hold without bringing away at the same time a limb of their antagonist. These little creatures are fed and kept apart in bamboo cages, and the custom of making them devour each other is so common that, during the summer months, scarcely a boy is to be seen without his cage of grasshoppers."[1]   The boys in Washington City, who call the Mantis the "Rear-horse," are also fond of this amusement.

Among the legends of St. Francis Xavier, the following is found.   Seeing a Mantis moving along in its solemn way, holding up its two fore-legs, as in the act of devotion, the Saint desired it to sing the praises of God, whereupon the insect caroled forth a fine canticle.[2]

The *Mantis religiosa* of America is said to make a most interesting pet when tamed, which can be done in a very short time and with but little pains.   Professor Glover, of the Maryland Agricultural College, tells me he once knew a lady in Washington who kept a Mantis on her window which soon grew so tame as to take readily a fly or other small insect out of her hand.   But Mrs. Taylor, in her Orthopterian Defense, has given us the particulars in full of a Mantis which she had petted.   She speaks of it under the name of "Queen Bess," and in her most interesting style, as follows :

"Queen Bess, of famous memory, would alight on my shoulder and take all her food from me half a dozen times a day.   When she omitted her visit I knew she had been hunting on her own account.   All night long she would keep watch and guard under the mosquito-net.   The silk (the thread with which the insect was bound) was fastened to the post of the bed ; and woe betide an unfortunate mosquito who fancied for his supper a drop of claret.   It was the drollest, the most laughter-moving sensation, to feel one of these trumpeters saluting your nose or forehead, and hear Queen Bess approaching with those long claws, creeping slowly, softly, nearer and nearer ; to feel the fine prick of the lancet setting in for a tipple ; then you would sup-

---

[1] *Trav. in China*, p. 159.   Cf. Williams' *Middle Kingdom*, i. 273.
[2] Ins. Arch., 63.

pose a dozen fine needles had been suddenly drawn across the part; then, *presto!* Bess's strong, saber-like claws had the jolly trumpeter tucked into her capacious jaws before you could open your eyes to ascertain the state of affairs.

"These creatures very seldom fly far," continues Mrs. Taylor, "but walk in a most stately and dignified manner. Queen Bess could not bear to be overlooked or slighted (!); and as sure as she saw me bending over the magnifier with an insect, and I thought she was ten yards off, the insect would be incontinently snapped out of my fingers. Many a valuable specimen disappeared in this way. I learned to put her at these times in the sounding-board of an Æolian harp, which was generally placed in the window. Her majesty liked music of this kind amazingly; as the vibration was *felt* though not heard. I presume she fancied she was serenaded by the singing leaves of the forest. I knew she would have remained there spell-bound until driven forth by hunger, if I did not remove her when I was not afraid of her company.

"As I have begun my 'experiences,'" continues the same writer, "I will go through with them and confess that I was obliged from circumstances to attach more than accident to her prophetic capacity—her fortune-telling. I have not a grain of superstition to contend against in other matters, having so much reverence for the Creator of all things that I certainly have no fear of anything earthly or spiritually conveyed to the senses. But I was taught by the saddest teacher, Experience, that whenever Queen Bess's refusal went unheeded I was the sufferer. The first time I ever tried it was to determine a vacillating presentiment I felt about trying a new horse whose reputation was far from good. I placed Queen Bess before me, held up my finger:

"'Attention! Queen Bess, would you advise me to try that horse?'

"She was standing on her hind legs, her antennæ erect, wings wide spread. I repeated the question. Antennæ fell; wings folded; and down she went, gradually, until her head and long thorax were buried beneath her front legs. I took her advice, and did not venture. Two days later the horse threw his rider and killed him.

"Here was the turning-point. Was I to allow such folly to master me? If French girls do take a Mantis at the junction of three roads, and ask her on which their lover will come, and watch the insect turning and examining each road with her weird sibyl head,[1]—if French girls commit such follies, should I, a staid American woman, follow their example—putting my faith in the caprices of an insect? Pshaw! I was above such folly. So the next time Queen Bess was consulted a more decided refusal was given; but I disregarded her warning, and most sorely did I repent it. Again she would approve, by standing more erect, if possible, spreading and closing her wings; then all was sunshine with me. So it went for many months. Many others have had the same experience, if they will confess it honestly. I learned to obey the hidden head more carefully than any other, I am sorry to say; and I never, in one single instance, knew her to refuse her opinion; and I never knew it to be wrong in whatever way she announced it."

This same superstitious woman says that boys and girls try their future expectations by making a mimic chariot, ballasting it with small pebbles, shot, or any such like thing, and harnessing the Mantis in with silk. Upon being freighted she rises immediately, as if to try the weight; if too heavy she will not fly. Lighten the chariot, and she will soar away to a tree or a field; then her owner is to be a lucky boy. If she will not go at all, or only a short distance, and soon come down, misfortune is to be his doom.[2]

Other superstitions among us, with respect to the Mantis are as follows:

When the Mantis (Rear-horse) kneels, it sees an angel in the way, or hears the rustle of its wings. When it alights on your hand, you are about to make the acquaintance of a distinguished person; if it alights on your head, a great honor will shortly be conferred upon you. If it injures you in any way, which it does but seldom, you will lose a valued friend by calumny. Never kill a Mantis, as it bears charms against evil.

From the great resemblance of many species of Mantis to the leaves of the trees upon which they feed, some travelers, who have observed them, have declared that they saw the leaves of trees become living creatures, and take flight.

---

[1] This superstition I have found in no other place.
[2] Harper's *New Monthly Mag.*, xxiv. 491, 2.

Madame Merian informs us of a similar opinion among the
Indians of Surinam, who believed these insects grew like
leaves upon the trees, and when they were mature, loosened
themselves and crawled, or flew away.

We find also in the works of Piso an account of insects
becoming plants. Speaking of the Mantis, that author
says: "Those little animals change into a green and tender
plant, which is of two hands breadth. The feet are fixed
into the ground first; from these, when necessary humidity
is attracted, roots grow out, and strike into the ground;
thus they change by degrees, and in a short time become a
perfect plant. Sometimes only the lower part takes the
nature and form of a plant, while the upper part remains as
before, living and movable; after some time the animal is
gradually converted into a plant. In this Nature seems to
operate in a circle, by a continual retrograde motion."[1]

There may be, however, much truth in this remarkable
metamorphosis; for, that an insect may strike root into the
earth, and, from the co-operation of heat and moisture, con-
genial to vegetation, produce a plant of the cryptogamic
kind, cannot be disputed. Westwood states that he has seen
a species of *Clavaria*, both of the undivided and branched
kinds, which had sprung from insects, and were four times
larger than the insects themselves. In truth, it cannot
then be denied that Piso may not have seen a plant of a
proportionate magnitude which had likewise grown out of
a Mantis. The pupæ of bees, wasps, and cicadas, have
been known to become the nidus of a plant, to throw up
stems from the front part of the head, and change in every
respect into a vegetable, and still retain the shell and ex-
terior appearance of the parent insect at the root. Specimens of these vegetated animals are frequently brought
from the West Indies. Mr. Drury had a beetle in the perfect state, from every part of which small stalks and fibers
sprouted forth; they were entirely different from the tufts
of hair that are observed in a few Coleopterous insects, such
as the *Buprestis fascicularius* of the Cape of Good Hope,
and were certainly a vegetable production.[2] Mr. Atwood,

---

[1] Donovan seems to think that Ovid's account of the Transformation of Phaeton's Sisters into trees, had its origin in some such idea
as this.—*Insects of China*, p. 18, note. See also Chamb. *Journal*, xi.
367, 2d Ser.

[2] Donovan's *Ins. of China*, p. 19.

in his account of Dominica, describes a "vegetable fly" as follows: "It is of the appearance and size of a small Cock-chafer, and buries itself in the ground, where it dies; and from its body springs up a small plant, resembling a young coffee-tree, only that its leaves are smaller. The plant is often overlooked, from the supposition people have of its being no other than a coffee plant, but on examining it properly, the difference is easily distinguished. . . . The head, body, and feet of the insect appearing at the foot as perfect as when alive."[1]

Dr. Colin, of Philadelphia, has mentioned, also, on the authority of a missionary, a "vegetable fly," similar to the last mentioned, on the Ohio River.[2]

The inhabitants of the Sechell Islands raise the *Mantis siccifolia*, or Dry-leaf Mantis, as an object of commerce and natural history.

Achetidæ—Crickets.

In the Island of Barbados, the natives look upon the creaking chirp of a species of Cricket, to which Hughes has given the name of the *Ash-colored* or *Sickly Cricket*, when heard in the house, as an omen of death to some one of the family.[4]

In England, also, is the Cricket's chirp sometimes looked upon as prognosticating death. "When Blonzelind expired," Gay, in his Pastoral Dirge, says,

> And shrilling Crickets in the chimney cry'd.[5]

So also in Reed's Old Plays is the Cricket's cry ominous of death:

> And the strange Cricket i' th' oven sings and hops.

The same superstition is found in the following line from the Œdipus of Dryden and Lee:

---

[1] Smith's *Nature and Art*, x. 240.
[2] *Amer. Phil. Trans.*, vol. iii. *Introd.*
[3] Cuvier, *An. Kingd.—Ins.*, ii. 173.
[4] *Nat. Hist. of Barbados*, p. 90.
[5] 4th Pastoral, line 102.

Owls, ravens, Crickets, seem the watch of death.

Gaule mentions, among other vain observations and superstitious ominations thereupon, "the Cricket's chirping behind the chimney stack, or creeping on the foot-pace."[1]

Dr. Nathaniel Horne, after saying that "by the flying and crying of ravens over their houses, especially in the dusk of evening, and when one is sick, they conclude death," adds, "the same they conclude of a Cricket crying in a house where there was wont to be none."[2]

"Some sort of people," says Mr. Ramsay, in his Elminthologia, "at every turn, upon every accident, how are they therewith terrified! If but a Cricket unusually appear, or they hear but the clicking of a Death-watch, as they call it, they, or some one else in the family, shall die!"[3]

Gilbert White, the accurate naturalist of Selborne, speaking of Crickets, says: "They are the house wife's barometer, foretelling her when it will rain; and are prognostics sometimes, she thinks, of ill or good luck, of the death of a near relation, or the approach of an absent lover. By being the constant companions of her solitary hours, they naturally become the objects of her superstition."[4]

The voice of the Cricket, says the Spectator, has struck more terror than the roaring of a lion.

Mrs. Bray also notices that the Cricket's chirp in England, which in almost all other countries, and in that too in some families, as will be shown hereafter, is considered a cheerful and a welcome note, the harbinger of joy,—is deemed by the peasantry ominous of sorrow and evil.[5]

"In Dumfries-shire," says Sir William Jardine, "it is a common superstition that if Crickets forsake a house which they have long inhabited, some evil will befall the family; generally the death of some member is portended. In like manner the presence or return of this cheerful little insect is lucky, and portends some good to the family."[6]

Melton also says,—"17. That it is a sign of death to

---

[1] *Mag-astromancers Posed and Puzzel'd*, p. 181.
[2] *Dæmonologia*, 1650, p. 59.
[3] *Elminth.*, 8vo. Lond., 1668, p. 271.
[4] *Nat. Hist. of Selborne*, p. 255.
[5] *Tamar and Tavy*, i. 321.
[6] *The Mirror*, xix. 180.

some in that house where Crickets have been many years, if on a sudden they forsake the chimney."[1]

The departure of Crickets from a hearth where they have been heard, is, at the present time, in England, considered an omen of misfortune.[2]

From the above statements of Mr. White, Mrs. Bray, and Sir William Jardine, we learn that in England the Cricket's chirp is not always ominous of evil, but sometimes also of good luck, of joy, and of the approach of an absent lover.

A correspondent of the "Notes and Queries" mentions the Cricket's cry as foreboding good luck.[3] So also a writer for "The Mirror," remarking, it is singular that the House-cricket should by some persons be considered an unlucky, by others a lucky, inmate of the mansion. Those who hold the latter opinion, he adds, consider the destruction of these insects the means of bringing misfortunes on their habitations.[4] Grose thus expresses this last superstition: Persons killing these insects (including the Lady-bird, before mentioned) will infallibly, within the course of the year, break a bone, or meet with some other dreadful misfortune.[5]

That the belief that the appearance of Crickets in a house is a good omen, and prognosticates cheerfulness and plenty, is pretty generally entertained in England, may be inferred also from the manner in which it has been embodied by Cowper, in his address to a Cricket

Chirping on his kitchen hearth.

His words are:

Whereso'er be thine abode,
Always harbinger of good.

And again in that admirable little tale of Charles Dickens, entitled "The Cricket on the Hearth," this good and happy superstition is embodied. "It's sure to bring us good fortune, John! It always has been so. To have a Cricket on the hearth is the luckiest thing in the world," says its heroine.

---

[1] *Astrologaster*, p. 45.
[2] *Notes and Queries*, iii. 3.
[3] *Ibid.*
[4] *The Mirror*, xix. 180.
[5] Grose, *Antiq. Prov. Gloss.*, p. 121.

All these superstitions are more or less entertained in America, brought here by the English themselves, and retained by their descendants. That the Cricket is the "harbinger of good," it gives me pleasure to say, is the most common.

Another superstition obtaining in this country, and particularly in Maryland and Virginia, is that Crickets are old folks and ought not therefore to be destroyed. This probably arose from Crickets being found about the kitchen hearth where the old folks were accustomed to sit.

Milton chose for his contemplative pleasures a spot where Crickets resorted:

> Where glowing embers through the room
> Teach light to counterfeit a gloom,
> Far from all resort of mirth,
> Save the Cricket on the hearth.[1]

The learned Scaliger is said to have been particularly delighted with the chirping of these animals, and was accustomed to keep them in a box for his amusement in his study.[2]

Mrs. Taylor, the writer of a very interesting series of papers on insects for Harper's Magazine, relates that in her travels through Wales, she obtained several House-crickets in the old Castle of Caernarvon. These she carried with her, in her journeyings to and fro over the Kingdom, for several years, and at last brought them to this country, where they were liberated in the snuggest corner of a Southern hearth. Again a wanderer for many years, she went back to the old house to see how her chirping friends were coming on, but, alas! she was told by the then residents, with the utmost calmness, "they had had great difficulty in *scalding* them out, and they hoped there was not one left on the premises!"[3]

In certain countries of Africa, Crickets are reported to constitute an article of commerce. Some persons rear them, feed them in a kind of iron oven, and sell them to the natives, who are very fond of their music, thinking it induces sleep.[4]

---

[1] *Il Penserosa.*
[2] Mouffet, *Theat. Insect.*, p. 136.
[3] Harper's *Mag*, xxvi. 497.
[4] Mouff. *Theat. Ins.*, p. 136.

De Pauw finds some traces of the Egyptian worship of the Scarabæus in this fondness for the music of the "holy Crickets," as he calls them, of Madagascar! By the rearing of which insects, he tells us, the Africans make a living, and the rich would think themselves at enmity with heaven, if they did not preserve whole swarms in ovens constructed expressly for that purpose.[1]

The youth of Germany, Jaeger says, are extremely fond of Field-crickets, so much so, that there is scarcely a boy to be seen who has not several small boxes made expressly for keeping these insects in. So much delighted are they, too, with their music, that they carry these boxes of Crickets into their bed-rooms at night, and are soothed to sleep with their chirping lullaby.[2]

On the contrary, others, as has been before mentioned, think there is something ominous and melancholy in the Cricket's cry, and use every endeavor to banish this insect from their houses. "Lidelius tells us," says Goldsmith, "of a woman who was very much incommoded by Crickets, and tried, but in vain, every method of banishing them from her house. She at last accidentally succeeded; for having one day invited several guests to her house, where there was a wedding, in order to increase the festivity of the entertainment, she procured drums and trumpets to entertain them. The noise of these was so much greater than what the little animals were accustomed to, that they instantly forsook their situation, and were never heard in that mansion more."[3] Like many other noisy persons, Crickets like to hear nobody louder than themselves.

In the Island of Sumatra, Capt. Stuart tells me, a black Cricket is looked upon with great respect, amounting almost to adoration. It is deemed a grievous sin to kill it.

Baskets full of Field-crickets, Lopes de Gomara says, were found among the provisions of the Indians of Jamaica when they were first discovered.[4]

"The Criquet called Gryllus," says Pliny in the words of Holland, "doth mitigat catarrhs and all asperities offending the throat, if the same bee rubbed therewith: also if a man

---

[1] De Pauw. ii. 106.
[2] *Life of Amer. Ins.*, p. 114.
[3] *Earth and Animat. Nat.*, iv. 216.
[4] Sloane's *Nat. Hist. of Jamaica*, ii. 204.

doe but touch the amygdals or almonds of the throat, with the hand wherewith he hath bruised or crushed the said Cricquet, it will appease the inflamation thereof."[1] Again, "The Cricket digged up and applied to the plase, earth and all where it lay, is very good for the ears. Nigridius," continues Pliny, "attributeth many properties to this poore creature, and esteemeth it not a little: but the Magicians much more by a faire deale: and why so? Forsooth because it goeth, as it were, reculing backward, it pierceth and boreth a hole into the ground, and never ceaseth all night long to creake very shrill.

"The manner of hunting and catching them is this, They take a flie and tie it above the middest at the end of a long haire of one's head, and so put the said flie into the mouth of the Cricquet's hole; but first they blow the dust away with their mouth, for fear lest the flie should hide herself therein; the Cricket spies the sillie flie, seaseth upon her presently and claspeth her round, and so they are both drawne foorth together by the said haire."[2]

At the present time, children in France practice the same method of capturing Crickets for amusement; substituting, however, an ant for the "sillie flie," and a long straw for "the haire of one's head." Hence comes the common proverb in France, *il est sot comme un grillon.* A ruse for capturing the larva of the Cicindela, now commonly practiced by entomologists, is founded on the same principle.

Pliny further says: "The Cricquets above rehearsed, either reduced into a liniment, or else bound too, whole as they be, cureth the accident of the lap of the eare, wounds, contusions, bruises," etc.[3]

Dr. James, quoting Schroder and Dale, says: "The ashes of the Cricket (*Gryllus domesticus*) exhibited, are said to be diuretic; the expressed juice, dropped into the eyes, is a remedy for weakness of the sight, and alleviates disorders of the tonsils, if rubbed on them."[4]

The English name *Cricket,* the French *Cri-cri,* the Dutch *Krekel,* and the Welsh *Cricell* and *Cricella,* are evidently derived from the *creak*-ing sounds of these insects.

---

1 *Nat. Hist.*, xxx. 4. Holland, p. 378. II.
2 *Ibid.*, xxix. 6. Holland, p. 370. K.
3 Pliny, *Nat. Hist.*, xxix. 6. Holl., p. 371. A.
4 *Med. Dict.*

Gryllidæ—Grasshoppers.

Mr. Hughes, after describing an ash-colored Grasshopper (which may be his ash-colored cricket before mentioned),[1] remarks that the superstitious of the inhabitants of Barbados are very apprehensive of some approaching illness to the family, whenever this insect flies into their houses in the evening or in the night.[2]

Athenæus tells us the ancient Greeks used to eat the common Grasshopper and the Monkey-grasshopper as provocatives of the appetite. Aristophanes says:

> How can you, in God's name, like Grasshoppers,
> Catching them with a reed, and Cercopes?[3]

Turpin tells us there is a kind of brown Grasshopper in Siam, which the natives consider a delicate food.[4]

"Fernandus Oniedus declareth furthermore," says Peter Martyr in his History of the West Indies, "that in a certain region called Zenu, lying fourescore and tenne miles from Darrina Eastwarde, they exercise a strange kinde of marchaundize: For in the houses of the inhabitantes they found great chests and baskets, made of twigges and leaves of certaine trees apt for that purpose, being all ful of Grasshoppers, Grilles, Crabbes, Crefishes, Snails also, and Locustes, which destroie the fields of corne, all well dried and salted. Being demanded why they reserved such a multitude of these beastes: they answered, that they kept them to be sowlde (sold) to the borderors, which dwell further within the lande, and that for the exchange of these pretious birdes, and salted fishes, they received of them certayne straunge thinges, wherein partly they take pleasure, and partly use them for the necessarie affaires."[5]

In the account of the voyages of J. Huighen Linschoten, it is stated that the inhabitants of Cumana eat "horse-

---

[1] The Grasshopper, however, according to Mr. Hughes' description, is twice as large as the cricket; it being two inches, the cricket but one inch, in length.—P. 85 and 90.

[2] *Nat. Hist. of Barb.*, p. 85.

[3] Athen. *Deipnos.* L. 4, c. 12. The Cercope, or Monkey-grasshopper, was so called from having a long tail like a monkey, *cercops.*

[4] Pinkert. *Col of Voy. and Trav.*, ix. 612.

[5] *Hist. of West Indies*, p. 121–2.

leeches, bats, Grasshopers, spiders, bees, and raw, sodden, and roasted lice. They spare no living creature whatsoever, but they eat it."[1]

"Among the choice delicacies with which the California Digger Indians regale themselves during the summer season," says the Empire County Argus, "is the Grasshopper roast. Having been an eye-witness to the preparation and discussion of one of their feasts of Grasshoppers, we can describe it truthfully. There are districts in California, as well as portions of the plains between Sierra Nevada and the Rocky Mountains, that literally swarm with Grasshoppers, and in such astonishing numbers that a man cannot put his foot to the ground, while walking there, without crushing great numbers. To the Indian they are a delicacy, and are caught and cooked in the following manner: A piece of ground is sought where they most abound, in the center of which an excavation is made, large and deep enough to prevent the insect from hopping out when once in. The entire party of Diggers, old and young, male and female, then surround as much of the adjoining grounds as they can, and each with a green bough in hand, whipping and thrashing on every side, gradually approach the center, driving the insects before them in countless multitudes, till at last all, or nearly all, are secured in the pit. In the mean time smaller excavations are made, answering the purpose of ovens, in which fires are kindled and kept up till the surrounding earth, for a short distance, becomes sufficiently heated, together with a flat stone, large enough to cover the oven. The Grasshoppers are now taken in coarse bags, and, after being thoroughly soaked in salt water for a few moments, are emptied into the oven and closed in. Ten or fifteen minutes suffice to roast them, when they are taken out and eaten without further preparation, and with much apparent relish, or, as is sometimes the case, reduced to powder and made into soup. And having from curiosity tasted, not of the soup, but of the roast, really, if one could divest himself of the idea of eating an insect as we do an oyster or shrimp, without other preparation than simple roasting, they would not be considered very bad eating, even by more refined epicures than the Digger Indians."[2]

---

[1] Voy., ii. 239. Wanley's *Wonders*, ii. 373.
[2] Quoted in Simmond's *Curios. of Food*, p. 304.

An item dated Tuesday, Aug. 21st, 1742, in the Gentleman's Magazine, states: "Great damage has been done to the pastures in the country, particularly about Bristol, by swarms of Grasshoppers; the like has happened in Pennsylvania to a surprising degree."[1]

A common species in Sweden, the *Decticus verrucivorus*, is employed by the native peasants to bite the warts on their hands; the black fluid which it emits from its mouth being supposed to possess the power of making these excrescences vanish.[2] This black fluid, from whatever Grasshoppers it may be emitted, is called by our boys "tobacco spit," which it much resembles; and they attribute to it also a wart-curing quality. When they catch one, they hold it between the thumb and fore-finger, and cry out,—

> Spit, spit tobacco spit,
> And then I'll let you go.

The exuviæ of a Grasshopper called *Semmi* or *Sebi*, Kempfer tells us, are preserved for medicinal uses, and sold publicly in shops both in Japan and China.[3]

Dr. James, quoting Dioscorides, says: "Grasshoppers (*Locusta Anglica minor, vulgatissima*, Raii *Ins.* 60.) in a suffumigation relieve under a dysury, especially such as is incident to the female sex. The Locusta Africanus is a very good antidote against the poison of the Scorpion."[4]

After describing the Grasshopper of Italy, Brookes says: "It is often an amusement among the children of that country to catch this animal; and, by tickling the belly with their finger, it will whistle as long as they chuse to make it."[5]

In France, Grasshoppers are called *Sauterelles*, Hoppers; and in Germany, *Heupferde*, Hay-horses, because they generally feed on grasses, and their head has something of the form of a horse's head.

If Grasshoppers appear early in the summer in great numbers, they foretell famine and drouth,—a superstition obtaining in Maryland.

---

[1] *Gent. Mag.*, xii. 442.
[2] Good, *Study of Med.*, iv. 515.
[3] Pinkerton's *Voy. and Trav.*, vii. 705.
[4] *Med. Dict.*
[5] *Nat. Hist. of Ins.*, p. 67.

Locustidæ—Locusts.

Moufet says : " That Locusts should be generated of the carkasse of a mule or asse (as Plutarch reports in the life of Cleonides) by putrefaction, I cannot with philosophers determine; first, because it was permitted to the Jewes to feed on them ; secondly, because no man ever yet was an eyewitness of such a putrid and ignoble generation of Locusts."[1]

The first record of the ravages of the Locusts, which we find in history, is the account in the Book of Exodus of the visitation to the land of Egypt. "And the Locusts went up over all the land of Egypt, and rested in all the coasts of Egypt—very grievous were they. . . . . For they covered the face of the whole earth, so that the land was darkened ; and they did eat every herb of the land, and all the fruit of the trees which the hail had left; and there remained not any green thing in the trees, or in the herbs of the field, through all the land of Egypt."[2]

It is to the Bible, too, we go to find the best account, for correctness and sublimity, of the appearance and ravages of these terrific insects. It is thus given by the prophet Joel : "A day of darkness and of gloominess, a day of clouds and of thick darkness, as the morning spread upon the mountains: a great people and a strong ; there hath not been ever the like, neither shall be any more after it, even to the years of many generations. A fire devoureth before them ; and behind them a flame burneth; the land is as the garden of Eden before them, and behind them a desolate wilderness ; yea, and nothing shall escape them. Like the noise of chariots[3] on the tops of mountains shall they leap, like the noise of a flame of fire that devoureth the stubble, as a strong people set in battle array. Before their faces the people shall be much pained : all faces shall gather blackness. They shall run like mighty men ; they shall climb the wall like men of war, and they shall march every one on his ways,

---

[1] *Theatr. Ins.*, p. 120.   Topsel's *Hist. of Beasts*, p. 984.

[2] Exod , chap. x.

[3] Of the symbolical Locusts in the Apocalypse it is said—"And the sounds of their wings was as the sound of chariots, of many horses running to battle."—ix. 9.

and they shall not break their ranks; neither shall one thrust another, they shall walk every one in his path; and when they fall upon the sword they shall not be wounded. They shall run to and fro in the city; they shall run upon the wall, they shall climb up upon the houses; they shall enter in at the windows like a thief. The earth shall quake before them, the heavens shall tremble; the sun[1] and the moon shall be dark, and the stars shall withdraw their shining." The usual way in which they are destroyed is also noticed by the prophet. "I will remove far off from you the northern army, and will drive him into a land barren and desolate, with his face towards the east sea, and his hinder part towards the utmost sea, and his stink shall come up, because he hath done great things."[2]

Paulus Orosius tells us that in the year of the world 3800, during the consulship of M. Plautius Hypsæus, and M. Fulvius Flaccus, such infinite myriads of Locusts were blown from the coast of Africa into the sea and drowned, that being cast upon the shore in immense heaps they emitted a stench greater than could have been produced by the carcasses of one hundred thousand men. A general pestilence of all living creatures followed. And so great was this plague in Numidia, where Micipsa was king, that eighty thousand persons died; and on the sea-coast, near Carthage and Utica, about two hundred thousand were reported to have perished. Thirty thousand soldiers, appointed as the garrison of Africa, and stationed in Utica, were among the number. So violent was the destruction that the bodies of more than fifteen hundred of these soldiers, from one gate of the city, were carried and buried in the same day.[3]

St. Augustine also mentions a plague to have arisen in Africa from the same cause, which destroyed no less than eight hundred thousand persons (*octigenta hominum millia*) in the kingdom of Masanissa alone, and many more in the territories bordering upon the sea.[4]

Blown from that quarter of the globe, Locusts have occasionally visited both Italy and Spain. The former country was severely ravaged by myriads of these desolating in-

---

[1] Cf. Ex. x. 15; Jer. xlvi. 23; Judg. vi. 5, viii. 12; Nah. iii. 15.
[2] Joel, ii. 2–10, 20.
[3] Oros, *Contra Pag.*, l. 5, c. 2.
[4] Kirb. and Sp. *Introd.*, i. 217; Cuv. *An. Kingd.—Ins.*, ii. 206.

truders, in the year 591.   These were of a larger size than
common, as we are informed by Mouffet, who quotes an
ancient historian ; and from their stench, when cast into the
sea, arose a pestilence which carried off near a million of
men and cattle.[1]

In A.D. 677, Syria and Mesopotamia were overrun by
Locusts.[2]

"About the year of our Lord 872," we read in Wanley's
Wonders, "came into France such an innumerable company
of Locusts, that the number of them darkened the very
light of the sun; they were of extraordinary bigness, had a
sixfold order of wings, six feet, and two teeth, the hardness
whereof surpassed that of stone.   These eat up every green
thing in all the fields of France.   At last, by the force of
the winds, they were carried into the sea (the Baltic) and
there drowned; after which, by the agitation of the waves,
the dead bodies of them were cast upon the shores, and from
the stench of them (together with the famine they had made
with their former devouring) there arose so great a plague,
that it is verily thought every third person in France died of
it."[3] These Locusts devoured in France, on an average every
day, one hundred and forty acres ; and their daily marches, or
distances of flight, were computed at twenty miles.[4]

In 1271, all the cornfields of Milan were destroyed ; and
in the year 1339, all those of Lombardy.[5]   We read in
Bateman's Doome, that in 1476, "grasshoppers and the
great rising of the river Isula did spoyle al Poland."   A
famine took place in the Venetian territory in 1478, occasioned by these terrific scourges, in which thirty thousand
persons are reported to have perished.   Mouffet mentions
many other instances of their devastations in Europe,—in
France, Spain, Italy, and Germany.[6]

A passage of Locusts in France, in 1613, entirely cut up,
even to the very roots, more than fifteen thousand acres of
corn in the neighborhood of Arles, and had even penetrated
into the barns and granaries, when, as it were by Providence, many hundreds of birds, especially starlings, came to

---

[1] Mouff., *Theat. Ins.*, p. 123.
[2] Shaw, *Zool.*, vi. 187.
[3] *Wonders*, ii. 507.
[4] Shaw, *Zool.*, vi. 187.   [5] *Ibid.*
[6] *Theatr. Insect.*, p. 123.

diminish their numbers. Notwithstanding this, nothing could be more astonishing than their multiplication, for the fecundity of the Locust is very remarkable. Upon an order issued by government, for the collection of their eggs, more than three thousand measures were collected, from each of which, it was calculated, would have issued nearly two millions of young ones.[1] In 1650, they entered Russia, in immense divisions, in three different places; thence passed over into Poland and Lithuania, where the air was darkened by their numbers. In many parts they lay dead to the depth of four feet. Sometimes they covered the surface of the earth like a dark cloud, loaded the trees, and the destruction which they produced exceeded all calculation.[2] In 1645, immense swarms visited the islands of Formosa and Tayowan, and caused such a famine that eight thousand persons died of hunger.[3]

"In 1649," says Sir Hans Sloane, "the Locusts destroyed all the products of the island of Teneriffe. They came from the coast of Barbary, the wind being a Levant thence. They flew as far as they could, then one alighted in the sea, and another on it, so that one after another they made a heap as big as the greatest ship above water, and were esteemed almost as many under. Those above water, next day, after the sun's refreshing them, took flight again, and came in clouds to the island, whence the inhabitants had perceived them in the air, and had gathered all the soldiers of the island and of Laguna together, being 7 or 8000 men, who laying aside their arms, some took bags, some spades, and having notice by their scouts from the hills when they alighted, they went straight thither, made trenches, and brought their bags full, and covered them with mould. . . . . After two months fruitless management of them in this manner, the ecclesiastics took them in hand by penances, etc. But all would not do: the Locusts staid their four months; cattle eat them and died, and so did several men, and others stuck out in botches. The other Canary islands were so troubled, also, that they were forced to bury their provisions. They were troubled forty years before with the like calamity."[4]

---

[1] Cuvier, *An. Kingd.—Ins.*, ii. 212.
[2] Bingley, *Anim. Biog.*, iii. 258.
[3] *Hist. of Ins.* (Murray, 1838), ii. 188.
[4] *Nat. Hist. of Jam.*, quot. in *Gent. Mag.*, xviii. 302.

Barbot, after mentioning a famine that happened in North Guinea in 1681, which destroyed many thousands of the inhabitants of the Continent, and forced many to sell themselves for slaves, to only get sustenance, says these fearful famines are also some years occasioned by the dreadful swarms of Locusts, which come from the eastward and spread over the whole country in such prodigious multitudes, that they darken the very air, passing over head like mighty clouds. They leave nothing that is green wheresoever they come, either on the ground or trees, and fly so swiftly from place to place, that whole provinces are devastated in a very short time. Barbot adds, terrific storms of hail, wind, and such like judgments from Heaven, are nothing to compare to this, which when it happens, there is no question to be made but that multitudes of the natives must starve, having no neighboring countries to supply them with corn, because those round about them are no better husbands than themselves, and are no less liable to the same calamities.[1]

Of a swarm, which in the year 1693 covered four square miles of ground, a German author has made the following estimate. Observing that, when he trod on the ground, at least three were crushed, and that in a square German measure, less than an English foot, ten were destroyed; and after determining the number of these square measures in the four miles, he concluded that ninety-two billions, one hundred and sixty millions of Locusts were congregated on the surface. This is altogether a very moderate calculation, for not only is their number more compact in breadth, but they are often piled knee-high on the earth.[2]

In 1724, Dr. Shaw was a witness of the devastations of these insects in Barbary. He has given us a description of their habits.[3] For four successive years, from 1744 to 1747, Locusts ravaged the southern provinces of Spain and Portugal.[4] In a letter from Transylvania, dated August 22d, 1747, a graphic description is given of two vast columns that overswept that country. "They form," says the writer, "a close compact column about fifteen yards deep, in breadth about four musket-shot, and in length about four leagues;

---

[1] Churchill's *Col. of Voy. and Trav.*, v. 33.
[2] *Ins.* (Murray, 1838), ii. 188.   [3] *Ibid.*, ii. 197.
[4] *Gent. Mag.*, lxx. 989.

they move with such force, or rather precipitation, that the air trembles to such a degree as to shake the leaves upon the trees, and they darkened the sky in such a manner, that when they passed over us I could not see my people at twenty feet distance."[1] This flight was four hours in passing over the Red Tower. The guards here attempted to stop them, by firing cannon at them; and where, indeed, the balls and shot swept through the swarm, they gave way and divided; but, having filled up their ranks in a moment, they proceeded on their journey.[2] In an item dated Hermanstadt, July 24, 1748, it is stated that on the day before, a hussar, coming from the plague committee, saw such a host of these insects near Szanda, that they covered the country for a mile round, and were so thick, that he was obliged to dismount from his horse, and halt for three hours, until the inhabitants of the district, coming with all sorts of instruments, beat about and forced with loud cries these pests to quit the spot.[3] In another item, dated Warsaw, August 15, 1748, it is stated that a certain prince sent out soldiers against the Locusts, who fired upon them not only with small arms, but with cannons. They succeeded in dividing the Locusts, but unluckily with the noise frightened away the storks and cranes which daily consume many of these insects.[4] Some stragglers from these swarms which so desolated Wallachia, Moldavia, Transylvania, Hungary, and Poland, in the years 1747 and '48, made their way into England, where they caused some alarm.[5] During this grand invasion of Europe, they even crossed the Baltic, and visited Sweden in 1749. Charles the Twelfth, in Bessarabia, imagined himself, it is said, assailed by a hurricane, mingled with tremendous hail, when a cloud of these insects suddenly falling, and covering both men and horses, arrested his entire army in its march.[6]

During the devastations committed by the Locusts in Spain in 1754, '55, '56, and '57, a body of them entered the

---

[1] *Phil. Trans.*, vol. xlvi., and *Gent. Mag.*, xvii. 435.   [2] *Ibid.*
[3] *Ins.* (Murray, 1838), ii. 190.
[4] *Ibid.*, 191. Dr. Shaw says, Governors of particular provinces of the East oftentimes command a certain number of the military to take the field against armies of Locusts, with a train of artillery.—*Zool.*, vi. 131, note.
[5] *Phil. Trans.*, vol. xlvi.
[6] Cuv. *An. King.*—*Ins.*, ii. 211.

church of Almaden, and devoured the silk garments that adorned the images of the saints, not sparing even the varnish on the altars.[1]

In 1750 and '53 Poland was again devastated by Locusts.[2] In June, 1772, there were several swarms of "large black flies of the Locust kind," that did incredible damage to the fruits of the earth, seen in England. Salt water, it is said, was found effectual in destroying them.[3]

From 1778 to 1780 the empire of Morocco was terribly devastated by Locusts: every green thing was eaten up, not even the bitter bark of the orange and pomegranate escaping—a most dreadful famine ensued. The poor wandered over the country, in search of a wretched subsistence from the roots of plants. They picked, from the dung of camels, the undigested grains of barley, and devoured them with eagerness. Vast numbers perished, and the streets and roads were strewed with the unburied carcasses. On this sad occasion, fathers sold their children, and husbands their wives. When they visit a country, says Mr. Jackson, from whom we have gathered the above facts, speaking of the same empire, it behooves every one to lay in provision for a famine, for they stay from three to seven years. When they have devoured all other vegetables, they attack the trees, consuming first the leaves and then the bark.[4]

To prevent the fatal consequences which would have resulted from a passage of Locusts in 1780 near Bontzhida, in Transylvania, fifteen hundred persons were ordered each to gather a sack full of the insects, part of which were crushed, part burned, and part interred. Notwithstanding this, very little diminution was remarked in their numbers, so astonishing was their multiplication, until very cold and sharp weather had come on. In the following spring there were millions of eggs disinterred and destroyed by the people, who were levied "en masse" for the operation; but notwithstanding all this, many places of tolerable extent were still to be found, in which the soil was covered with young Locusts, so that not a single spot was left naked. These

---

[1] Dillon's *Trav. in Spain*, quot. in *Ins.* (Murray, 1838), ii. 205.

[2] *Gent. Mag.*, xx. 382; xxiii. 387.

[3] *Ibid.*, xlii. 293.

[4] Jackson's *Trav. in Morocco*, p. 105. Cf. Lempriere, Pinkerton's *Col. of Voy. and Trav.*, xv. 709.

were finally, however, swept into ditches, the opposite sides of which were provided with cloths tightly stretched, and crushed.[1]

When the provincial governors of Spain are informed in the spring that Locusts have been seen, they collect the soldiers and peasants, divide them into companies and surround the district. Every man is furnished with a long broom, with which he strikes the ground, and thus drives the young Locusts toward a common center, where a vast excavation, with a quantity of brushwood, is prepared for their reception, and where the flame destroys them. Three thousand men were thus employed, in 1780, for three weeks, at Zamora; and it was reckoned that the quantity collected exceeded 10,000 bushels.[2] In 1783, 400 bushels more were collected and destroyed in the same way.[3]

Mr. Barrow informs us that in South Africa, in 1784 and 1797, two thousand square miles were literally covered by Locusts, which, being carried into the sea by a northwest wind, formed, for fifty miles along shore, a bank three or four feet high; and when the wind was in the opposite point, the horrible odor which they exhaled was perceptible a hundred and fifty miles off.[4]

The immense column of Locusts which ravaged all the Mahratta territory, and was thought to have come from Arabia, extended, Mr. Kirby's friend told him, five hundred miles, and was so dense as thoroughly to hide the sun, and prevent any object from casting a shadow. This horde was not composed of the migratory Locust, but of a red species, which imparted a sanguine color to the trees on which they settled.[5]

Mr. Forbes describes a flight of Locusts which he saw soon after his arrival at Baroche in 1779. It was more than a mile in length, and half as much in breadth, and appeared, as the sun was in the meridian, like a black cloud at a distance. As it approached, its density obscured the solar rays, causing a gloom like that of an eclipse, over the gar-

---

[1] Cuv. *An. King.*—*Ins.*, ii. 212.
[2] *Gent. Mag.*, lxii. 543.
[3] *Ibid.*, liii. 526, Pt. 1.
[4] *Trav.*, etc., 257.
[5] K. and S. *Introd.*, i. 219.

dens, and causing a noise like the rushing of a torrent. They were almost an hour in passing a given point.[1]

In another place, this traveler states that, in one considerable tract near the confines of the Brodera district, he witnessed a mournful scene, occasioned by a scourge of Locusts. They had, some time before he came, alighted in that part of the country, and left behind them, he says, "an awful contrast to the general beauty of that earthly paradise." The sad description of Hosea, he adds, was literally realized: "That which the palmer-worm hath left, hath the caterpillar eaten. They have laid waste the vine, and barked the fig-tree; they have made it clean bare, and the branches thereof are made white: the pomegranate-tree, the palm-tree also, and the apple-tree, even all the trees of the field are withered. Howl, O ye husbandmen! for the wheat and for the barley; because the harvest of the field is perished. How do the beasts groan! The herds of cattle are perplexed, because they have no pasture; yea, the flocks of sheep are made desolate!"[2]

On the 16th of May, 1800, Buchanan met with in Mysore a flight of Locusts which extended in length about three miles. He compares the noise they made to the sound of a cataract.[3] This swarm was very destructive to the young crops of jola.[4]

In 1811, at Smyrna, at right angles to a flight of Locusts, a man rode forty miles before he got rid of the moving column. This immense flight continued for three days and nights, apparently without intermission. It was computed that the lowest number of Locusts in this swarm must have exceeded 168,608,563,200,200! Captain Beaufort determined that the Locusts of this flight, which he himself saw, if framed into a heap, would have exceeded in magnitude more than a thousand and thirty times the largest pyramid of Egypt; or if put on the ground close together, in a band of a mile and an eighth in width, would have encircled the globe! This immense swarm caused such a famine in the district of Marwar, that the natives fled for subsistence in a living torrent into Guzerat and Bombay; and out of every

---

[1] *Orient. Mem.*, ii. 273.
[2] *Ibid*, iii. 338.
[3] Pinkerton's *Col. of Voy. and Trav.*, viii. 595.
[4] *Ibid.*, viii. 613.

hundred of these Marwarees, Captain Carnac estimates, ninety-nine died that year! Near the town of Baroda, these poor people perished at the rate of five hundred a day; and at Ahmedabad, a large city of two hundred thousand inhabitants, one hundred thousand died from this awful visitation![1]

In 1816, Captain Riley met with a flight of Locusts in the north of Africa, which extended in length about eight miles, and in breadth three. He tells us, also, he was informed that several years before he came to Mogadore, nearly all the Locusts in the empire, which at that time were very numerous, and had laid waste the country, were carried off in one night, and drowned in the Atlantic Ocean: that their dead carcasses a few days afterward were driven by winds and currents on shore, all along the western coast, extending from near Cape Spartel to beyond Mogadore, forming in many places immense piles on the beach: that the stench arising from their remains was intolerable, and was supposed to have produced the plague which broke out about that time in various parts of the Moorish dominions.[2] Before this plague in 1799, Mr. Jackson tells us, from Mogadore to Tangier the face of the earth was covered by them, and relates the following singular incident which occurred at El Araiche: The whole region from the confines of the Sahara was ravaged by the Locusts; but on the other side of the river El Kos not one of them was to be seen, though there was nothing to prevent their flying over it. Till then they had proceeded northward; but upon arriving at its banks they turned to the east, so that all the country north of El Araiche was full of pulse, fruits and grain, exhibiting a most striking contrast to the desolation of the adjoining district. At length they were all carried by a violent hurricane into the Western Ocean; the shore, as in former instances, was covered by their carcasses, and a pestilence (confirming the statement, and verifying the supposition of Captain Riley) was caused by the horrid stench which they emitted: but when this evil ceased, their devastations were followed by a most abundant crop.[3]

In 1825 the Russian empire was overrun to a very alarm-

---

[1] *Penny Mag.*, 1843, p. 231.
[2] *Narrative*, p. 234, and p. 238.
[3] *Trav. in Morocco*, p. 105.

ing extent by young Locusts. About Kiew, as far as the eye could reach, they lay piled up one upon another to the height of two feet. Through the government of Ekatharinoslaw and Cherson to the Black Sea, a distance of about 400 miles, they covered the ground so thickly that a horse could not walk fast through them. The sight of such an immense number, says an eye-witness, Mr. Jaeger, of the most destructive and rapacious insects, justly occasioned a melancholy foreboding of famine and pestilence, in case they should invade the cultivated and populous countries of Russia and Poland. It was at this juncture, however, that the Emperor Alexander sent his army of thirty thousand soldiers to destroy them. These forming a line of several hundred miles, and advancing toward the south, attacked them with shovels, and collected them, as far as possible, in sacks and burned them. This is the largest army of soldiers sent against Locusts we have any record of.[1]

In 1824, Locusts made their appearance at the Glen-Lynden Colony in South Africa, being the first time they had been seen there since 1808. In 1825, they continued to advance from the north; in 1826, the corn crops at Glen-Lynden were totally destroyed by them; and in 1827, 1828, and 1829, they extended their ravages through the whole of the northern and southern districts of the colony. In 1830, they again disappeared.[2]

The following graphic description of the swarm that visited Glen-Lynden in 1825 is from the pen of Mr. Pringle. He says: "In returning to Glen-Lynden, we passed through a flying swarm, which had exactly the appearance, as it approached, of a vast snow-cloud hanging on the slope of a mountain from which the snow was falling in very large flakes. When we got into the midst of them, the air all around and above was darkened as by a thick cloud; and the rushing sound of the wings of the millions of these insects was as loud as the dash of a mill-wheel. . . . . . The column that we thus passed through was, as nearly as I could calculate, about half a mile in breadth, and from two to three miles in length."[3]

---

[1] Jaeg. on Ins., p. 103.

[2] Pringle's S. Africa, p. 54. The Missionary Moffat has written the history of the scourge of 1826.—Miss. Lab , p. 447-9.

[3] Ibid.

In 1835, a plague of Locusts made their appearance in China, in the neighborhood of Quangse, and in the western departments of Quangtung. The military and people were ordered out to exterminate them, as they had done two years before. A more rational mode, however, was adopted by the authorities, of offering a bounty of twelve or fifteen cash per catty of the insects. They were gathered so fast for this price, that it was immediately lowered to five or six cash per catty. A strike followed, and the Locusts were left in quiet to do as much damage as they could.[1]

Nieuhoff tells us, Locusts in the East Indies are so destructive that the inhabitants are oftentimes obliged to change their habitations, for want of sustenance. He adds that this has frequently happened in China and the Island of Tojowac.[2]

In 1828–9, in the provinces lying between the Black and Caspian Seas, Locusts appeared in such vast numbers as were never seen in that country before.[3]

In 1839, Kaffraria was again visited by Locusts, which, together with the war at that time, caused so great a famine that many persons perished for want of subsistence.[4] Again in 1849–50, this country was visited by this dreadful scourge. The whole country, says the Rev. Francis Fleming, was covered with them; and when they arose, the cloud was so dense that this gentleman was obliged to dismount, and wait till they passed over.[5]

Mr. Jules Remy says, that at his arrival at Salt Lake, he observed upon the shore, on the top of the salt, a deposit of a foot deep which was entirely composed of dead Locusts— *Œdipoda corallipes*. These insects, driven by a high wind in prodigiously thick clouds, had been drowned in the lake, after having, during the course of the summer (of 1855), destroyed the rising crops, and even the prairie grass. A famine ensued; but the Mormons, continues Mr. Remy, only saw in this scourge a fresh proof of the truth of their religion, because it had happened, as among the Israelites, in the seventh year after their settlement in the country.[6]

---

[1] *Chinese Repository.*

[2] Churchill's *Col. of Voy. and Trav.*, ii. 317.

[3] *Penny Mag.* 1843.

[4] Backhouse, p. 264.

[5] *Kaffraria*, p. 79.

[6] Remy & Brenchley's *Voy. to G. Salt Lake City*, iv. 440, note; Burton's *City of the Saints*, p. 345.

According to Lieutenant Warren, whose graphic description is here borrowed, these devastating insects of our great western plains are "nearly the same as the Locusts of Egypt; and no one," continues this officer, "who has not traveled on the prairie, and seen for himself, can appreciate the magnitude of the swarms. Often they fill the air for many miles in extent, so that an inexperienced eye can scarcely distinguish their appearance from that of a shower of rain or the smoke of a prairie fire. The height of their flight may be somewhat appreciated, as Mr. Evans saw them above his head, as far as their size would render them visible, while standing on the top of a peak of the Rocky Mountains 8500 feet above the plain, and an elevation of 14,500 above that of the sea, in the region where the snow lies all the year. To a person standing in one of the swarms as they pass over and around him, the air becomes sensibly darkened, and the sound produced by their wings resembles that of the passage of a train of cars on a railroad, when standing two or three hundred yards from the track. The Mormon settlements have suffered more from the ravages of these insects than probably all other causes combined. They destroyed nearly all the vegetables cultivated last year at Fort Randall, and extended their ravages east as far as Iowa."[1]

The Mormons, in their simple and picturesque descriptions, say that these insects ("Crickets"—*Œdipoda corallipes*, Haldemars) are the produce of "a cross between the Spider and the Buffalo."[2]

In Egypt, in 1843, the popular idea was that the hordes of Locusts, which were then ravaging the land, were sent by the comet observed about that time for twelve days in the southwest.[3]

Pliny, in the words of his translator, Holland, says: "Many a time have the Locusts been knowne to take their flight out of Affricke, and with whole armies to infest Italie: many a time have the people of Rome, fearing a great famine and scarcity toward, beene forced to have recourse unto Sybil's bookes for remedie, and to avert the ire of the gods.

---

[1] Quot. by Burton, *City of the Saints*, p. 86.   Cf. Long's *Exped.*, ii. 31.

[2] Remy and Brenchley's *Voy. to G. S. Lake City*, i. 440, note; Burton's *City of the Saints*. p. 345.

[3] Lepsius, *Disc. in Egypt*, p. 50.

In the Cyrenaick region within Barbarie, ordained it is by law, every three years to wage warre against them, and so to conquer them. . . . . Yea, and a grievous punishment lieth upon him that is negligent in this behalf, as if hee were a traitour to his prince and countrey. Moreover, within the Island Lemnos there is a certaine proportion and measure set down, how many and what quantity every man shall kill; and they are to exhibit unto the magistrate a just and true account thereof, and namely, to shew what measure full of dead Locusts. And for this purpose they make much of Iaies, Dawes, and Choughs, whom they do honour highly, because they doe flie opposite against the Locusts, and so destroy them. Moreover in Syria, they are forced to levie a warlike power of men against them, and to make ridance by that means."[1]

Democritus says, if a cloud of Locusts is coming forward, let all persons remain quiet within doors, and they will pass over the place; but if they suddenly arrive before they are observed, they will hurt nothing, if you boil bitter lupines, or wild cucumbers, in brine, and sprinkle it, for they will immediately die. They will likewise pass over the subjacent spot, continues Democritus, if you catch some bats and tie them on the high trees of the place; and if you take and burn some of the Locusts, they are rendered torpid from the smell, and some indeed die, and some drooping their wings, await their pursuers, and they are destroyed by the sun. You will drive away Locusts, continues this same writer, if you prepare some liquor for them, and dig trenches, and besprinkle them with the liquor; for if you come there afterward, you will find them oppressed with sleep; but how you are to destroy them is to be your concern. A Locust will touch nothing, he concludes, if you pound absinthium, or a leek, or centaury with water, and sprinkle it.[2]

Didymus says, to preserve vines from that species of Locusts called by the ancients *Bruchus*, set three grains of mustard around the stem of the vine at the root; for these being thus set, have the power of destroying the Bruchus.[3]

Nieuhoff tells us that when a swarm of Locusts is seen in China, the inhabitants, to prevent their alighting, "march to and again the fields with their colors flying, shouting and

---

[1] *Nat. Hist.*, xi. 29; Holland, Pt. I. p. 327, F-II.
[2] Owen's *Geoponika*, ii. 187–8.   [3] *Ibid.*, 188.

hallooing all the while; never leaving them till they are driven into the sea, or some river, where they fall down and are drowned."[1]

Volney says, that when the Locusts first make their appearance on the frontiers of Syria, the inhabitants strive to drive them off by raising large clouds of smoke; and if, as it too frequently happens, their herbs and wet straw fail them, they dig trenches, in which they bury them in great numbers. The most efficacious destroyers of these insects are, however, he adds, the south and southeasterly winds, and the bird called the Samarmar.[2]

Capt. Riley tells us, it is said at Mogadore, and believed by the Moors, Christians, and Jews, that the Bereberies inhabiting the Atlas Mountains have the power to destroy every flight of Locusts that comes from the south, and from the east, and thus ward off this scourge from all the countries north and west of this stupendous ridge, merely by building large fires on the parts of the mountains over which the Locusts are known always to pass, and in the season when they are likely to appear, which is at a definite period, within a certain number of days in almost every year. The Atlas being high, and the peaks covered with snow, these insects become chilled in passing over them, when, seeing the fires, they are attracted by the glare, and plunge into the flame. What degree of credit ought to be attached to this opinion, Capt. Riley says he does not know, but is certain that the Moorish Sultan used to pay a considerable sum of money yearly to certain inhabitants of the sides of the Atlas, in order to keep the Locusts out of his dominions. He also adds, the Moors and Jews affirmed to him, that during the time in which the Sultan paid the said yearly stipend punctually, not a Locust was to be seen in his dominions; but that when the Emperor refused to pay the stipulated sum, because no Locusts troubled his country, and thinking he had been imposed upon, that the very same year the Locusts again made their appearance, and have continued to lay waste the country ever since.[3]

An impostor, who is believed to have been a French adventurer, at one time, it is said, endeavored to persuade the

---

[1] Pinkerton's *Col. of Voy. and Trav.*, vii. 257.
[2] Volney's *Trav.*, i. 387.
[3] Riley's *Narrative*, p. 236-7.

people of Morocco that he could destroy all the Locusts by a chemical process.[1]

The superstitious Tartars of the Crimea, in order to rid their country of its most destructive enemy, the Locusts, at one time sent over to Asia Minor, whence these insects had come, to procure Dervises to drive them away by their incantations, etc. These divines prayed around the mosques, and, as a charm, ordered water to be hung out on the minarets, which, with the prayers, were meant to entice a species of blackbird to come in multitudes and devour the Locusts! The water thus hung out is said to be still preserved in the mosques. On this occasion, the Dervises collected eighty thousand rubles, the poorest shepherd giving half a ruble.[2]

We read in "Purchas's Pilgrims," of Locusts being exorcised and excommunicated, so that they immediately flew away![3] From this interesting collection the following is clipped: "In the yeere 1603, at Fremona, great misery happened by Grasse-hoppers, from which Paez freed the Catholikes, by Letanies and sprinkling the Fields with Holy-water; when as the Fields of Heretikes, seuered only by a Ditch, were spoyled by them. Yea, a Heretike vsing this sacred sprinkling, preserued his corne, which, to a Catholike neglecting in one Field, was lost, and preserued in another by that coniured aspersion (so neere of kinne are these Locusts to the Deuill, which is said to hate Holy-water)."[4]

In the south of Europe rewards are offered for the collection both of the Locusts and their eggs; and at Marseilles, it is on record that, in the year 1613, 20,000 francs were paid for this purpose. In 1825, the same city paid a sum of 6200 francs for destroying these pests to agriculture.[5] We read in the eighty-first volume of the Gentleman's Magazine, that most of the Agricultural Societies of Italy have offered premiums for the best method of destroying Locusts: that in many districts several thousand persons are employed in searching for the eggs; that in four days the inhabitants of the .district of Ofanto collected at one time 80,000 sacks full, which were thrown into the river.[6]

---

[1] Richardson's *Sahara*, i. 338.
[2] *The Mirror*, xv. 429.
[3] *Pilgr.*, ii. 1047.
[4] *Ibid.*, ii. 1186.
[5] Baird's *Cyclop. of Nat. Sci.*
[6] *Gent. Mag.*, lxxxi. (Pt. II.) 273.

The noise Locusts make when engaged in the work of destruction has been compared to the sound of a flame of fire driven by the wind, and the effect of their bite to that of fire.[1] Volney says: "The noise they make, in browsing on the trees and herbage, may be heard at a great distance, and resembles that of an army foraging in secret." His following sentence may also be introduced here: "The Tartars themselves are a less destructive enemy than these little animals."[2] Robbins compares their noise to that of small pigs when eating corn.[3] The noise produced by their flight and approach, the poet Southey has strikingly described:

> Onward they came a dark continuous cloud
> Of congregated myriads numberless,
> The rushing of whose wings was as the sound
> Of a broad river headlong in its course
> Plunged from a mountain summit, or the roar
> Of a wild ocean in the autumn storm,
> Shattering its billows on a shore of rocks![4]

Another comparison may be introduced here, to give some idea of the infinite numbers of these insects. Dr. Clarke compares a cloud of them to a flight of snow when the flakes are carried obliquely by the wind. They covered his carriage and horses, and the Tartars assert that people are sometimes suffocated by them. The whole face of nature might have been described as covered with a living veil. They consisted of two species—*Locusta tartarica* and *L. migratoria;* the first is almost twice the size of the second, and, because it precedes it, is called by the Tartars the herald or messenger.[5]

In the Account of the admirable Voyage of Domingo Gonsales, the little Spaniard, to the World of the Moon, by Help of several Gansa's, or large Geese, we find the following: "One accident more befel me worth mention, that during my stay, I say, I saw a kind of a reddish cloud coming toward me, and continually approaching nearer, which, at last, I perceived, was nothing but a huge swarm of Locusts. He that reads the discources of learned men concerning them

---

[1] Vide Bochart, *Hierozoic*, L. IV. c. 5, 474–5.
[2] Volney, *Trav.*, i. 304.
[3] Robbins' *Journal*, p. 228.
[4] Southey's *Thalaba*, i. 171.
[5] Clarke's *Travels*, i. 348.

(as John Leo, of Africa, and others, who relate that they are seen for several days in the air before they fall on the earth), and adds thereto this experience of mine, will easily conclude that they can come from no other place than the globe of the moon.'"

To accompany this piece of satire, the following suits well:

A Chinese author, quoted by Rev. Thomas Smith, observes, that Locusts never appear in China but when great floods are followed by a very dry season; and that it is his opinion that they are hatched by the sun from the spawn of fish left by the waters on the ground![2]

So far the history of the Locust has been but a series of the greatest calamities which human nature has suffered—famine, pestilence, and death. No wonder that, in all ages and times, these insects have so deeply impressed the imagination, that almost all people' have looked on them with superstitious horror. We have shown how that their devastations have entered into the history of nations. Their effigies, too, like those of other conquerors of the earth, have been perpetuated in coins.

We are the army of the great God, and we lay ninety-and-nine eggs; were the hundredth put forth, the world would be ours—such is the speech the Arabs put into the mouth of the Locust. And such is the feeling the Arabs entertain of this insect, that they give it a remarkable pedigree, and the following description of its person: It has the head of the horse, the horns of the stag, the eye of the elephant, the neck of the ox, the breast of the lion, the body of the scorpion, the hip of the camel, the legs of the stork, the wings of the eagle, and the tail of the dragon.[3]

The Mohammedans say, that after God had created man from clay, of that which was left he made the Locust: and

---

[1] *Harleian Miscel.*, ii. 523.

[2] *Nature and Art*, vi. 109.

[3] Bochart, *Hierozoic*, Pt. II L. iv. c. 5, 475.—Much of this description is quite oriental, but such is the general resemblance to some of the animals mentioned, that in Italy it still bears the name of "Cavalletta." A German name for this Locust, as well as the Grasshopper (before mentioned), is the "Hay-horse." About the Locust's neck, too, the integuments have some resemblance to the trappings of a horse: some species, however, have the appearance of being hooded. In the Bible. Locusts are compared to horses.—Joel, ii. 4: Rev. ix. 7. Ray says, "*Caput oblongum, equi instar prona spectans.*"

in utter despair, they look upon this devastating scourge as a just chastisement from heaven for their or their nation's sins, or as directed by that fatality in which they all believe.[1]

The wings of some Locusts being spotted, were thought by many to be leaves from the book of fate, in which letters announcing the destiny of nations were to be read. Paul Jetzote, professor of Greek literature at the Gymnasium of Stettin, wrote a work on the meaning of three of these letters, which were, according to him, to be seen on the wings of those Locusts which visited Silesia in 1712. These letters were B. E. S., and formed the initials of the Latin words "Bella Erunt Sæva," or "Babel Est Solitudo;" also the German words, "Bedeutet Erschreckliche Schlacten," portending frightful battles, "Bedeutet und Erfreuliche Siege," portending happy victories. There are Greek and Hebrew sentences likewise, in which, no doubt, the professor showed as much learning, judgment, and spirit of prophecy as in those already quoted.[2]

A quite common belief in our own country is, that every Locust's wing is marked with either the letter W, portending War, or the letter P, portending Peace.

Not content with the dreadful presence of this plague, the inhabitants of most countries took that opportunity of adding to their present misery by prognosticating future evils. The direction of their flight pointed out the kingdom doomed to bow under the divine wrath. The color of the insect designated the national uniform of such armies as were to go forth and conquer.[3]

Aldrovandus states, on the authority of Cruntz, that Tamerlane's army being infested by Locusts, that chief looked on it as a warning from God, and desisted from his designs on Jerusalem.[4]

Mouffet says: "If any credit may be given to Apomasaris, a man most learned in the learning of the Indians, Persians, and Egyptians, to dream of the coming of Locusts is a sign of an army coming against us, and so much as they shall seem to hurt or not hurt us, so shall the enemy."[5]

---

[1] Riley's *Narrative*, p. 234.
[2] *Ins.* (Murray, 1838), ii. 186.
[3] *Ibid.*, 187.   [4] *Ibid.*
[5] *Theatr. Ins.*, p. 125. Topsel's *Hist. of Beasts*, p. 988.

We now turn to the history of the Locust as an article of food—a striking benefit directly derived from insects. For as they are the greatest destroyers of food, so as some recompense they furnish a considerable supply of it to numerous nations—as they cause, they are frequently the means of preventing famines. They are recorded to have done this from the remotest antiquity.

In the curious account given by Alexis of a poor Athenian family's provisions, mention of this insect is found:

> For our best and daintiest cheer,
> Through the bright half of the year,
> Is but acorns, onions, peas,
> Ochros, lupines, radishes,
> Vetches, wild pears nine and ten,
> With a Locust now and then.[1]

Diodorus Siculus, who lived about threescore years before our Saviour's birth, first, if I mistake not, described the Acridophagi, or Locust-eaters, of Ethiopia. He says they are smaller than other men, of lean and meager bodies, and exceeding black: that in the spring the south winds rise high, and drive an infinite number of Locusts out of the desert, of an extraordinary bigness, furnished with most dirty and nasty colored wings; and these are plentiful food and provision for them all their days. This historian has also given us an account of their peculiar mode of catching these insects: In their country there is a large and deep vale, extending far in length for many furlongs together: all over this they lay heaps of wood and other combustible material, and when the swarms of Locusts are driven thither by the force of the winds, then some of the inhabitants go to one part of the valley, and some to another, and set the grass and other combustible matter on fire, which was before thrown among the piles; whereupon arises a great and suffocating smoke, which so stifles the Locusts as they fly over the vale, that they soon fall down dead to the ground. This destruction of them, he continues, is continued for many days together, so that they lie in great heaps; and the country being full of salt, they gather these heaps together, and season them sufficiently with this salt, which gives them an excellent relish, and preserves them a

---

[1] St. John's *Man. and Cust. of Anct. Greeks*, iii. 95.

long time sweet, so that they have food from these insects all the year round.

Diodorus concludes his history of this people, with an account of the strange and wonderful death that comes to them at an early age, the result of eating this kind of food: They are exceeding short-lived, never living to be over forty; and when they grow old, winged lice breed in their flesh, not only of divers sorts, but of horrid and ugly shapes; that this plague begins first at the abdomen and breast, and in a short time eats and consumes the whole body. (*Phthiriasis.*)[1]

Strabo, most probably quoting from the above passage from Diodorus, speaks of a nation bordering on that of the Struthophagi, or Bird-eaters, whose food consisted entirely of Locusts, and who were carried off by the same most horrible disease.[2]

Pliny remarks: "The people of the East countries make their food of grasshoppers, even the very Parthians, who otherwise abound in wealth."[3]

The Arabs, who are compelled at the present day to inhabit the desert of Sahara, welcome the approach of Locusts as the means, oftentimes, of saving them from famishing with hunger. Robbins tells us their manner of preparing these insects for food is, by digging a deep hole in the ground, building a fire at the bottom, and filling it with wood. Then, after the earth is heated as hot as possible, and the coals and embers taken out, they prepare to fill the cavity with the live Locusts, confined in a bag holding about five bushels. Several hold the bag perpendicularly over the hole with the mouth near the surface of the ground, while others stand round with sticks. The bag is then opened, and the Locusts shaken with great force into the hot pit, while the surrounding persons immediately throw sand upon them to prevent their flying off. The mouth of the hole is now completely covered with sand, and another fire built upon the top of it. When the Locusts are thoroughly roasted and become cool, they are picked out with the hand, thrown upon tent-cloths, or blankets, and placed in the

---

[1] Diod. Sic. *Hist.*, L. III. c. 2.   Booth's Trans., 170–1.

[2] Strabo. *Geog.*, L. XVI. c. 4, § 13.

[3] *Nat. Hist.*, xi. 26.   Holl. Pt. I. p. 325. E.   Cf. Pliny, *Nat. Hist.*, xi. 29.

sun to dry. During this process, which requires two or three days, they must be watched with the utmost care, to prevent the live Locusts from devouring them, if a flight should happen to be passing at the time. When perfectly dry, they are pounded slightly, pressed into bags, or skins, and are ready for transportation. To prepare them now for present eating, they are pulverized in mortars, and mixed with water sufficient to make a kind of dry pudding. They are, however, sometimes eaten singly without pulverizing, after breaking off the head, wings, and legs. Mr. Robbins considers them nourishing food.[1]

Locusts are sometimes boiled at Wadinoon for food for men and beasts.[2]

The Arabs of Morocco, we learn from Mr. Jackson, esteem Locusts a great delicacy; and, during the summer of 1799 and the spring of 1800, after the plague had almost depopulated Barbary, dishes of them were served up at the principal repasts. Their usual way of dressing these insects, was to boil them in water half an hour, then sprinkle them with salt and pepper, and fry them, adding a little vinegar. The body of the insect is only eaten, and resembles, according to this gentleman, the taste of prawns. For their stimulating qualities, the Moors prefer them to pigeons. A person may eat a plateful of them containing two or three hundred without any ill effects.[3] In another place, however, Mr. Jackson says the poor people, when obliged to live altogether on this kind of food, become meager and indolent.[4]

In Morocco, the price of provisions falls when the Locusts have entered the neighborhood.[5]

The authority of Capt. Riley is, that Locusts are esteemed very good food by the Moors, Arabs, and Jews of Barbary, who catch large numbers of them in their season, and throw them, while alive and jumping, into a pan of boiling argan oil, where they are allowed to remain, hissing and frying, till their wings are burned off and their bodies sufficiently cooked; they are then poured out and eaten. Riley says

---

[1] Rob. *Journal*, p. 172.
[2] *Ibid.*, p. 228.
[3] Jackson's *Morocco*, p. 104.
[4] *Ibid.*, p. 106.
[5] *Wand. and Adv. in S. Afr.*, i. 137.

they resemble, in consistence and flavor, the yelks of hard-boiled hens' eggs.[1]

Capt. Beechey tells us he saw many asses, heavily laden with Locusts for food, driven into the town of Mesurata, in Tripoli.[2]

Barth, in Central Africa, saw whole calabashes filled with roasted Locusts, which, he says, occasionally form a considerable part of the food of the natives, particularly if their grain has been destroyed by this plague, as they can then enjoy not only the agreeable flavor of the dish, but also take a pleasant revenge for the ravages of their fields.[3]

Adanson, after describing an immense swarm of Locusts that covered an extent of several leagues which he saw, says the negroes of Gambia eat these insects, and have different ways of dressing them—some pounding and boiling them in milk, others only boiling them on coals.[4]

Dr. Sparrman says the Hottentots rejoice greatly upon the arrival of the Locusts, although they never fail to destroy every particle of verdure on the ground. But, continues the doctor, they make themselves ample amends for this loss, for, seizing these marauding animals, they eat them in such numbers as, in the space of a few days, to get visibly fatter and in a better condition. The females are principally eaten, especially when about to migrate, before they are able to fly, when their wings are short and their bodies heavy and distended with eggs. The soup prepared of these is of a brown coffee color, and, when cooled, from the eggs has a fat and greasy appearance.[5]

Dr. Sparrman also relates a curious notion which the Hottentots about the Visch River have with respect to the origin of the Locusts : that they proceed from the good will of a great master-conjurer a long way to the north, who, having removed the stone from the mouth of a certain deep pit, lets loose these insects in order to furnish them with food.[6] This is not unlike the account, given by the author of the Apocalypse, of the origin of the symbolical Locusts,

---

[1] Riley's *Narrat.*, p. 237.
[2] *Exped. to Africa*, p. 107.
[3] *Cent. Africa*, ii 30.
[4] *Pinkerton's Col. of Voy. and Trav.*, xvi. 634.
[5] *Travels to C. of Good Hope*, i. 263.
[6] *Ibid.*

which are said to ascend upon an angel's opening the pit of the abyss.[1]

The Korannas and Bushmen of the Cape save the Locusts in large quantities, and grind them between two stones into a kind of a meal, which they mix with fat and grease, and bake in cakes. Upon this fare, says Mr. Fleming, they live for months together, and chatter with the greatest joy as soon as the Locusts are seen approaching.[2]

Locusts in Madagascar are greatly esteemed by the natives as food.[3]

The account of the missionary Moffat differs somewhat from and is much more complete than Mr. Fleming's and Dr. Sparrman's. He says the natives of S. Africa embrace every opportunity of gathering Locusts, which can be done during the night. Whenever the cloud alights at a place not very distant from a town, the inhabitants turn out with sacks, and often with pack-oxen, gather loads, and return next day with millions. The Locusts are then prepared for eating by simple boiling, or rather steaming, as they are put into a large pot with a little water, and covered closely up; after boiling for a short time, they are taken out and spread on mats in the sun to dry, when they are winnowed, something like corn, to clear them of their legs and wings; and, when perfectly dry, are put into sacks, or laid upon the house floor in a heap. The natives eat them whole, adding a little salt when they can obtain it, or pound them in a wooden mortar; and, when they have reduced them to something like meal, they mix them with a little water and make a cold stir-about.

When Locusts abound, the natives become quite fat, and would even reward any old lady who would say that she had coaxed them to alight within reach of the inhabitants.

Mr. Moffat thinks the Locust not bad food, and, when well fed, almost as good as shrimps.[4]

The plan of gathering Locusts by night is occasionally attended with danger. "It has happened that in gathering them people have been bitten by venomous reptiles. On one occasion a woman had been traveling for several miles

[1] *Revel.* ix. 2, 3.
[2] Fleming's *Kaffraria*, p. 80.
[3] Holman's *Travels,* p. 487.
[4] *Miss. Lab.*, p. 448-9.

with a large bundle of Locusts on her head, when a serpent, which had been put into the sack with them, found its way out. The woman, supposing it to be a thong dangling about her shoulders, laid hold of it with her hand, and, feeling that it was alive, instantly precipitated the bundle to the ground and fled."[1]

Pringle, in his song of the wild Bushman, has the following lines :

> Yea, even the wasting Locust-swarm,
> Which mighty nations dread,
> To me nor terror brings nor harm ;
> I make of them my bread.[2]

Flights of Locusts are considered so much of a blessing in South Africa, that, as Dr. Livingstone states, the *rain-doctors* sometimes promised to bring them by their incantations.[3]

Carsten Niebuhr says that all Arabians, whether living in their own country or in Persia, Syria, and Africa, are accustomed to eat Locusts. They distinguish several species of insect, to which they give particular names. The red Locust, which is esteemed fatter and more succulent than any other, and accordingly the greatest delicacy, they call *Muken;* another is called *Dubbe,* but they abstain from it because it has a tendency to produce diarrhœa. A light-colored Locust, as well as the Muken, is eaten.

In Arabia, Locusts, when caught, are put in bags, or on strings, to be dried; in Barbary, they are boiled, and then dried upon the roofs of the houses. The Bedouins of Egypt roast them alive, and devour them with the utmost voracity. Niebuhr says he saw no instance of unwholesomeness in this article of food; but Mr. Forskal was told it had a tendency to thicken the blood and bring on melancholy habits. The former gentleman also says the Jews in Arabia are convinced that the fowls, of which the Israelites ate so largely of in the desert, were only clouds of Locusts, and laugh at our translators, who have supposed that they found quails where quails never were.[4]

The wild Locusts upon which St. John fed have given rise

---

[1] Quot. in Anderson's *L. Ngami,* p. 284.
[2] *Ibid.,* p. 283.
[3] *Trav. and Res. in S. Africa,* p. 48.
[4] Pinkerton's *Col. of Voy. and Trav.,* x. 189.

to great discussion—some authors asserting them to be the
fruit of the carob-tree, while others maintain they were the
true Locusts, and refer to the practice of the Arabs in Syria
at the present day. "They who deny insects to have been
the food of this holy man," says Hasselquist, "urge that
this insect is an unaccustomary and unnatural food; but
they would soon be convinced of the contrary, if they would
travel hither, to Egypt, Arabia, or Syria, and take a meal
with the Arabs. Roasted Locusts are at this time eaten by
the Arabs, at the proper season, when they can procure
them; so that in all probability this dish has been used in
the time of St. John. Ancient customs are not here subject to many changes, and the victuals of St. John are not
believed unnatural here; and I was assured by a judicious
Greek priest that their church had never taken the word in
any other sense, and he even laughed at the idea of its being a bird or a plant."[1]

Mr. Forbes incidentally remarks that in Persia and Arabia, roasted Locusts are sold in the markets, and eaten with
rice and dates, and sometimes flavored with salt and spices.[2]

The *Acridites lincola* (*Gryllus Ægypticus* of Linnæus)
is the species commonly sold for food in the markets of
Bagdad.

In fact, Locusts have been eaten in Arabia from the remotest antiquity. This is evinced by the sculptured slabs
found by Layard at Kouyunjic; for, among other attendants carrying fruit, flowers, and game, to a banquet, are seen
several bearing dried Locusts fastened on rods. And being
thus introduced in this bas-relief among the choicest delicacies, it is most probable they were also highly prized by
the Assyrians. Layard has figured one of these Locust
bearers, who upon the sculptured slab is about four and a
half feet in height.[3]

The Chinese regard the Locust, when deprived of the
abdomen, and properly cooked, as passable eating, but do
not appear to hold the dish in much estimation.[4]

Mr. Laurence Oliphant, in Tientsin, China, saw bushels
of fried Locusts hawked about in baskets by urchins in the

---

[1] Hasselq. *Trav.*, p. 419.
[2] *Orient. Mem.*, i. 46.
[3] Layard's *Nin. and Bab.*, p. 289.
[4] *Chinese Repository*.

streets. Locust-hunting, he asserts, was a favorite and profitable occupation among the juvenile part of the community. He thought the taste not unlike that of periwinkle.[1]

Williams says: "The insect food (of the Chinese) is confined to Locusts and Grasshoppers, Ground-grubs and Silkworms; the latter are fried to a crisp when cooked."[2]

Dampier says in the Bashee (Philippine) Islands, Locusts are eaten as a regular food. The natives catch them in small nets, when they come to devour their potato-vines, and parch them over the fire in an earthen pan. When thus prepared the legs and wings fall off, and the heads and backs, which before were brownish, turn red like boiled shrimps. Dampier once ate of this dish, and says he liked it well enough. When their bodies were full they were moist to the palate, but their heads cracked in his teeth.[3]

Ovalle states that in the pampas of Chili, bread is made of Locusts and of Mosquitos.[4]

According to Mr. Jules Remy, our Western Indians eat in great quantities what are generally there called *Crickets*, the *Œdipoda corallipes*.[5]

In the southern parts of France, M. Latrielle informs us, the children are very fond of the fleshy thighs of Locusts.[6]

The Arabs believe the Locusts have a government among themselves similar to that of the bees and ants; and when "Sultan Jeraad," King of the Locusts, rises, the whole mass follow him, and not a solitary straggler is left behind to witness the devastation. Mr. Jackson himself evidently believed this from the manner he has narrated it.[7] An Arab once asserted to this gentleman, that he himself had seen the great "Sultan Jeraad," and described his lordship as being larger and more beautifully colored than the ordinary Locust.[8]

Capt. Riley also mentions that each flight of Locusts is said to have a king which directs its movements with great regularity.[9]

---

[1] *Lord Elgin's Miss. to China and Japan*, p. 273.
[2] *Middle Kingdom*, ii. 50.
[3] *Voy.*, i. 430. Pinkerton's *Col. of Voy. and Trav.*, xi. 49.
[4] *Ibid.*, xiv. 128.　　　　　　[5] Vol. ii. p. 525.
[6] Cuvier, *An. King.—Ins.*, ii. 205.
[7] Jackson's *Morocco*, p. 103.　　　　　[8] *Ibid.*, p. 106.
[9] *Narrative*, p. 235.

The Chinese believe the same, and affirm that this leader is the largest individual of the whole swarm.[1]

Benjamin Bullifant, in his observations on the Natural History of New England, says: "The Locusts have a kind of regimental discipline, and as it were commanders, which show greater and more splendid wings than the common ones, and arise first when pursued by fowls, or the feet of a traveler, as I have often seriously remarked."[2]

The truth, however, is found in the Bible. They have no king.[3]

The Saharawans, or Arabs of the desert, "whose hands are against every man,"[4] and who rejoice in the evil that befalls other nations, when they behold the clouds of Locusts proceeding toward the north are filled with the greatest gladness, anticipating a general mortality, which they call *El-khere*, the good, or the benediction; for, when Barbary is thus laid waste, they emerge from their arid recesses in the desert and pitch their tents in the desolated plains.[5]

Pausanias tells us, that in the temple of Parthenon there was a brazen statue of Apollo, by the hand of Phidias, which was called Parnopius, out of gratitude for that god having once banished from that country the Locusts, which greatly injured the land. The same author asserts that he himself has known the Locusts to have been thrice destroyed by Apollo in the Mountain Lipylus, once exterminating them by a violent wind; at another time by vehement heat; and the third time by unexpected cold.[6]

At a time when there were great swarms of Locusts in China, as we learn from Navarette, the Emperor went out into his gardens, and taking up some of these insects in his hands, thus spoke to them: The people maintain themselves on wheat, rice, etc., you come to devour and destroy it, without leaving anything behind; it were better you should devour my bowels than the food of my subjects. Having concluded his speech, the monarch was about to put them in a fair way of "devouring his bowels" by swallowing them, when some that stood by telling him they were venomous,

---

[1] *Chinese Repository.*
[2] *Phil. Trans.* for 1698.
[3] *Prov.* xxx. 27.
[4] *Genes.* xvi. 12.
[5] Jackson's *Travels in Morocco*, p. 105–6.
[6] *Hist. of Greece*, b. i. c. 24.

he nobly answered, "I value not my life when it is for the good of my subjects and people to lose it," and immediately swallowed the insects. History tells us the Locusts that very moment took wing, and went off without doing any more damage; but whether or not the heroic Emperor recovered leaves us in ignorance.[1]

Mr. J. M. Jones gives the following ludicrous account of the capture of a Locust in the Bermudas. While walking one hot day in the vicinity of the barracks at St. George's, with his lamented friend, the late Col. Oakly (56th Regt.), on the lookout for insects, a very fine specimen of the Locust sprung up before them. The former chased it for a while unavailingly, but determined not to be balked of his prey; the colonel then joined in the pursuit, and after a sharp and hot chase, bagged his game right before a sentry-box; the sentry, as in duty bound, standing with arms presented, in the presence of a field officer, who was, however, in a rather undignified position to receive the salute. They had gained their prize, however, and had a hearty laugh, in which we fancy the sentry could scarcely help joining.[2]

Capt. Drayson, in his South African Sporting, tells the following anecdote: A South African, riding through a flock of Locusts, was struck in the eye by one of them, and, though blinded momentarily in the injured eye, he still kept the other on the insect, which sought to escape by diving among the crowd on the ground. So, dismounting, he captured it, passed a large pin through its body, and thrust it in his waistcoat pocket; and whenever the damaged eye smarted, he pulled it out again, and stuck the pin through it in a fresh place.[3]

Darwin tells us that when the "Beagle" was to windward of the Cape de Verd Islands, and when the nearest point of land, not directly opposed to the trade-wind, was Cape Blanco on the coast of Africa, 370 miles distant, a large Grasshopper—*Acrydium*—flew on board![4] But Sir Hans Sloane mentions a much more remarkable flight in his History of Jamaica; for when the Assistance frigate was about

---

[1] *Hist. Acct. of China*, b. ii. c. 15, and Church *Col. of Voy. and Trav.*, i. 95.

[2] *Naturalist in Bermuda*, p. 112.

[3] *S. African Sport.*, p. 220.

[4] Darwin's *Res.*, p. 159.

300 leagues to windward of Barbados, he says a Locust alighted on the forecastle among the sailors![1]

Several species of Locusts are beautifully marked; these were sought after by young Jewish children as playthings.[2]

The eggs of the *Chargol* Locust, *Truxalis nasuta?*, the Jewish women used to carry in their ears to preserve them from the earache.[3]

The word *Locust*, Latin *Locusta*, is derived by the old etymologists from *locus*, a place, and *ustus*, burned,—"quod tactu multa *urit* morsu vero omnia erodat." True Locusts are the *Acridium*, or *Criquets*, of Geoffroy, and the *Gryllus* of Fabricius. The Migratory-locust, *Locusta migratoria*, a rather small insect, is the most celebrated species of the family. To it almost all the devastations before mentioned have been attributed. It is most probable, however, many species have been confounded under the same name.

In Spain, as we are told by Osbeck, the people of fashion keep a species of Locust—called there *Gryllo*—in cages—*grillaria*,—for the sake of its song.[4] De Pauw says that, like Canary birds, they were kept in cages to sing during the celebration of mass.[5]

The song of a Spanish Gryllo on one occasion, if we may credit the historian, was the means of saving a vessel from shipwreck. The incident evinces the perilous situation of Cabeza de Vara, in his voyage toward Brazil, and is related by Dr. Southey in his history of that country as follows:

"When they had crossed the Line, the state of the water was inquired into, and it was found, that of a hundred casks there remained but three, to supply four hundred men and thirty horses. Upon this, the Adelantado gave orders to make for the nearest land. Three days they stood toward it. A soldier, who had set out in ill health, had brought a Gryllo, or ground cricket, with him from Cadiz, thinking to be amused by the insect's voice; but it had been silent the whole way, to his no little disappointment. Now, on the fourth morning, the Gryllo began to sing its shrill rattle, scenting, as it was immediately supposed, the land. Such

---

[1] *Hist. of Jam.*, ii. 201.
[2] Smith's *Bib. Dict.*
[3] *Ibid.*
[4] *Travels*, i. 71.
[5] *Egypt and China*, ii. 106.

was the miserable watch which had been kept, that upon looking out at the warning, they perceived high rocks within bowshot; against which, had it not been for the insect, they must inevitably have been lost. They had just time to drop anchor. From hence they coasted along, the Gryllo singing every night, as if it had been on shore, till they reached the Island of St. Catalina."[1]

To account for the singular sound produced by the *Platyphyllon concavum*, which much resembles the expression *Katy did*, so much so that the insect is now called the Katydid,—a curious legend is told in this country, and particularly in Virginia and Maryland. Mrs. A. L. Ruyter Dufour has kindly embodied it in the following verses for me:

> Two maiden sisters loved a gallant youth,
>   Once in the far-off days of olden time:
> With all of woman's fervency and truth ;—
>   So runs a very ancient rustic rhyme.
>
> Blanche, chaste and beauteous as a Fairy-queen,
>   Brave Oscar's heart a willing captive led;
> Lovely in soul as was her form and mien,
>   While guileless love its light around her shed.
>
> A Juno was the proud and regal Kate,—
>   Her love thus scorn'd, her beauty thus defied,
> Like Juno's turn'd her love to vengeful hate :—
>   Mysteriously the gallant Oscar died.
>
> Bereft of reason, faithful Blanche soon lay ;—
>   The mystery of this fearful fate none knew,
> Save proud, revengeful Kate, who would not say
>   It was her hand had dared the deed to do.
>
> Justice and pity then to Jove appealed,
>   That the dark secret be no longer hid :
> Young Oscar's spirit he at once concealed,
>   That cries, each summer night, *Kate, Katy-did!*

Rose Hill, D.C., June 24, 1864.

If a Katy-did enters your house, an unlooked-for visitor will speedily come. If it sings there, some of your family will be noted for fine musical powers. These superstitions obtain in Maryland.

---

[1] *Hist. of Brazil*, i. 105.

ORDER IV.

NEUROPTERA.

Termitidæ—White-ants.

THE Termites or White-ants (which are *ants* only by a misnomer) are found in both the Indies, in Africa, and in South America, where they do vast damage, in consequence of their eating and perforating wooden buildings, utensils, furniture, and indeed all kinds of household stuff, which are utterly destroyed by them if not timely prevented. They are found also in Europe, and, about thirty years ago, from the extent of their ravages in the West of France, and particularly at Rochelle, caused considerable alarm.[1]

There is a story commonly told, if not commonly credited throughout India, of the Termites demolishing a chest of dollars at Bencoolen, which is in a great degree cleared up by the following anecdote introduced by Mr. Forbes in his Memoirs: A gentleman having charge of a chest of money, unfortunately placed it on the floor in a damp situation; and, as a matter of course in that climate, the box was speedily attacked by the Termites, which had their burrow just under the place the treasure stood. Soon annihilating the bottom, these devouring insects were not any more ceremonious in respect to the bags containing the specie; which, being thus let loose, fell piece by piece gradually into the hollows in the Termites' burrow. When the cash was demanded, and not to be found, all were greatly amazed at the wonderful powers, both of teeth and stomachs, of the little marauders, which were supposed to have consumed the silver and gold as well as the wood. But, after some years, however, the house requiring repair, the whole sum was found several feet deep in the earth; and, thanks, the

---

[1] Baird's *Cyclop. of Nat. Sci.* The species here referred to was the *Termes lucifuga.*

(132)

Termites were rescued from that obloquy which the supposed power of feasting on precious metals had cast on their whole race.[1]

Kempfer, during his stay at a Dutch fort on the coast of Malabar, one morning discovered some peculiar marks like arches upon his table, about the size of his little finger. Suspecting they were the work of Termites, he made an accurate examination, and, much to his surprise, found not only what he expected to be true, but that these voracious insects had pierced a passage of that thickness up one leg of the table, then across the table, and so down again through the middle of another leg into the floor! What made it the more wonderful was that it had all been done in the few hours that intervened between his retiring to rest and his rising.[2]

Mr. Forbes, on surveying a room which had been locked up during an absence of a few weeks, observed a number of advanced works in various directions toward some prints and drawings in English frames; the glasses appeared to be uncommonly dull, and the frames covered with dust. "On attempting," says he, "to wipe it off, I was astonished to find the glasses fixed on the wall, not suspended in frames as I left them, but completely surrounded by an incrustation cemented by the White-ants, who had actually eaten up the deal frames and back-boards, and the greater part of the paper, and left the glasses upheld by the incrustation, or covered way, which they had formed during their depredation."[3]

It is even asserted, says Kirby and Spence, that the superb residence of the Governor-general at Calcutta, which cost the East India Company such immense sums, is now going rapidly to decay in consequence of the attacks of these insects. But not content with the dominions they have acquired, and the cities they have laid low on Terra Firma, encouraged by success, the White-ants have also aimed at the sovereignty of the ocean, and once had the hardihood to attack even a British ship of the line—the Albion; and, in spite of the efforts of her commander and his valiant crew, having boarded they got possession of her, and handled her

---

[1] *Orient. Mem.*, i. 303-4.

[2] Kempf. *Japan*, ii. 127; also Pinkerton's *Col. of Voy. and Trav.*, vii 701.

[3] *Orient. Mem.*, i. 362.

so roughly, that when brought into port, being no longer fit for service, she was obliged to be broken up.[1]

Lutfullah, in his Autobiography, relates the following: "I returned the couch kindly sent to me by a friend, with my thanks, and made my bed on the ground, placing my new desk of Morocco leather at the head to serve as a pillow, and went to bed. In the morning, when roused by the bugle, I found my bed strewed with damp dust, my skin excoriated in some parts, and my back irritated in others. I called my servant, who was saddling my horse. 'Mahdilli,' said I angrily, 'you have been throwing dust all over my bed and self, in shaking the trappings of the horse near my bed in the tent.'—'No, sir, I have done no such thing,' was his reply. When I took up my cloak it fell to pieces in my hand ; the blanket was in the same state, and the bottom of my desk, with some valuable papers, were destroyed. 'What misfortune is this?' cried I to Mahdilli, who immediately brought a burning stick to examine the cause, and coolly observed, 'It is the White-ants, sir, and no misfortune, but a piece of bad luck, sir.' Poor man ! in all mishaps, I always found him attaching blame to destiny, and never to his own or my imprudence."[2]

The Caffres, as we are informed by Mr. Latrobe, when first permitted to settle at Guadenthal, before they could build ovens, according to the custom of their country, availed themselves of the Ant-hills found in that neighborhood; for, having destroyed the inhabitants by fire and smoke, they scooped them out hollow, leaving a crust of a few inches in thickness, and used them for baking, putting in three loaves at a time.[3]

Mr. Southey says that in Brazil the Spaniards hollow out the nests of the Termites, and use them for ovens.[4] The authority of Messrs. Kidder and Fletcher is, that in Brazil, "the Termites' dwelling is sometimes overturned by the slaves, the hollow scooped wider, and is then used as a bake-oven to parch Indian-corn."[5]

Mr. Latrobe also tells us that the clay of which these

[1] *Introd.*, i. 247.
[2] *Autobiog.*, Lond., 1858, p. 222–3.
[3] *Latr. S. Africa*, p. 315.
[4] *Hist. of Brazil*, i. 319.
[5] Kid. and Fletch., *Brazil*, p. 443.

Ant-hills are formed, is so well prepared by the industrious Termites, *Termes bellicosus*, that it is used for the floors of rooms in South Africa both by the Hottentots and farmers.[1]

Mr. Southey states that in Brazil "the Spaniards pulverize the nests of the Termites, and with the powder form a flooring for their houses, which becomes as hard as stone, and on which it is said no fleas or other insects will harbor."[2] The early Spanish settlers built the walls of their houses of the same earth ; and some of which, which were erected in the seventeenth century, are said to be still in existence.[3]

Ant-hills, or rather the Termites which inhabit them, have also been used as an instrument of perhaps the most infernal torture the ingenuity of man has ever invented. For, in South Africa, at one time, the wretched victim, whether prisoner of war or offending subject, having been smeared with some oily substance, was partially interred in one of these heaps, and, if not first roasted to death by the burning sun, was literally devoured alive by the myriads of insects which have their habitation there. It has been asserted that even some Englishmen have met this dreadful fate.[4]

At Unyamwezi, in the lake regions of Central Africa, the natives chew the clay of Ant-hills as a substitute when their tobacco fails. They call this clay "sweet earth." It is said the Arabs have also tried it without other effects than nausea.[5]

The goldsmiths of Ceylon employ the powdered clay of Ant-hills in preference to all other substances in the preparation of crucibles and moulds for their fine castings, for so delicate is the trituration to which the Termites subject this material;[6] and Knox says, "the people use this finer clay to make their earthen gods of, it is so pure and fine."[7]

Termites, as an article of food, are eaten by the inhabitants of many countries. Mr. Kœnig, in his essay on the history of these insects, read before the Society of Natural-

---

[1] *S. Africa*, p. 315.
[2] *Hist. of Brazil*, i. 319.
[3] Kidder and Fletcher, *Brazil*, p. 442.
[4] Barter's *Dorp and Veld*, p. 81.
[5] Burton's *Central Africa*, i. 202.
[6] Tennent, *Nat. Hist. of Ceylon*, p. 412.
[7] Knox, *Ceylon*, Pt. I. ch. vi. p. 24.

ists of Berlin, tells us, that to catch the Termites before
their emigration, the natives of the East Indies make two
holes in the nest, one to windward, and the other to leeward ;
at the latter aperture, they place a pot, rubbed with aromatic
herbs.   On the windward side they make a fire, the smoke
of which drives these insects into the pots.   By this method
they take a great quantity, of which they make, with flour,
a variety of pastry, which they sell to the poorer people.
This author adds, that in the season in which this aliment
is abundant, the abuse of it produces an epidemic colic and
dysentery, which carries off the patient in two or three
hours.[1]

The Africans, says Mr. Smeatham, are less ingenious in
catching and preparing them.   They content themselves in
collecting those which fall into the water at the time of
emigration.   They skim them off the surface with calabashes,
filling large caldrons with them, then grill them in iron
pots, over a gentle fire, stirring them as coffee is stirred.
They thus eat them by handfuls, without sauce, or any other
preparation, and find them delicious.   This gentleman has
several times eaten them cooked in this manner, and thinks
them delicate, nourishing, and wholesome, being sweeter than
the grub of the palm-tree weevil (*Calandra palmarum*),
and resembling in taste sugared cream or sweet almond
paste.[2]

The Hottentots, Dr. Sparrman informs us, eat them
greedily boiled and raw, and soon grow fat and plump upon
this food.[3]

An idea may be formed of this dish by what once occurred
to Dr. Livingstone on the banks of the Zouga, in South
Africa.   The Bayeiye chief Palani visiting this traveler while
eating, he gave him a piece of bread and preserved apricots ;
and as the chief seemed to relish it much, he asked him if
he had any food equal to that in his country.   "Ah !" said
the chief, "did you ever taste White-ants?"   As the doctor
never had. he replied, "Well, if you had, you never could
have desired to eat anything better."[4]

In the lake regions of Central Africa, says Burton, man '

---

[1] *Phil. Trans.*, lxxi. 167–8, note.   [2] *Ibid.*
[3] *Voy. to Cape of Good Hope*, i. 261 ; Cf. Alexander's *Exped. into
Africa.* i. 52.
[4] *Trav. in S. Africa*, p. 501.

revenges himself upon the White-ant, and satisfies his craving for animal food, which in those regions oftentimes becomes a principle of action,—a passion,—by boiling the largest and fattest species, and eating them as a relish with his insipid porridge.[1]

Buchanan says the Termes, or White-ant, is a common article of food among one of the Hindoo tribes; Mr. Forbes says, of the low castes in Mysore, and the Carnatic.[2] Captain Green relates that, in the ceded districts of India, the natives place the branches of trees over the nests, and then by means of smoke drive out the insects; which attempting to fly, their wings are broken off by the mere touch of the branches.[3]

The female Termite, in particular, is supposed by the Hindoos to be endowed with highly nutritive properties, and, we are told by Mr. Broughton, was carefully sought after and preserved for the use of the debilitated Surjee Rao, Prime-minister of Scindia, chief of the Mahrattas.[4]

The Hottentots not only eat the Termites in their perfect state, but also, when their corn is consumed and they are reduced to the necessity, in their pupa. These pupæ, which they call "rice," on account of their resemblance to that grain, they usually wash, and cook with a small quantity of water. Prepared in this way they are said to be palatable; and if the people find a place where they can obtain them in abundance, they soon become fat upon them, even when previously much reduced by hunger. A large nest will sometimes yield a bushel of pupæ.[5]

Termite queens in the East Indies are given alive to old men for strengthening the back.[6]

---

[1] Burton's *Cent. Africa.* i. 202.

[2] Buchanan. i. 7: Forbes, *Orient. Mem.*, i. 305.

[3] Kirb. and Sp. *Introd.*. i. 308, note.

[4] *Letters written in a Mahratta Camp in* 1809.

[5] Backhouse, p. 584.

[6] *Phil. Trans.*, lxxi. 167–8, note.

Ephemeridæ—Day-flies.

The name of Ephemeridæ has been given to the insects, so called, in consequence of the short duration of their lives, when they have acquired their final form. There are some of them which never see the sun; they are born after it is set, and die before it reappears on the horizon.

These insects, indifferently called also Day-flies and May-flies, usually make their appearance in the districts watered by the Seine and the Marne, in the month of August; and in such countless myriads, that the fishermen of these rivers believe they are showered down from heaven, and accordingly call the living cloud of them *manna*—manna for fish, not men. Reaumur once saw them descend in this region so fast, that the step on which he stood by the river's bank was covered by a layer four inches thick in a few minutes. He compares their falling to that of snow with the largest flakes.[1]

Scopoli assures us that such swarms are produced every season in the neighborhood of some particular spots in the Duchy of Carniola, that the countrymen think they obtain but a small portion, unless every farmer can carry off about twenty cartloads of them into his fields for the purpose of a manure.[2]

Libellulidæ—Dragon-flies.

On account of the long and slender body, peculiar to the insects of this family, they are with us sometimes called *Devil's Darning-needles*, but more commonly *Dragon-flies*. In Scotland they are known by the name of *Flying Adders*, for the same reason. The English, from an erroneous belief that they sting horses, call them *Horse-stingers*. In France, from their light and airy motions, and brilliant, variegated dress, they are called *Demoiselles;* and in Germany, for the same reason, and that they hover over, and lived during

---

[1] *Memoirs*, vi. 485. Quot. by K. and S. *Introd.*, i. 284. Cuv. *An. Kingd.—Ins.*, ii. 315. *Ins. Trans.*, p. 373.
[2] Quot. by Shaw, *Zool.*, vi. 250.

their first stages in, water, *Wasser-jungfern*—Virgins of the Water. Another German name for them is *Florfliegen*—Gauze-flies, in allusion to their net-like wings. Our boys also call them *Snake-feeders* and *Snake-doctors*, in the belief that they wait upon snakes in the capacity of feeders and doctors; and so firm are they in this belief, that frequently I have been laughed at for asserting the contrary to them. The belief probably arose from the manner in which the Dragon-fly sometimes falls a prey to the snakes. Hovering over ponds, they are fond of alighting on little sticks and twigs just out of the water, and mistaking the heads of snakes, which probably swam there for the purpose, for such twigs, they are instantly caught by the snakes.

On the 30th and 31st of May, 1839, immense cloud-like swarms of Dragon-flies passed in rapid succession over the German town of Weimar and its neighborhood. They were the *Libellula depressa*, a species which, in general, is rather scarce in that part of Germany. The general direction of this migration was from south by west to north by east. The insects were in a vigorous state, and some of the flocks flew as high as 150 feet above the level of the River Ilm.

At Göttingen on June the 1st, at Eisenach on May the 30th and 31st of the same year, swarms of the same species were seen flying from east to west; and at Calais, June 14th, similar clouds, though of a different species, were noticed on their way toward the Netherlands. At Halle, also, on May 30th, a short time before a thunder-storm, swarms of the Dragon-fly, *L. quadrimaculata*, were seen by Dr. Buhle, flying very rapidly from south to north. The *L. quadrimaculata* is not generally found in the neighborhood of Halle.

This wonderful migration, for it is a phenomenon of rare occurrence, extended from the 51st to the 52d degree of latitude, and was observed within 27° 40′ and 30° east of Ferro. But the instance of Calais renders it probable that it extended over a great part of Europe.

Another migration of Dragon-flies was observed at Weimar on the 28th of June, 1816. The insects, in this instance, belonged also to the *L. depressa*. They were taken then, as were they also in 1839, for locusts by the common people, and looked upon as the harbingers of famine and war.

In these migrations they followed the direction of the

rivers, with the currents. They did not, however, always keep close by them, since they must spread over wide districts in order to subsist.

To account for the great multiplication of these insects, in the year 1839, is by no means difficult. From the beginning to the 21st of May (in the latter part of which month, it will be remembered, they appeared), the weather had been exceedingly rainy; rivers and lakes overflowed their banks and inundated immense areas of low grounds, whereby myriads of the *larvæ* and *pupæ* (which live entirely in water) of the *Libellulæ*, which, under other circumstances, would have remained in deep water, and become the prey of their many enemies, fish, etc., were brought into shallow water, and hot weather following, from May 21st to May 29th, converted these shallows and swamps into true hot-beds for them. Their development into perfect insects was thus rendered rapid, so that, somewhat earlier than usual, they appeared, and in far greater, their undiminished, numbers; and, being very voracious in their appetite, as well in the imago as the pupa state, they were obliged to migrate immediately to satisfy it.[1]

Mr. Gosse observed in Jamaica, Oct. 8th, 1845, a swarm of Dragon-flies in the air, about twenty feet from the level of the ground. They floated and danced about, over the stream of water that runs through Blue-fields, much in the manner of gnats, which they resembled also in their immense numbers.[2] And Rev. T. J. Bowen, on one occasion, in descending the Ogun River (in the Yoruba country, Africa), met millions of Dragon-flies, about one-fourth of an inch in length, making their way up the country by following the course of the stream.[3]

It is commonly said among us, that if a Dragon-fly be killed, there will soon be a death in the family of the killer.

---

[1] *Mag. of Nat. Hist.*, iii. 516–8.

[2] Gosse's *Jamaica*, p. 251.

[3] *Gram. and Dict. of the Yoruba Language.* Smithson. Public., p. xiii.

Myrmeleonidæ—Ant-lions.

When children meet with the funnel-shaped pitfalls of the larva of the Ant-lion, *Myrmeleon formicales*, they are wont to put their heads close to the ground and softly sing *ooloo-ooloo-ooloo*, till the larva, mistaking the sound for that of a fly escaping his trap, throws up a shower of sand to bring its supposed victim down again.

Ant-lions are held in great esteem in many sections of our country, so much so that they are not suffered to be in any way injured.

HYMENOPTERA.

Uroceridæ—Sirex.

IN a work called "*Ephemerides des curieux de la nature*," is an observation apparently relative to this family of insects, which, if true, would be very extraordinary indeed. It is there said, that in the town of Czierck and its environs, there were seen in 1679 some unknown winged insects which, with their stings, mortally wounded both men and beasts. They fell abruptly upon men without provocation, and attached themselves to the naked parts of the body: the sting was immediately followed by a hard tumor, and if care was not taken of the wound within the first three hours, by hastily extracting the poison from it, the patient died in a few days after. These insects killed five and thirty men in this diocese, and a great number of oxen and horses. Toward the end of September, the winds brought some of them into a small town on the confines of Silesia and Poland; but they were so feeble on account of the cold, that they did but little mischief there. Eight days after, they all disappeared. These animals have all of them four wings, six feet, and carry under the belly a long sting provided with a sheath, which opens and separates in two. They make a very sharp noise in attacking men. Some of them are ornamented with yellow circles (*Sirex gigas*, or *S. fusicornis?* M. Latreille), and others are similar to them in all respects, but they have the back altogether black, and their stings are more venomous (*S. spectrum* or *juvencus?*). The author of these observations gives an extended description of the species with the yellow circles, which he accompanies with figures, in which the character of *Sirex* may be clearly distinguished.[1]

---

[1] Cuv. *An. King.—Ins.*, ii. 404.

(142)

Cynipidæ—Gall-flies.

In the spring of 1694, some Galls hung down like chains upon the oaks in Germany, and the common people, who had never observed them before, imagined them to be magical knots.[1]

A very old and common superstition is, that every oak-apple contains either a maggot, a fly, or a spider: the first foretelling famine, the second war, and the third, the spider, pestilence. Matthiolus gravely affirms this conceit to be true;[2] and the learned Sir Thomas Browne, in his Pseudo-doxia Epidemica, has thought it worth his while, with much gravity, to explode it. He, however, while combating one popular error, falls himself into another, for want of that philosophical knowledge of insects which later times have succeeded in obtaining. We pass this by, and hurry to his conclusion: "We confess the opinion may hold some verity in analogy, or emblematical phancy; for pestilence is properly signified by the spider, whereof some kinds are of a very venomous nature: famine by maggots, which destroy the fruits of the earth; and war not improperly by the fly, if we rest in the phancy of Homer, who compares the valiant Grecian unto a fly. Some verity it may also have in itself, as truly declaring the corruptive constitution in the present sap and nutrimental juice of the tree; and may consequently discover the disposition of the year according to the plenty or kinds of those productions; for if the putrefying juices of bodies bring forth plenty of flies and maggots, they give forth testimony of common corruption, and declare that the elements are full of the seeds of putrefaction, as the great number of caterpillars, gnats, and ordinary insects do also declare. If they run into spiders, they give signs of higher putrefaction, as plenty of vipers and scorpions are confessed to do; the putrefying materials producing animals of higher mischief according to the advance and higher strain of corruption."[3]

---

[1] They were produced by that species of Gall-fly. *Cynips*, delineated by Reaumur in his *Hist. of Ins.*, vol. iii. tabl. 40. *The Mirror*, xxx. 234.

[2] K. and S. *Introd.*, i. 33.

[3] Browne's *Works*, ii. 376.

Moufet says: "In oak acorns and spongy apples some-times worms breed, and astrologers presage that year to be likely to produce a great famine and dearth. . . . . . It is strange that Ringelbergius writes, *lib. de experiment*, that these worms may be fed to be as big as a serpent, with sheep's milk; yet Cardanus confirms the same, and shewes the way to feed them, *Lib. de rer. varietat.*"[1]

There is a very curious operation performed at the present day in the Levant with one of these Gall-flies, which is termed *caprification*. The object of it is to hasten the maturity of figs; and the species employed for that purpose is the *Cynips ficus caricæ*, or *Cynips psenes* of Linnæus; it consists in placing on a fig-tree, which does not produce flowers or early figs, some of these last strung together with a thread. The insects which issue from them, full of fecundating dust, introduce themselves through the eye into the interior of the second figs, fecundate by this means all the grains, and provoke the ripening of the fruit.

This operation, of which some authors have spoken with admiration, appeared to Hasselquist and Olivier, both competent observers, who have been on the spot, to be of no advantage whatsoever in fertilizing the fig;[2] and scientific men of the present day generally hold that it cannot be of any utility, for each fig contains some small flowers toward the eye, capable of fecundating all the female flowers in the interior, and moreover this fruit will grow, ripen, and become excellent to eat even when the grains are not fecundated.[3]

A curious kind of gall, produced on the rose-trees by the *Cynips rosæ*, which is known by the name of *Bedeguar*, has been placed among the remedies which may be successfully employed against diarrhœa and dysentery, and useful in cases of scurvy, stone, and worms.[4]

The galls of commerce, commonly called *Nut-galls*, are found on the *Quercus infectoria*, a species of oak growing in the Levant, and are produced by the *Cynips Gallæ tinctorum*. When gathered before the insects quit them, the nut-galls contain more astringent matter, and are then known as Black, Blue, or Green-galls. When the insects

---

[1] *Theatr. Ins.*, 252. Topsel's *Hist. of Beasts*, p. 1085.
[2] Hasselquist's *Travels*, p. 253.
[3] Cuv. *An. King.—Ins.*, ii. 424.
[4] *Ibid.*, p. 427.

have escaped, they are less astringent, and are called White-galls. They are of great importance in the arts, being very extensively used in dyeing and in the manufacture of ink and leather. They are the most powerful of all the vegetable astringents, and are sometimes used, both internally and externally, with great effect in medicine. Those imported from Syria are the most esteemed, and, of these, those found in the neighborhood of Moussoul are considered the best.[1]

The gall of the field cirsium formerly enjoyed a very great reputation, for it was considered, when carried simply in the pocket, as a sovereign remedy against hemorrhages. It, no doubt, owed this virtue to its resemblance to the principal sign of this disease, the swelling of the vein.[2]

The galls of the ground-ivy, produced by the *Cynips glecome*, have been eaten as food in France; they have an agreeable taste, and to a high degree the odor of the plant which bears them. Reaumur, however, is doubtful whether they will ever rank with good fruits.[3]

The galls of the sage (*Salvia pomifera*, *S. triloba*, and *S. officinalis*), which are very juicy, like apples, and crowned with rudiments of leaves resembling the calyx of that fruit, are gathered every year, as an article of food, by the inhabitants of the Island of Crete. This is the statement of Poumefort. Olivier confirms it, and adds: They are esteemed in the Levant for their aromatic and acid flavor, especially when prepared with honey and sugar, and form a considerable article of commerce from Scio to Constantinople, where they are regularly exposed in the market.[4]

The celebrated "Dead Sea Fruits," often called *Poma insana*, or Mad-apples, *Mala Sodomitica*, etc., which have given rise to great controversy among Oriental scholars and Biblical commentators, are produced by the *Cynips insana* on the low oaks (*Quercus infectoria*) growing on the borders of the Dead Sea.[5]

---

[1] Baird's *Cyclop. of Nat. Sci.* Cf. Cuv.—*Ins.*, ii. 428; K. and S. *Introd.*, i. 318. Medict. Virt. Cf. Geoffroy's *Treatise on Subs. used in Physic*, p. 369.

[2] Cuv. *An. Kingd.*—*Ins.*, ii. 428. Cf. Geoffroy's *Subs. used in Phys.*, p. 369.

[3] Reaum. iii. 416. Cf. Cuv. *Ibid.* ii. 429. K. and S. *Introd.*, i. 310.

[4] Smith's *Introd. to Bot.*, p. 346. Olivier's *Trav.*, i. 139. Cf. *Ibid.*

[5] Baird's *Cyclop. of Nat. Sci.*

Formicidæ—Ants.

Herodotus, who wrote in the fifth century before the birth of Christ, tells the following fabulous story without the slightest trace of diffidence or disbelief: There are other Indians bordering on the City of Caspatyrus and the country of Pactyica, settled northward of the other Indians, whose mode of life resembles that of the Bactrians. They are the most warlike of the Indians, and these are they who are sent to procure the gold; for near this part is a desert by reason of the sand. In this desert then, and in the sand, there are Ants in size somewhat less indeed than dogs, but larger than foxes. Some of them are in the possession of the King of the Persians, which were taken there. These Ants, forming their habitations under ground, heap up the sand as the Ants in Greece do, and in the same manner; and they are very like them in shape. The sand that is heaped up is mixed with gold. The Indians, therefore, go to the desert to get this sand, each man having three camels, on either side a male one harnessed to draw by the side, and a female in the middle; this last the man mounts himself, having taken care to yoke one that has been separated from her young as recently as possible; for camels are not inferior to horses in swiftness, and are much better able to carry burdens. . . . . The Indians then, adopting such a plan and such a method of harnessing, set out for the gold, having before calculated the time, so as to be engaged in their plunder during the hottest part of the day, for during the heat the Ants hide themselves under ground. . . . . When the Indians arrive at the spot, having sacks with them, they fill these with the sand, and return with all possible expedition; for the Ants, as the Persians say, immediately discovering them by the smell, pursue them, and they are equaled in swiftness by no other animal, so that if the Indians did not get the start of them while the Ants were assembling, not a man of them could be saved. Now the male camels (for they are inferior in speed to the females) slacken their pace, dragging on, not both equally; but the females, mindful of the young they have left, do not slacken their pace. Thus the Indians, as the Persians say, obtain the greatest part of their gold.[1]

---

[1] Herod., B. 3, 102–5. Cary's *Trans.*, p. 214.

Concerning these remarkable Ants, Strabo and Arrian have preserved the statement of Megasthenes, who traveled in India about two centuries later than the time of Herodotus. As given by Strabo, who is somewhat more particular in his story than Arrian, it is as follows: Megasthenes, speaking of the Myrmeces (or Ants), says, among the Derdæ, a populous nation of the Indians, living toward the East and among the mountains, there was a mountain plain of about 3000 stadia in circumference; that below this plain were mines containing gold, which the Myrmeces, in size not less than foxes, dig up. They are excessively fleet, and subsist on what they catch. In winter they dig holes and pile up the earth in heaps, like moles, at the mouths of the openings. The gold dust which they obtain requires little preparation by fire. The neighboring people go after it by stealth with beasts of burden; for if it is done openly, the Myrmeces fight furiously, pursuing those that run away, and, if they seize them, kill them and the beasts. In order to prevent discovery, they place in various parts pieces of the flesh of wild beasts, and when the Myrmeces are dispersed in various directions, they take away the gold dust, and, not being acquainted with the mode of smelting it, dispose of it in its rude state at any price to merchants.[1]

Nearchus says he has himself seen several of the skins of these Ants, which were as large as the skins of leopards. They were brought by the Macedonian soldiers into Alexander's camp.[2]

Pliny, as a matter of course, believed this marvelous story, and has inserted it in brief in his compilation of natural history. He adds, too, that in his time there were suspended in the temple of Hercules, at Erythræ, this Ant's horns, which were looked upon as quite miraculous for their size. He also informs us it was of the color of a cat.[3]

Strabo and Arrian, from the manner in which they refer to the statements of Megasthenes and Nearchus, no doubt disbelieved them;[4] not so, however, Pomponius Mela.[5]

---

[1] Strabo, *Geog.*, B. xv. c. 1, § 44. Hamilton's *Trans.*, iii. 101. Cf. Arrian's *Ind. Hist.*, c. 15. Rooke's *Trans.*, ii. 211.

[2] *Ibid.*

[3] Pliny, *Nat. Hist.*, B. xi. c. 31. Bost. and Riley's *Trans.*, iii. 39.

[4] *Ubi supra*, and Strabo, B. xv. c. 1, § 37.

[5] Pomp., *Vita Apollon. Tyan.*, B. vi. c. 1.

M. de Veltheim thinks this animal, which, as Pliny says, "has the color of a cat, and is in size as large as an Egyptian wolf," is nothing more than, and really is, the *Canis corsac*, the small fox of India, and that by some mistake it was represented by travelers as an ant. It is not improbable, Cuvier says, that some quadruped, in making holes in the ground, may have occasionally thrown up some grains of the precious metal. Another interpretation of this story has also been suggested. We find some remarks of Mr. Wilson, in the *Transactions of the Asiatic Society*, on the Mahabharata, a Sanscrit poem, that various tribes on the mountains Meru and Mandara (supposed to lie between Hindostan and Thibet) used to sell grains of gold, which they called *paippilaka*, or *Ant-gold*, which, they said, was thrown up by Ants, in Sanscrit called *pippilaka*. In traveling westward, this story (in itself, no doubt, untrue) may very probably have been magnified to its present dimensions.[1]

The laborious life and foresight of the Ant have been celebrated throughout all antiquity, and from the wise Solomon down to the amiable La Fontaine, the sluggard has been referred to this insect to "learn her ways and be wise."[2] The Arabians held the wisdom of these animals in such estimation, that they used to place one of them in the hands of a newly-born infant, repeating these words: "May the boy turn out clever and skillful."[3] But their wisdom is magnified by all, and in the panegyrics of their providence we always find the following curious notion. Plutarch, in his Land and Water Creatures Compared, thus mentions it: "But that which surpasseth all other prudence, policy, and wit, is their (the Ants') caution and prevention which they use, that their wheat and other corn may not spurt and grow. For this is certain, that dry it cannot continue alwayes, nor sound and uncorrupt, but in time will wax soft, resolve into a milky juice, when it turneth and beginneth to swell and chit; for fear, therefore, that it become not a generative seed, and so by growing, loose the nature and property of food for their nourishment, *they gnaw that end thereof or head where it is wont to spurt and bud forth.*"[4]

---

[1] Bostick and Riley's *Trans. of Pliny*, iii. 39, note.
[2] Prov. vi. 6. Cf. Prov. xxx. 23.
[3] Smith's *Bib. Dict.*
[4] Holland's *Trans.*, p. 787.

The ancients, observing the Ants carry their pupæ, which in shape, size, and color very much resemble a grain of corn, and the ends of which they sometimes pull open to let out the inclosed insect, no doubt mistook the one for the other, and this action for depriving the grain of the embryo of the plant.

Some modern writers, as Addison[1] and Pluche,[2] it is curious to observe, have fallen into this ancient error; so ancient, in fact, it is that some have supposed the Hebrew name of the Ant to be derived from it.[3]   Among the poets, Prior asks:

> Tell me, why the *Ant*
> In *summer's plenty thinks of winter's want?*
> By constant journey *careful to prepare*
> *Her stores*, and *bringing home the corny ear,*
> By what instruction *does she bite the grain?*
> Lest, hid in earth, and taking root again,
> It might elude the foresight of her care.[4]

Thus Watts, also:

> They don't wear their time out in sleeping or play;
> But *gather up corn* in a sunshiny day,
>   And *for winter they lay up their stores:*
> They manage their work in such regular forms,
> One would think they *foresaw* all the frosts and the storms,
>   And so *brought their food within doors.*[5]

And Smart:

> The *sage, industrious Ant*, the *wisest insect*,
> And *best economist* of all the field:
> For when as yet the favorable sun
> Gives to the genial earth th' enlivening ray,
> ——All her subterranean avenues,
> And storm-proof cells, with management most meet,
> And unexampled housewif'ry, she frames;
> Then to the field she hies, and *on her back*
> *Burden immense! brings home the cumbrous corn:*
> Then, many a weary step, and many a strain,
> And many a grievous groan subdued, at length
> Up the huge hill she hardly heaves it home;

---

[1] *Guardian*, No. 156–7.

[2] *Nat. Displ.*, i. 128.

[3] *Namahl a Namal Circumcidit.*—Browne's *Pseud. Epid.*—*Works*, ii. 531.

[4] *Poems: Solomon.*

[5] *Hymns: The Emmet.*

> Nor rests she here her providence, but *nips*
> *With subtle tooth the grain*, lest from *her garner*,
> In mischievous fertility, it steal,
> And back to daylight vegetate its way.[1]

Milton also entertained this erroneous opinion :

> First crept
> The *parsimonious Emmet, provident*
> *Of future*, in small room large heart inclos'd;
> Pattern of just equality perhaps
> Hereafter, join'd in her popular tribes
> Of commonalty.[2]

And also Dr. Johnson :

> Turn on the *prudent Ant* thy heedless eyes,
> Observe her labors, sluggard ! and be wise.
> No stern command, no monitory voice,
> Prescribes her duties or directs her choice;
> Yet *timely provident* she hastes away,
> To snatch the blessings of a plenteous day :
> When fruitful Summer loads the teeming plain,
> *She crops the harvest, and she stores the grain.*[3]

There is an old Eastern proverb, that "what the Ant *collects* in a year the monks eat up in a night," which seems to be founded on the supposition that the Ants provide themselves with stores of food. Juvenal, also, observes, in his Sixth Satire, that "after the example of the Ant, some have learned to *provide* against cold and hunger."[4]

"Since, therefore," says Moufet, "(to winde up all in a few words) they (the Ants) are so exemplary for their great piety, prudence, justice, valour, temperance, modesty, charity, friendship, frugality, perseverance, industry and art; it is no wonder that Plato, in Phædone, hath determined, that they who without the help of philosophy have lead a civill life by custom or from their own diligence, they had their souls from Ants, and when they die they are turned to Ants again. To this may be added the fable of the Myrmidons, who being a people of Ægina, applied themselves to diligent labour in tilling the ground, continual digging, hard toiling, and con-

---

[1] *On the Omnis. of God.*
[2] *Par. Lost*, B. vii. l. 484.
[3] *Saturday Mag..* xix. 190.
[4] Lawson's *Bible Cyclop.*, ii. 505.

stant sparing, joyned with virtue, and they grew thereby so rich, that they passed the common condition and ingenuity of men, and Theogonis knew not how to compare them better than to Pismires, that they were originally descended from them, or were transformed into them, and as Strabo reports they were therefore called Myrmidons. The Greeks relate the history otherwise than other men do; namely, that Jupiter was changed into a Pismire, and so deflowered Eurymedusa, the mother of the Graces, as if he could no otherwise deceive the best woman, then in the shape of the best creature. Hence ever after was he called Pismire Jupiter, or, Jupiter, King of Pismires. . . . . . .

"They do better, in my opinion, who observe the Pismire, and grow rich by following his manners in labor, industry, rest, and study. We read of Midas that he was the richest King of all the West, and when he was a boy, the Pismires carryed grains of wheat into his mouth while he slept, and so foreshowed without doubt that he should be endowed with the Pismire's prudence, and should by his labour and frugality, gain so much riches, that he should be called the Golden boy of fortune, and the Darling of prosperity. *Ælianus.* And when the Ants did devour and eat up the live serpent of Tiberius Cæsar, which he so dearly loved, did they not thereby give him sufficient warning that he should take heed to himself for fear of the multitude, by whom he was afterwards cruelly murthered? *Suetonius.*"[1]

Of the wars and battles of the Ants, now so familiar from the writings of Huber and others, one of the oldest records is that given by Æneas Sylvius, who afterward became Pope Pius II., of an engagement contested with obstinacy by a great and a small species, on the trunk of a pear-tree. "This action," he states, "was fought in the pontificate of Eugenius the Fourth, in the presence of Nicholas Pistoriensis, an eminent lawyer, who related the whole history of the battle with the greatest fidelity." Another engagement of the same description is recorded by Olaus Magnus, as having happened previous to the expulsion of Christiern the Second, of Sweden, and the smallest species, having been victorious, are said to have buried the bodies of their

---

[1] *Theatr. Ins.*, p. 245-6. Topsel's *Hist. of Beasts*, p. 1078. Vide Pierius' *Hieroglyph.*, p. 73-6.

own soldiers that had been killed, while they left those of their adversaries a prey to the birds.[1]

Alexander Ross, in his Appendix to the Arcana Microcosmi, p. 219, tells us: "That the cruel battels between the Venetians and Insubrians, and that also between the Liegeois and the Burgundians, in which about thirty thousand men were slain, were presignified by a great combat between two swarms of Emmets (Ants)."[2]

Ants were used in divination by the Greeks, and generally foretold good.[3] They were also considered an attribute of Ceres.[4]

The following extract is from an English North-Country chap-book, entitled the Royal Dream Book: "To dream of Ants or Bees denotes that you will live in a great town or city, or in a large family, and that you will be industrious, happy, well married, and have a large family."[5] The Ant and the Bee are common figures to express these predictions.

I heard a mother once say to her child, "Never destroy Ants, for they are fairies, and will so bewitch our cows that they will give no milk." This superstition prevails in particular about Washington and in Virginia.

Mrs. Meer Hassan Ali, in an interesting article on the Ants of India, remarks that she has often witnessed the Hindoos, male and female, depositing small portions of sugar near Ants' nests as acts of charity to commence the day with.

With the natives of India, this lady also tells us, it is a common opinion that wherever the Red-ants colonize, prosperity attends the owner of that house.[6]

We read in Purchas's *Pilgrims*, that "the natives of Cambaia and Malabar will go out of the path if they light on an Ant-hill, lest they might happily treade on some of them."[7]

Other insects, as will be noticed in the course of this

---

[1] Mouf. *Theatr. Ins.*, p. 242.
[2] Quot. in Brande's *Pop. Antiq.*, iii. 224.
[3] Harwood's Grec. *Antiq.*, p 200.
[4] Stosch. Cl., ii. 227–8. Fosbr. *Encycl. of Antiq.*, ii. 738.
[5] Quot. in Brande's *Pop. Antiq.*, iii. 134.
[6] *The Mirror*, xxx. 216.
[7] *Pilgrims*, v. 542.

volume, are looked upon by these people with the same respect.

Moufet says: "In Isthmus the priests sacrificed Pismires to the sun, either because they thought the sun the most beautiful, and therefore they would offer unto him the most beautiful creature, or the most wise, as seeing all things, and therefore they offered unto him the wisest creature."[1]

In the twenty-seventh chapter of the Koran, which was revealed at Mecca, and entitled the Ant, we find, among other strange things, an odd story of the Ant, which has therefore given name to the chapter. It is as follows: "And his armies were gathered together unto Solomon, consisting of genii, and men, and birds; and they were led in distant bands, until they came to the valley of Ants.[2] And an Ant, seeing the hosts approaching, said, O Ants, enter ye into your habitations, lest Solomon and his army tread you under foot, and perceive it not. And Solomon smiled, laughing at her words, and said, O Lord, excite me that I may be thankful for thy favour, wherewith thou hast favoured me, and my parents; and that I may do that which is right, and well pleasing unto thee: and introduce me, through thy mercy, into paradise, among my servants, the righteous."[3]

Thevenot mentions "Solomon's Ant" among the "Beasts that shall enter into Paradise" in the belief of the Turks, and gives the following reason: "Solomon was the greatest king that ever was, for all creatures obey'd him, and brought him presents, amongst others, an Ant brought him a Locust, which it had dragged along by main force: Solomon, perceiving that the Ant had brought a thing bigger than itself, accepted the present, and preferred it before all other creatures."[4]

Plutarch, speaking of the Ant, says: "Aratus in his prognostics setteth this down for a rain toward, when they bring forth their seeds and grains (pupæ), and lay them abroad to take the air:

---

[1] *Theatr. Ins.*, 246. Topsel's *Hist. of Beasts*, p. 1079.

[2] The valley seems to be so called from the great number of Ants which are found there. Some place it in Syria, and others in Tayeb.—*Al Beidawi, Jallalo'ddin.*

[3] *The Koran*, p. 310. Translated by Geo. Sale.

[4] *Trav. in the Levant*, Pt. I. p. 41.

14*

> ' When Ants make haste with all their eggs aload,
>   Forth of their holes to carry them abroad.' "[1]

In the Treasvrie of Avncient and Moderne Times, it is also asserted that "when Ants walk the thickest, and more than in vsuall numbers, meeting together confusedly, it is a manifest signe of raine."[2]

It is related of the celebrated Timour, that being once forced to take shelter from his enemies in a ruined building, he sat alone many hours; and, desirous of diverting his mind from his hopeless condition, at length fixed his observation upon an Ant which was carrying a grain of corn (probably a pupa) larger than itself, up a high wall. Numbering the efforts it made to accomplish this object, he found that the grain fell sixty-nine times to the ground; but the seventieth time it reached the top of the wall. "This sight," said Timour, "gave me courage at the moment, and I have never forgotten the lesson it conveyed."[3]

Plutarch, in his comparison between land and water creatures, narrates the following anecdote: "Gleanthus the Philosopher, although he maintaineth not that beasts have any use of reason, made report nevertheless that he was present at the sight of such a spectacle and occurrent as this. There were (quoth he) a number of Ants which went toward another Ant's hole, that was not their own, carrying with them the corpse of a dead Ant; out of which hole, there came certain other Ants to meet them on the way (as it were) to parl with them, and within a while returned back and went down again; after this they came forth a second, yea a third time, and retired accordingly until in the end they brought up from beneath (as it were a ransom for the dead body) a grub or little worm; which the others received and took upon their shoulders, and after they had delivered in exchange the aforesaid corpse, departed home."[4]

Of the ingenuity of the Ant in removing obstacles, the following anecdote is a very appropriate illustration: A gentleman of Cambridge one day observed an Ant dragging along what, with respect to the creature's size, might be denominated a log of wood. Others were severally employed,

---

[1] *Land and Water Creatures Compared*, Holland, p. 787.
[2] B. 7, c. 16, p. 665; printed 1613. ·
[3] Strong's *Nat. Hist.*, iii. 163.
[4] Holland's *Trans.*, p. 787.

each in its own way. Presently the Ant in question came to an ascent, where the weight of the wood seemed for a while to overpower him: he did not remain long perplexed with it; for three or four others, observing his dilemma, came behind and pushed it up. As soon, however, as he got it on level ground, they left it to his care, and went to their own work. The piece he was drawing happened to be considerably thicker at one end than the other. This soon threw the poor fellow into a fresh difficulty; he unluckily dragged it between two bits of wood. After several fruitless efforts, finding it would not go through, he adopted the only mode that even a man in similar circumstances would have taken: he came behind it, pulled it back again, and turned it on its edge; when, running again to the other end, it passed through without the slightest difficulty.[1]

Franklin was much inclined to believe Ants could communicate their thoughts or desires to one another, and confirmed his opinion by several experiments. Observing that when an Ant finds some sugar, it runs immediately under ground to its hole, where, having stayed a little while, a whole army comes out, unites and marches to the place where the sugar is, and carry it off by pieces; and that if an Ant meets with a dead fly, which it cannot carry alone, it immediately hastens home, and soon after some more come out, creep to the fly, and carry it away; observing this, he put a little earthen pot, containing some treacle, into a closet, into which a number of Ants collected, and devoured the treacle very quickly. He then shook them out, and tied the pot with a thin string to a nail which he had fastened in the ceiling, so that it hung down by the string. A single Ant by chance remained in the pot, and when it had gorged itself upon the treacle, and wanted to get off, it was under great concern to find a way, and kept running about the bottom of the pot, but in vain. At last it found, after many attempts, the way to the ceiling, by going along the string. After it was come there, it ran to the wall, and thence to the ground. It had scarcely been away half an hour, when a great swarm of Ants came out, got up to the ceiling, and crept along the string into the pot, and began to eat again. This they continued till the treacle was all

---

[1] Chamb. *Misc.*, x. 17.

eaten; in the mean time one swarm running down the string, and the other up.[1]

It has been suggested, that in such instances as the preceding, the Ants may have been led by the scent or trace of treacle likely to be left by the solitary prisoner; and the following case, related by Bradley, is quoted to favor the opinion: "A nest of Ants in a nobleman's garden discovered a closet, many yards within the house, in which conserves were kept, which they constantly attended till the nest was destroyed. Some, in their rambles, must have first discovered this depot of sweets, and informed the rest of it. It is remarkable that they always went to it by the same track, scarcely varying an inch from it, though they had to pass through two apartments; nor could the sweeping and cleaning of the rooms discomfit them, or cause them to pursue a different route "[2]

Dionisio Carli, of Piacenza, a missionary in Congo, lying sick at that place, was awakened one night by his monkey leaping on his head, and almost at the same time by his Blacks crying out, much to his surprise, "Out! Out! Father!" Thoroughly awake now, Carli asked them what was the matter? "The Ants," they cried, "are broke out, and there is no time to be lost!" Not being able to stir, he bid them carry him into the garden, which they did, four of them lifting him upon his straw bed; and yet though very quick about it, the Ants had already commenced crawling up his legs. After shaking them off their master, the Blacks took straw and fired it on the floor of four rooms, where these insects by this time were over half a foot thick. The pests being thus destroyed, Carli was conveyed back to his chamber, where he found the stench so great from the burnt bodies, that he was forced, he says, to hold his *monkey* close to his nose!

These Ants, Carli relates, ate up every living object within their reach; and of one cow, which was accidentally left over night in the stable through which they passed, nothing but the bones were found the next morning.[3] We need not wonder at this, if we believe what Bosman has said of the Black-ants of Guinea, which were so surprisingly rapacious

---

[1] Kalm in Pinkerton's *Col. of Voy. and Trav.*, xiii. 474.
[2] Chamb. *Misc.*, x. 22.
[3] Pinkerton's *Col. of Voy. and Trav.*, xvi. 174.

that no animal could stand before them.   He relates an instance where they reduced for him one of his live sheep in
one night to a perfect skeleton, and that so nicely that it
surpassed the skill of the best anatomists.[1]   Du Chaillu
says the elephant and gorilla fly before the attack of the
Bashikouay-ants, and the black men run for their lives.
Many a time has he himself, he says, been awakened out of
a sleep, and obliged to rush out of his hut and into the
water to save his life![2]   The Driver-ants[3] of Western Africa,
A. nomma arcens, have been known to kill the *Python nata-
lensis*, the largest serpent of that part of the world.[4]

Col. St. Clair, after a visit by a species of small Red-ants,
makes mention of the following instance, among others, of
their singular destructiveness: "I next discovered that a
little pet deer, which I had purchased from a negro, was
extremely ill.   I could not discover the cause of its malady,
until, placing it on its legs, I observed that it would not let
one foot touch the ground, and, on examining it, I found, to
my grief, that the Red-ants had absolutely eaten a hole into
the bone.   The poor little animal pined all that day and
died in the evening "[5]

Capt. Stedman relates that the Fire-ants of Surinam
caused a whole company of soldiers to start and jump about
as if scalded with boiling water; and its nests were so
numerous that it was not easy to avoid them.[6]   And Knox,
in his account of Ceylon, mentions a black Ant, called by
the natives *Coddia* or *Kaddiya*,[7] which, he says, "bites
desperately, as bad as if a man were burnt by a coal of fire;
but they are of a noble nature, and will not begin unless you
disturb them."   The reason the Singhalese assign for the
horrible pain occasioned by their bite is curious, and is thus
related by Knox: "Formerly these Ants went to ask a wife
of the *Noya*, a venomous and noble kind of snake;[8] and be-

---

[1] *Guinea*, p. 276; Astley's *Col. of Voy. and Trav.*, ii. 727.

[2] Du Chaillu, p. 312 and 108.

[3] Allied to the Stinger (*ota*) of Yoruba, and *Idzalco*, "the fighter
which makes one go."—*T. J. Bowen.*

[4] Livingstone's *Travels*, p. 468.

[5] St. Clair's *W. Indies*, i 167–8.

[6] Stedm. *Surinam*, ii. 94.

[7] Of similar size and ferocity as the great Red-ant of Ceylon, the
Dimiya, Formica smaragdina.—Tennent, *N. H. of Ceyl.*, p. 424.

[8] The Cobra de Capello, *Naja tripudians*, Merr.

cause they had such a high spirit to dare to offer to be related to such a generous creature, they had this virtue bestowed upon them, that they should sting after this manner. And if they had obtained a wife of the Noya, they should have had the privilege to sting full as bad as he."[1] Capt. Stedman has a story of a large Ant that stripped the trees of their leaves, to feed, as was supposed by the natives of Surinam, a blind serpent under ground,[2] which is somewhat akin to this: as is also another, related to Kirby and Spence by a friend, of a species of Mantis, taken in one of the Indian islands, which, according to the received opinion among the natives, was the parent of all their serpents.[3] But, the reverse: Among the harmless snakes of Mexico is a beautiful one about a foot in length, and of the thickness of the little finger, which appears to take pleasure in the society of Ants, insomuch that it will accompany these insects upon their expeditions, and return with them to their usual nest. From this peculiarity it is called by the Spaniards and Mexicans the "Mother of the Ants."[4]

When in Africa, Du Chaillu was told by the natives that criminals in former times were exposed to the path of the Bashikouay-ants, as the most cruel way of putting them to death.[5] This dreadful manner of torturing was at one time also practiced by the Singhalese, and I have heard that several British soldiers have thus met their fate. The Termites have been referred to before as having been employed for a similar purpose.

To check the ravages of the Coffee-bug, *Lecanium coffea*, Walker, which for several years was devastating some of the plantations of Ceylon, the experiment was made of introducing the Red-ants, *Formica smaragdina*, Fab., which feed greedily on the Coccus.[6] But the remedy threatened

---

[1] Knox, *Hist. Rel. of Ceylon*, Pt. I. ch. vi. p. 24.

[2] Stedm. *Surinam*, ii. 142.

[3] K. and S. *Introd.*, i. 123.

[4] Smith's *Nature and Art*, xii. 195. Clavigero supposes that all the attachment which the snake shows to the Ant-hills proceeds from its living on the Ants themselves.

[5] Du Chaillu, p. 312.

[6] The Swiss farmers, in order to rid their trees of caterpillars, allure the Ants to climb the trees, where, being confined by a circle of pitch round the holes, hunger soon causes them to attack the noxious larvæ.

to be attended with some inconvenience, for, says Tennent, the Malabar coolies, with bare and oiled skins, were so frequently and fiercely assaulted by the Ants as to endanger their stay on the estates.

The pupæ or cocoons of the Ants, during the day, are placed near the surface of the Ant-hills to obtain heat, which is indispensable to the growth of the inclosed insects. This is taken advantage of in Europe to collect the cocoons in large quantities as food for nightingales and larks. The cocoons of a species of Wood-ant, *Formica rufa*, are the only kind chosen. In most of the towns of Germany, one or more individuals make a living during summer by this business alone. "In 1832," says a contributor to the Penny Encyclopedia, "we visited an old woman at Dottendorf, near Bern, who had collected for fourteen years. She went to the woods in the morning, and collected in a bag the surfaces of a number of Ant-hills where the cocoons were deposited, taking Ants and all home to her cottage, near which she had a small tiled shed covering a circular area, hollowed out in the center, with a trench full of water around it. After covering the hollow in the center with leafy boughs of walnut or hazel, she strewed the contents of her bag on the level part of the area within the trench, when the Nurse-ants immediately seized the cocoons, and carried them into a hollow under the boughs. The cocoons were thus brought into one place, and after being from time to time removed, and black ones separated by a boy who spread them out on a table, and swept off what were bad with a strong feather, they were ready for market, being sold for about 4*d.* or 6*d.* a quart. Considerable quantities of these cocoons are dried for winter food of birds, and are sold in the shops."[1]

Ants not only furnish food to man for his birds, but also food for himself, in both the pupa and imago states. Nicoli Conti, who traveled in India in the early part of the fifteenth century, says the Siamese eat a species of Red-ant, of the size of a small crab, which they consider a great delicacy seasoned with pepper.[2] At the present day, the pupæ of a species of Ants are a costly luxury with these people. They are not much larger than grains of sand, and are sent to table curried, or rolled in green leaves, mingled with shreds

---

[1] *Penny Encycl.*, *sub* Ant.
[2] *Hakluyt Society*, ii. 13.

or very fine slices of fat pork.[1] And in the province of
Michuacan, Mexico, is a singular species of Ant, which
carries on its abdomen "a little bagful of a sweet substance,
of which the children are very fond: the Mexicans suppose
this to be a kind of honey collected by the insect; but
Clavigero thinks it rather its eggs."[2]

Piso, De Laet, Marcgrave, and other writers mention
their being an article of food in different parts of South
America. Piso speaks of yellow Ants called *Cupia* inhabiting Brazil, the abdomen of which many used for food, as
well as a large species under the name of *Tama-joura:*
"Alia præterea datur grandis species *Tama-ioura* dicta
digiti articulum adæquans. Quarum etiam clunes dessicantur
et friguntur pro bono alimento."[3] Says De Laet: "Denique
formicæ hic visuntur grandissimæ, quas indigenæ vulgo
comedunt; et in foris venales habent."[4] And again: "For-
micis vescebantur, easquæ studiose ad victum educabant."[5]
Lucas Fernandes Piedrahita, in his Historia General de
las Conquistas del Nuevo Regno de Granada, states that
cakes of Cazave and Ants were eaten in that country:
"Al tiempo de tostarlas para este efecto, dan el mismo olor
que los quesillos, que se labran para comer asados."[6] Herrera says, the natives of New Granada made their main food
of Ants, which they kept and reared in their yards.[7] Sloane
confirms this, and says they are publicly sold in the markets.[8]
Abbeville de Noromba tells us these great Ants are fricasseed.[9]
Schomburgk, in his journey to the sources of the Essequibo,
one evening saw all the boys of a village out shouting and
chasing with sticks and palm leaves a large species of winged
Ant, which they collected in great numbers in their calabashes for food. When roasted or boiled, he says, the natives considered these insects a great delicacy.[10] Humboldt
informs us that Ants are eaten by the Marivatanos and
Margueritares, mixed with resin for sauce.[11]

---

[1] *The Mirror*, xxxi. 342.
[2] Smith's *Nature and Art*, xii. 197.
[3] *Hist. Nat.*, i. 9, and v. 291. Cf. Sloane, *Hist. of Jam.*, ii. 221.
[4] *Amer. Utriusq. Desc.*, p. 333.   [5] *Ibid.*, p. 379.
[6] Southey's *Com. Place Book*, 3d S. p. 346-7.
[7] Herrera, vi. 5, 6.
[8] *Hist. of Jam.*, ii. 221.   [9] Quoted, *Ibid.*
[10] *Journ. of Geog. Soc.*, 1841, x. 175.
[11] Quot. by K. and S. *Introd.*, i. 309.

Mr. Consett, in his Travels in Sweden, makes mention of a young Swede who ate live Ants with the greatest relish imaginable.[1] This author states also, that in some parts of Sweden Ants are distilled along with rye, to give a flavor to the inferior kinds of brandy.[2]

The inhabitants of the Tonga Group have a superstitious belief that when their kings, and matabooles, or inferior chiefs, die, they are wafted to Bulotu—"the island of the blessed," but the spirits of the lower class remain in the world, and feed on Ants and lizards.[3]

Ants also furnish us with an acid, called by the chemists *Formic*, which is said to answer the same purposes as the acetous acid. It is obtained in two modes: 1st. By distillation; the insects are introduced into a glass retort, distilled by a gentle heat, and the acid is found in the recipient. 2d. By the process called lixiviation; the Ants are washed in cold water, spread out upon a linen cloth, and boiling water poured over them, which becomes charged with the acid part.[4]

Formic acid is shed so sensibly by the wood Ant, *Formica rufa*, when an Ant-hill is stirred, that it can occasion an inflammation. If a living frog, it is asserted, be fixed upon an Ant-hill which is deranged, the animal will die in less than five minutes, even without having been bitten by the Ants.[5]

We read in Purchas's Pilgrims that the large Ant of the West Indies is "so poysonfull that herewith the Indians infect their arrowes so remedilesse, that not foure of an hundred which are wounded escape."[6]

The medicinal virtues of the Ant are as follows: "Ants, *Formica minor* of Schroder, heat and dry, and incite to venery; their acid smell mightily refreshes the vital spirits. They are said to cure the Flora, Lepra, and Lentigo. The eggs (pupæ) are effectual against deafness, and correct the hairiness of the cheeks of children being rubbed thereon."

The Horse-ant, *Formica major*, Schrod., "provokes to

---

[1] *Trav. in Swed.*, p. 118, Lond. 1789, 4to.　　　　　[2] *Ibid.*
[3] Jenkin's *Voy. of U. S. Explor. Exped. Com. by Wilkes*, 8vo. Auburn, 1852, p. 319.
[4] Cuv. *An. Kingd.—Insects*, ii. 489.　　　　　[5] *Ibid.*
[6] *Pilgrims*, iii. 996.

venery, and the oil thereof, by infusion, is good for the gout and palsy."[1]

Sloane tells us the Spaniards in the West Indies have a very highly valued medicated earth called "Makimaki," which he thinks is made of the nests of Ants.[2]

There is a species of Ant in Cayenne, *Formica bispinosa*, which collects from the bombax and silk-cotton trees a sort of lint which the natives value much as a styptic in cases of hemorrhage.[3]

The magicians, as mentioned by Pliny, recommended that the parings of all the finger-nails should be thrown at the entrance of Ant-holes, and the first Ant to be taken which should attempt to draw one into the hole; for if this, they asserted, be attached to the neck of a patient, he will experience a speedy cure.[4]

The two following remarkable cures effected by Ants of themselves are worthy of being noticed: Schuman, a missionary among the negroes of Surinam, relates in one of his letters, that after a most dangerous attack of the acclimating fever, his body was covered with boils and painful sores. He lay in his cot as helpless as a child, and had no one to administer any relief or food but a poor old negro woman, who sometimes was obliged to follow the rest to the plantations in the woods. One morning while she was absent, after spending a most restless and painful night, he observed at sunrise an immense host of Ants entering through the roof, and spread themselves over the inside of his chamber; and expecting little else than that they would make a meal of him, he commended his soul to God, and hoped thus to be released from all suffering. They presently covered his bed, and entering his sores caused him the most tormenting pain. However, they soon quitted him, and continued their march, and from that time he gradually recovered his health.[5]

The second is a case of stiffness in the knee effectually cured: In 1798, Mrs. Jane Crabley, aged 56 years, began to complain of a most torturing pain, and considerable enlargement of the knee-pan, which she described as, and

---

[1] James's *Med. Dict.*
[2] *Hist. of Jam.*, ii. 221.
[3] Brande's *Encycl. of Sci. Lit.*, etc.
[4] Pliny, *Nat. Hist.*, xxviii. 7 (23).
[5] Southey's *Com. Place Book*, 3d S. p. 419.

which her neighbors believed to be, a smart paroxysm of gout. Early in February, 1799, the inflammation and pain entirely ceased, but the swelling continued, and rather increased. The joint of the knee, from disuse, became perfectly stiff, and, owing to the particular form and size of her breasts, no relief could be gained by the use of crutches. However, toward the end of May, the Ants became so strangely troublesome to her, that she was sometimes obliged to avail herself of the help of travelers to assist her in changing her station. Still, however, they followed her, and seemed entirely attracted by her now useless knee. She was at first considerably annoyed by these little torments, but, in a few days, became not only reconciled to their intrusion, but was desirous of having her chair placed where she imagined them most to abound, even giving them freer access to her knee by turning down her stocking; for, she said, "the cold numbness she suffered just around the patella was eased and relieved by their bite; and that it was even pleasurable;" and, strange to say, these insects bit her nowhere else. The skin at first was pale and sallow, but began now to assume a lively red color; a clear and subtile liquid oozed from every puncture the Ants had left; the swelling and stiffness of the joint gradually abated; and, on the 25th of July, she walked home with the help of a stick, and before winter perfectly recovered the use of her limb.[1]

Says Plutarch, as translated by Holland: "The bear finding herself upon fulness given to loth and distaste for food, she goes to find out Ants' nests, where she sits her down, lilling out her tongue, which is glib and soft with a kind of sweet and slimy humour, until it be full of Ants and their egges, then draweth it she in again, swalloweth them down, and thereby cureth her lothing stomack."[2]

Also, in the Treasurie of Avncient and Moderne Times, we find: "The Bear, being poysoned by the Hearbe named *Mandragoras*, or *Mandrake*, doth purge his bodie by the eating of Ants or Pismires."[3]

M. Huber, initiated in the mysteries of the life of these

---

[1] *Gent. Mag.*, Pt. II. lxxiii. 704–5, and Kirby's *Wond. Museum*, i. 353–5.

[2] *Land and Water Creatures Compared*, Holl. *Trans..* p. 793.

[3] B. 7, c. xv. p. 664. Printed 1613.

insects, and whose observations can be most relied on, has made us acquainted with two of their maladies: one is a species of vertigo, occasioned, as he thinks, by a too great heat of the sun, and which transforms them for two or three minutes into a sort of bacchantes; the other malady, much more severe, causes them to lose the faculty of directing themselves in a right line. These Ants turn in a very narrow circle, and always in the same direction. A virgin female, inclosed in a sand-box, and attacked by this mania, made a thousand turns by the hour, describing a circle of about an inch in diameter; it continued this operation for seven days, and even during the night.[1]

Immense swarms of winged Ants are occasionally met with, and some have been recorded of such prodigious density and magnitude as to darken the air like a thick cloud, and to cover the ground or water for a considerable extent where they settled. We find in the memoirs of the Berlin Academy a description of a remarkable swarm, observed by M. Gleditch, which from afar produced an effect somewhat similar to that of an Aurora Borealis, when, from the edge of the cloud, shoot forth by jets many columns of flame and vapor, many rays like lightning, but without its brilliancy. Columns of Ants were coming and going here and there, but always rising upward, with inconceivable rapidity. They appeared to raise themselves above the clouds, to thicken there, and become more and more obscure. Other columns followed the preceding, raised themselves in like manner, shooting forth many times with equal swiftness, or mounting one after the other. Each column resembled a very slender net-work, and exhibited a tremulous, undulating, and serpentine motion. It was composed of an innumerable multitude of little winged insects, altogether black, which were continually ascending and descending in an irregular manner.[2] A similar kind of Ants is spoken of by Mr. Accolutte, a clergyman of Breslau, which resembled columns of smoke, and which fell on the churches and tops of the houses, where they could be gathered by handfuls. In the German *Ephemerides*, Dr. Chas. Rayger gives an account of a large swarm which crossed over the town of Posen, and was directing its course toward the Danube. The whole

---

[1] Cuv. *An. Kingd.—Ins.*, ii. 472.
[2] *Mem. Berlin Acad.* for 1749.

town was strewed with Ants, so that it was impossible to walk without crushing thirty or forty at every step. And more recently, Mr. Dorthes, in the *Journal de Physique* for 1790, relates the appearance of a similar phenomenon at Montpellier. The shoals moved about in different directions, having a singular intestine motion in each column, and also a general motion of rotation. About sunset all fell to the ground, and, on examining them, they were found to belong to the *Formica nigra* of Linnæus.[1]

"In September, 1814," says Dr. Bromley, surgeon of the Clorinde, in a letter to Mr. MacLeay, "being on the deck of the hulk to the Clorinde (then in the river Medway), my attention was drawn to the water by the first lieutenant observing there was something black floating down the tide. On looking with a glass, I discovered they were insects. The boat was sent, and brought a bucketful of them on board; they proved to be a large species of Ant, and extended from the upper part of Salt-pan Reach out toward the Great Nore, a distance of five or six miles. The column appeared to be in breadth eight or ten feet, and in height about six inches, which I suppose must have been from their resting one upon another."[2] Purchas seems to have witnessed a similar phenomenon on shore. "Other sorts of Ants," says he, "there are many, of which some become winged, and fill the air with swarms, which sometimes happens in England. On Bartholomew, 1613, I was in the island of Foulness, on our Essex shore, where were such clouds of these flying pismires, that we could nowhere flee from them, but they filled our clothes; yea, the floors of some houses where they fell were in a manner covered with a black carpet of creeping Ants, which they say drown themselves about that time of the year in the sea."[3]

When Colonel Sir Augustus Frazer, of the British horse-artillery, was surveying, on the 6th of October, 1813, the scene of the battle of the Pyrenees from the summit of the mountain called Pena de Aya, or Les Quartres Couronnes, he and his friends were enveloped by a swarm of Ants, so numerous as entirely to intercept their view, so that they were

---

[1] *Penny Encycl.*, sub. Ant.
[2] K. and S. *Introd.*, ii. 54.
[3] *Pilgrimage*, p. 1090.

obliged to remove to another station in order to get rid of them.[1]

"Not long since," says Josselyn in his Voyage to New England, London, 1674, "winged Ants were poured down upon the Lands out of the clouds in a storm betwixt *Black-point* and *Saco*, where the passenger might have walkt up to the Ancles in them."[2]

Wingless Ants, in swarms or armies, also migrate at particular seasons; but for what purpose is not clear, except to obtain better forage. In Guiana, Mr. Waterton says he has met with a colony of a species of small Ant marching in order, each having in its mouth a leaf; and the army extended three miles in length, and was six feet broad.[3]

It is recorded by Oviedo and Herrera, that the whole island of Hispaniola was almost abandoned in consequence of the Sugar-Ant, *Formica omnivora* of Linnæus, which, in 1518 and the two succeeding years, overran in such countless myriads that island, devouring all vegetation, and causing a famine which nearly depopulated the Spanish colony. A tradition, says Schomburgk, prevails in Jamaica that the town of Sevilla Nueva, which was founded by Esquivel in the beginning of the sixteenth century, was entirely deserted for a similar reason. Herrera relates that, in order to get rid of this fearful scourge in Hispaniola, the priests caused great processions and vows to be made in honor of their patron saint, St. Saturnin, and that the day of this saint was celebrated with great solemnities, and the Ants in consequence began to disappear. How this saint was chosen, we read in Purchas's Pilgrims: "This miserie (caused by the Ants) so perplexed the *Spaniards*, that they sought as strange a remedie as was the disease, which was to chuse some Saint for their Patron against the Autes. *Alexander Giraldine*, the Bishop, having sung a solemne and Pontifical Masse, after the consecration and Elevation of the Sacrament, and devout prayers made by him and the people, opened a Booke in which was a Catalogue of the Saints, by lot to chuse some he or she Saint, whom God should please to appoint their Advocate against the Calamitie. And the Lot fell vpon Saint *Saturnine*, whose Feast is on the nine

<hr>

[1] K. and S. *Introd.*, ii 54.
[2] Joss. *Voy*, p. 118.
[3] Baird's *Cyclop. of Nat. Sci.*

and twentieth of Nouember; after which the Ant damage became more tolerable, and by little and little diminished, by God's mercie and intercession of that Saint."[1]

These devouring Ants showed themselves about the year 1760 in Barbados, and caused such devastations that, in the words of Dr. Coke, "it was deliberated whether that island, formerly so flourishing, should not be deserted." In 1763, Martinique was visited by these devastating hordes; and about the year 1770 they made their appearance in the island of Granada. Barbados, Granada, and Martinique suffered more than any other islands from this plague. Granada especially was reduced to a state of the most deplorable desolation; for, it is said, their numbers there were so immense that they covered the roads for many miles together; and so crowded were they in many places that the impressions made by the feet of horses, which traveled over them, would remain visible but for a moment or two, for they were almost instantly filled up by the surrounding swarms. Mr. Schomburgk assures us that calves, pigs, and chickens, when in a helpless state, were attacked by such large numbers of these Ants that they perished, and were soon reduced to skeletons when not timely assisted. It is asserted by Dr. Coke that the greatest precaution was requisite to prevent their attacks on men who were afflicted with sores, on women who were confined, and on children that were unable to assist themselves. Mr. Castle, from his own observation, states that even burning coals laid in their way, were extinguished by the amazing numbers which rushed upon them.

Notwithstanding the myriads that were destroyed by fire, water, poison, and other means, the devastations continued to such an alarming extent, that in 1776 the government of Martinique offered a reward of a million of their currency for a remedy against this plague; and the legislature of Granada offered £20,000 for the same object; but all attempts proved ineffectual until the hurricane in 1780 effected what human power had been unable to accomplish.

In 1814, the Ants again made their appearance in the island of Barbados, doing considerable injury; but happily they did not continue long.[2]

---

[1] Purchas's *Pilgrims*, iii. 998.

[2] Schomburgk's *Hist. of Barbados*, 640-3; and Coke's *West Indies*, ii. 313.

Malouet, in visiting the forests of Guiana, of which he has spoken in his travels into that part of the globe, perceived in the midst of a level savanna, as far as the eye could reach, a hillock which he would have attributed to the hand of man, if M. de Prefontaine, who accompanied him, had not informed him that, in spite of its gigantic construction, it was the work of black Ants of the largest species (most probably of the genus *Ponera*). He proposed to conduct him, not to the Ant-hill, where both of them would infallibly have been devoured, but to the road of the workers. M. Malouet did not approach within more than forty paces of the habitation of these insects. It had the form of a pyramid truncated at one-third of its height, and he estimated that its elevation might be about fifteen or twenty feet, on a basis of from thirty to forty. M. de Prefontaine told him that the cultivators were obliged to abandon a new establishment, when they had the misfortune to meet with one of these fortresses, unless they had sufficient strength to form a regular siege. This even occurred to M. de Prefontaine himself on his first encampment at Kourva. He was desirous of forming a second a little farther on, and perceived upon the soil a mound of earth similar to that which we have just described. He caused a circular trench to be hollowed, which he filled with a great quantity of dry wood, and, after having set fire to it in every point of its circumference, he attacked the Ant-hill with a train of artillery. Thus every issue was closed to the hostile army, which, to escape from the invasion of the flames and the shaking and plowing of the ground by the cannon-balls, was obliged to traverse, in its retreat, a trench filled with fire, where it was entirely cut off.[1]

The Portuguese found such prodigious numbers of Ants upon their first landing at Brazil, that they called them Rey de Brazil, King of Brazil, a name which they now there bear.[2]

Mr. Southey states, on the authority of Manoel Felix, that the Red-ants devoured the cloths of the altar in the Convent of S. Antonio, or S. Luiz (Maranham, Brazil), and also brought up into the church pieces of shrouds from the graves; whereupon the friars prosecuted them according to

---

[1] *Cuv. An. Kingd.—Ins.*, ii. 471.
[2] *Pinkerton's Col. of Voy. and Trav.*, xiv. 716.

due form of ecclesiastical law. What the sentence was in this case, we are unable to learn. A similar case, however, the historian informs us, had occurred in the Franciscan Convent at Avignon, where the Ants did so much mischief that a suit was instituted against them, and they were excommunicated, and ordered by the friars, in pursuance of their sentence, to remove within three days to a place assigned them in the center of the earth. The Canonical account gravely adds, that the Ants obeyed, and carried away all their young, and all their stores.[1]

Annius writes, that an ancient city situate near the Volscian Lake, and called Contenebra, was in times past overthrown by Ants, and that the place was thereupon commonly called to his day, "the camp of the Ants."[2]

Ctesias makes mention "of a horse-pismire (*i.e.* the bigger kind of them in hollow trees) which was fed by the Magi, till hee grew to such a vast bulke as to devour two pound of flesh a daye."[3]

Martial has written the following beautiful epigram on an Ant inclosed in amber: "While an Ant was wandering under the shade of the tree of Phaeton, a drop of amber enveloped the tiny insect; thus she, who in life was disregarded, became precious by death.

> "A drop of amber from the weeping plant,
> Fell unexpected and embalmed an Ant;
> The little insect we so much contemn
> Is, from a worthless Ant, become a gem."[4]

It has been said, remarks Mr. Southey, and regarded as a vulgar error, that Ants cannot pass over a line of chalk: the fact, however, is certain. Mr. Coleridge tried the experiment at Malta, he continues, and immediately discovered the cause: The formic acid is so powerful, that it acts upon the chalk, and the legs of the insect are burnt by the instantaneous effervescence![5]

Paxamus says, that if you take some Ants and burn them, you will drive away the others, as experience has taught us.

---

[1] Southey's *Hist. of Brazil*, iii. 334, note.
[2] Wanley's *Wonders*, ii. 507.
[3] Thom. Browne's *Works*, ii. 337, note.
[4] Martial, B. iv. 15.
[5] Southey, *Hist. of Brazil*, i. 645.

Ants also, he continues, will not touch a vessel with honey, although the vessel may happen to be without its cover, if you wrap it in white wool, or if you scatter white earth or ruddle round it. If a person, continues Paxamus, takes a grain of wheat carried by an Ant with the thumb of his left hand, and lays it in a skin of Phœnician dye, and ties it round the head of his wife, it will prove to be the cause of abortion in a state of gestation.[1]

Pliny says the proper remedy for the venom of the *Solipuga* or *Solpuga* Ant, and for that of all kinds of Ants, is a bat's heart.[2]

Callicrates used to make Ants, and other such little creatures, out of ivory, with so much skill and ingenuity that other men could not discern the counterfeits from the originals even with the help of glasses.[3]

Vespidæ—Wasps, Hornets.

Concerning the generation of the Wasp, Topsel and Moufet have the following: "Isidore affirms that Wasps come out of the putrefied carkasses of asses, although he may be mistaken, for all agree that the Scarabees are procreated from them: rather am I of opinion with Pliny, l. ii. c. 20, and the Greek authors, that they are sprung from the dead bodies of horses, for the horse is a valiant and warlike creature, hence is that verse frequently and commonly used among the Greeks:

Wasps come from horses, Bees from bulls are bred.

And indeed their more than ordinary swiftnesse and their eagernesse in fight, are sufficient arguments that they can take their original from no other creature (much less from an asse, hart, or oxe) since that Nature never granted to any creatures else, to excel both in swiftness and valour. And surely that I may give another sense of that proverb of Aristotle,

---

[1] Owen's *Geoponika*, ii. 148–9.
[2] *Nat. Hist.*, xxix. 29.
[3] Wanley's *Wonders*, i. 378.

> Hail the daughters of the wing-footed steed:

this would I suppose fit to be spoken in way of jest and scorn to scolding women, which do imitate the hastiness and froward disposition of the Wasp. Other sorts of them are produced out of the putrid corps of the Crocodiles, if Horus and the Ægytians be to be believed, for which reason when they mean a Wasp, they set it forth by an horse or crocodile. Nicander gives them the name *lukosnoadon*, because they sometimes come from the dead carkasses of wolves. Bellenacensis and Vicentius say, that Wasps come out of the putrefaction of an old deer's head, flying sometimes out of the eyes, sometimes out of the nostrils. . . . There are those also that affirm that Wasps are begotten of the earth and rottenness of some kind of fruits, as Albertus and the Arabick scholiast."

Of the Hornet, likewise, these writers tell the following fabulous stories: "The Latins call the Hornets *Crabrones*, perchance from the village Crabra in the countrey of Tusculum (where there are great store of them), or from the word *Caballus*, *i.e.* a horse, who is said to be their father. According to that of Ovid, *Met.* 15:

> The warlike horse if buried under ground,
> Shortly a brood of Hornets will be found.

Albertus calls it a yellow Bee. Cardanus will needs have them to arise from the dead mule. Plutarch, in the life of Cleomedes, saith they come out of horse flesh, as the Bees do out of the oxe his paunch. Virgil saith they are produced of the asse. . . . I conceive that those are produced of the harder flesh of the horse, and the Wasps of the more tender flesh.'"

The Hornet (but whether or not it was the common species, *Vespa crabro*, Linn., is uncertain), we learn from Scriptures was employed by Providence to drive out the impious inhabitants of Canaan, and subdue them under the hand of the Israelites.—"And I sent the Hornet before you, which drave them out before you, even the two kings of the Amorites."[2]

In the second volume of Lieutenant Holman's Travels,

---

[1] *Theatr. Ins.*, p. 40-50. Topsel's *Hist. of Beasts*, p. 921-7. Vide Pierius' *Hieroglyph.*, p. 267-8; Pernicies summota; Pugnacitas; Imperfecti mores civiles: Perturbator.

[2] *Josh.* xxiv. 12; *Deut.* vii. 20.

the following anecdote is related: "Eight miles from Gran-
die——, the muleteers suddenly called out 'Marambundas!
Marambundas!' which indicated the approach of Wasps.
In a moment all the animals, whether loaded or otherwise,
lay down on their backs, kicking most violently; while the
blacks, and all persons not already attacked, ran away in
different directions, all being careful, by a wide sweep, to
avoid the swarms of tormentors that came forward like a
cloud. I never witnessed a panic so sudden and complete,
and really believe that the bursting of a water-spout could
hardly have produced more commotion. However, it must
be confessed that the alarm was not without good reason,
for so severe is the torture inflicted by these pigmy assail-
ants, that the bravest travelers are not ashamed to fly, the
instant they perceive the host approaching, which is of common occurrence on the Campos."[1]

Dr. Fairfax, in the Philosophical Transactions, mentions
a lady, who had such a horror of Wasps, that during the
season in which they abound in houses, she always confined
herself to her apartment.[2]

Dr. James tells us: "The combs (of the Hornet) are
recommended in a drench for that disorder in horses, which
Vigetius, L. 2, c. 23, calls scrofula, meaning, I believe, what
we call the strangles."[3]

Hornets'-nest is smoked under horses' noses for distemper,
cold in the head, and such like diseases. It is also given to
horses in their feed for thick-windedness.

The nests of Hornets are gathered by the country people
to clean spectacles.

Topsel, in his History of Four-footed Beasts and Serpents,
has the following prognostications of the weather from the
appearances of Hornets: "They serve instead of good
almanacks to countrey people, to foretel tempests and
change of weather, as hail, rain, and snow: for if they flie
about in greater numbers, and be oftner seen about any
place, then usually they are wont, it is a signe of heat and
fair weather the next day. But if about twilight they are
observed to enter often their nests, as though they would
hide themselves, you must the next day expect rain, winde,

---

[1] Kirby's *Bridgewater Treatise.—Saturday Mag.*, ix. 239.
[2] *Phil. Trans.*, i. 201.
[3] *Med. Dict.*

or some stormy, troublesome or boysterous season: whereupon Avienus hath these verses :

> So if the buzzing troups of Hornets hoarse to flie,
> In spacious air 'bout Autumn's end you see,
> When Virgil star the evening lamp espie,
> Then from the sea some stormy tempest sure shall be."[1]

"In the year 190, before the birth of Christ," say Moufet
and Topsel, "as Julius witnesseth, an infinite multitude of
Wasps flew into the market at Capua, and sate in the temple of Mars, they were with great diligence taken and burnt
solemnly, yet they did foreshew the coming of the enemy
and the burning of the city."[2]

The first Wasp seen in the season should always be killed.
By so doing, you secure to yourself good luck and freedom
from enemies throughout the year.[3] This is an English
superstition, and it prevails in parts of America. We have
one, also, directly opposed to it, namely, that the first Wasp
seen in the season should not be killed if you wish to secure
to yourself good luck. Many of our people, too, will kill a
Wasp at no time, for, if killed, they say, it will bring upon
them bad luck.

If a Wasp stings you, our superstitious think that your
foes will get the advantage of you.

If the first Wasp seen in the season be seen in your house,
it is a sign that you will form an unpleasant acquaintance.
If the first Bee seen in the season be seen in your house, it
is a sign you will form a pleasant and useful acquaintance.
This arose doubtless from the apparent uselessness of the
former, and worth of the latter insect.

Wasps building in a house foretell the coming to want of
the family occupying it. Likewise arose from the unthrifti-
ness of this insect.

If Hornets build high, the winter will be dry and mild;
if low, cold and stormy. This is firmly believed in Virginia;
and the idea seems to be, that if the nest is built high it will
be more exposed to the wind than if built low.

That a person may not be stung by Wasps, Paxamus says:

---

[1] *Hist. of Beasts*, p. 660.
[2] *Theatr. Ins.*, p. 49. Topsel's *Hist. of Beasts*, p. 657, 927.
[3] *Notes and Queries*, ii. 165.

"Let the person be rubbed with the juice of wild-mallow, and he will not be stung."[1]

The Creoles of Mauritius eat the larræ of Wasps, which they roast in the combs. In taking the nests, they drive off the Wasps by means of a burning rag fastened to the end of a stick. The combs are sold at the bazaar of Port Louis.[2]

The following story, of the cunning of the fox in killing the Wasps to obtain their combs, is told by Ælian: "The fox (a subtile creature) is said to prey upon the Wasp in this manner: he puts his tail into the Wasps' nest so long till it be all covered with Wasps, which he espying, pulls it out and beats them against the next stone or tree he meets withall till they be all dead, this being done again and again till all the Wasps be destroyed, he sets upon their combs and devours them."[3]

The Chinese Herbal contains a singular notion, prevalent also in India, concerning the generation of the Sphex, or solitary Wasp. When the female lays her eggs in the clayey nidus she makes in houses, she incloses the dead body of a caterpillar in it for the subsistence of the worms when they are hatched. Those who observed her entombing the caterpillar did not look for the eggs, and immediately concluded that the Sphex took the worm for the progeny, and say, that as she plastered up the hole of the nest, she hummed a constant song over it, saying, *"Class with me! class with me!"*—and the transformation gradually took place, and was perfected in its silent grave by the next spring, when a winged Wasp emerged, to continue its posterity the coming autumn in the same mysterious way.[4]

Apidæ—Bees.

Concerning the piety of Bees, we find the following legends:

"A certaine simple woman having some stals of Bees which

----

[1] Owen's *Geoponika*, ii. 211.

[2] Backhouse's *Mauritius*, p. 32.

[3] Moufet, *Theatr. Insect.*, p. 47. Topsel's *Hist. of Four-footed Beasts and Serpents*, p. 925, 655.

[4] William's *Middle Kingdom; or Chinese Empire*, i. 274.

yeelded not vnto her hir desired profit, but did consume and die of the murraine; made her mone to another woman more simple than hir selfe: who gave her councel to get a consecrated host or round Godamighty and put it among them. According to whose advice she went to the priest to receive the host; which when she had done, she kept it in hir mouth, and being come home againe she tooke it out and put it into one of hir hives. Wherevpon the murraine ceased, and the honey abounded. The woman therefore lifting vp the hive at the due time to take out the honie, sawe there (most strange to be seene) a chapel built by the Bees with an altar in it, the wals adorned by marvelous skil of architecture with windowes conveniently set in their places: also a dore and a steeple with bels. And the host being laid vpon the altar, the Bees making a sweet noise flew round about it."[1]

Mr. Hawker's legend is to this effect: A Cornish woman, one summer, finding her Bees refused to leave their "cloistered home" and had "ceased to play around the cottage flowers," concealed a portion of the Holy Eucharist which she obtained at church:

> She bore it to her distant home,
>   She laid it by the hive
> To lure the wanderers forth to roam,
>   That so her store might thrive;—
> 'Twas a wild wish, a thought unblest,
> Some evil legend of the west.
>
> But lo! at morning-tide a sign
>   For wondering eyes to trace,
> They found above that Bread, a shrine
>   Rear'd by the harmless race!
> They brought their walls from bud and flower,
> They built bright roof and beamy tower!
>
> Was it a dream? or did they hear
>   Float from those golden cells
> A sound, as of a psaltery near,
>   Or soft and silvery bells?
> A low sweet psalm, that grieved within
> In mournful memory of the sin![2]

The following passage, from Howell's *Parley of Beasts,*

---

[1] Thom. Bozius *de signis Eccles.*, B. 14, c. iii. Quot. by Butler, *Fem. Monarchie*, c. i. 48.

[2] Quot. in *Notes and Queries*, ix. 167.

furnishes a similar legend of the piety of Bees. Bee speaks:

" Know, sir, that we have also a religion as well as you, and so exact a government among us here; our hummings you speak of are as so many hymns to the Great God of Nature; and there is a miraculous example in *Cæsarius Cisterniensis*, of some of the Holy Eucharist being let fall in a meadow by a priest, as he was returning from visiting a sick body; a swarm of Bees hard-by took It up, and in a solemn kind of procession carried It to their hive, and their erected an altar of the purest wax for it, where it was found in that form, and untouched."[1]

Butler, quoting Thomas Bozius, tells us the following:

" Certaine theeves (thieves) having stolen the silver boxe wherein the wafer-Gods vse to lie, and finding one of them there being loath, belike, that he should lie abroad all night, did not cast him away, but laid him under a hive: whom the Bees acknowledging advanced to a high roome in the hive, and there insteade of his silver boxe made him another of the whitest wax: and when they had so done, in worshippe of him, and set howres they sang most sweetly beyond all measure about it: yea the owner of them took them at it at midnight with a light and al. Wherewith the bishop being made acquainted, came thither with many others: and lifting vp the hive he sawe there neere the top a most fine boxe, wherein the host was laid, and the quires of Bees singing about it, and keeping watch in the night, as monkes do in their cloisters. The bishop therefore taking the host, carried it with the greater honour into the church: whether many resorting were cured of innumerable diseases."[2]

Another legend, from the School of the Eucharist, is as follows:

"A peasant swayed by a covetous mind, being communicated on Easter-Day, received the Host in his mouth, and afterwards laid it among his bees, believing that all the Bees of the neighborhood would come thither to work their wax and honey. This covetous, impious wretch was not wholly disappointed of his hopes; for all his neighbors' Bees came indeed to his hives, but not to make honey, but to render there the honours due to the Creator. The issue of their

---

[1] *Parley of Beasts*, p. 144. London, 1660.
[2] Bozius, *ubi supra*. Butler, *ubi supra*.

arrival was that they melodiously sang to Him songs of praise as they were able; after that they built a little church with their wax from the foundations to the roof, divided into three rooms, sustained by pillars, with their bases and chapiters. They had there also an Altar, upon which they had laid the precious Body of our Lord, and flew round about it, continuing their musick. The peasant . . . . coming nigh that hive where he had put the H. Sacrament, the Bees issued out furiously by troops, and surrounding him on all sides, revenged the irreverence done to their Creator, and stung him so severely that they left him in a sad case. This punishment made this miserable wretch come to himself, who, acknowledging his error, went to find out the parish priest to confess his fault to him . . . . " etc.[1]

We quote also another from the School of the Eucharist:

"A certain peasant of Auvergne, a province in France, perceiving that his Bees were likely to die, to prevent this misfortune, was advised, after he had received the communion, to reserve the Host, and to blow it into one of the hives. As he tried to do it, the Host fell on the ground. Behold now a wonder! On a sudden all the Bees came forth out of their hives, and ranging themselves in good order, lifted the Host from the ground, and carrying it in upon their wings, placed it among the combes. After this the man went out about his business, and at his return found that this advice had succeeded ill, for all his Bees were dead . . . ."[2]

We will close this series of legends with one from the Lives of the Saints:

"When a thief by night had stolen St. Medard's Bees, they, in their master's quarrel, leaving their hive, set upon the malefactor, and eagerly pursuing him which way soever he ran, would not cease stinging of him until they had made him (whether he would or no) to go back again to their master's house; and there, falling prostrate at his feet, submissly to cry him mercy for the crime committed. Which being done, so soon as the Saint extended unto him the hand of benediction, the Bees, like obedient servants, did forthwith stay from persecuting him, and evidently yielded

---

[1] Vicentius in *Spec. Moral.*, B. 2, D. 21, p. 3. N. and Q., x. 499.
[2] Pet. Cluniac, B. 1, c. i. N. and Q., x. 499.

themselves to the ancient possession and custody of their master."[1]

By the Greeks, Bees were accounted an omen of future eloquence;[2] the soothsayers of the Romans, however, deemed them always of evil augury.[3] They afforded also to the Romans presages of public interest, "clustering, as they do, like a bunch of grapes, upon houses or temples; presages, in fact, that are often accounted for by great events."[4] The instances of happy omens afforded by swarms of Bees are the following:

"It is said of Pindar," we read in Pausanias' History of Greece, "that when he was a young man, as he was going to Thespia, being wearied with the heat, as it was noon, and in the height of summer, he fell asleep at a small distance from the public road; and that Bees, as he was asleep, flew to him and wrought their honey on his lips. This circumstance first induced Pindar to compose verses."[5]

A similar incident is mentioned in the life of Plato:

"Whilst *Plato* was yet an infant carried in the arms of his mother *Perictione*, *Aristo* his father went to *Hymettus* (a mountain in *Attica* eminent for abundance of Bees and Honey) to sacrifice to the Muses or Nymphs, taking his Wife and Child along with him; as they were busied in the Divine Rites, she laid the Child in a Thicket of Myrtles hard by; to whom, as he slept (*in cunis dormienti*) came a Swarm of Bees, Artists of Hymettian Honey, flying and buzzing about him, and (as it is reported) made a Honeycomb in his mouth. This was taken for a presage of the singular sweetness of his discourse; his future Eloquence foreseen in his infancy."[6]

From Butler's Lives of the Saints we have the following:

"The birth of St. Ambrose happened about the year 340 B.C., and whilst the child lay asleep in one of the courts of

---

[1] Quot. in *Notes and Queries*, x. 499.

[2] Harwood, *Grec. Antiq.*, p. 200.

[3] Pliny, *Nat. Hist.*, ix. 18   [4] *Ibid.*

[5] Paus. *Hist. of Greece*, B. ix. c. xxiii. 3.

[6] Stanley's *Hist. of Philos.*, Pt. V. c. ii. p. 157, Lond. 1701. Cf. Pliny, *Nat. Hist.*, xi. 18.

Vide Pierius, *Hieroglyph.*, p. 261–5. Populus regi suo obsequens: Rex: Regnum: Grata eloquentia; Poeticæ amœnitas; Futuri seculi beatitudo: Dulcium appetitus; Diuturnæ valetudinis prosperitas; Meretrix: Exoticæ disciplinæ; Prophetarum oracula, etc.

his father's palace, a swarm of Bees flew about his cradle, and some of them even crept in and out at his mouth, which was open; and at last mounted up into the air so high, that they quite vanished out of sight. This," concludes the Reverend Alban, "was esteemed a presage of future greatness and eloquence."[1]

Another instance is mentioned in the Feminine Monarchie, printed at Oxford in 1634, p. 22.

"When *Ludovicus Vives* was sent by Cardinal Wolsey to Oxford, there to be a public professor of Rhetoric, being placed in the College of Bees, he was welcomed thither by a swarm of Bees; which sweet creatures, to signifie the incomparable sweetnesse of his eloquence, settled themselves over his head, under the leads of his study, where they have continued to this day. . . . . How sweetly did all things then accord, when in this neat μουσαίον newly consecrated to the Muses, the Muses' sweetest favorite was thus honoured by the Muses' birds."[2]

Moufet, in his Theater of Insects, and Topsel, in almost the same words in his History of Four-footed Beasts and Serpents, gives the following list of remarkable omens drawn from Bees:

"Whereas the most high God did create all other creatures for our use; so especially the Bees, not only that as mistresses they might hold forth to us a patern of politick and œconomic vertues, and inform our understanding; but that they might be able as extraordinary foretellers, to foreshew the success and event of things to come; for in the years 90, 93, 113, 208, before the birth of Christ, when as mighty huge swarms of Bees did settle in the chief marketplace, and in the beast-market upon private citizens' houses, and on the temple of Mars, there were at that time stratagems of enemies against Rome, wherewith the whole state was like to be surprised and destroyed. In the reign of Severus, the Bees made combes in his military ensigns, and especially in the camp of Niger. Divers wars upon this ensued between both the parties of Severus and Niger, and battels of doubtful event, while at length the Severian faction prevailed. The statues also of Antonius Pius placed

---

[1] *Lives of the Saints*, xii. 106.

[2] Quot. in N. and Q., x. 500. This story is not in the *Fem. Monarchie* of 1609, printed for Jos. Barnes.

here and there all over Hetruria, were all covered with swarms of Bees; and after that settled in the camp of Cassius; what great commotions after followed Julius Capitolinus relates in his history. At what time also, through the treachery of the Germans in Germany, there was a mighty slaughter and overthrow of the Romans. P. Fabius, and Q. Elius being consuls in the camp of Drusus in the tent of Hostilius Rutilus, a swarm of Bees is reported to have sate so thick, that they covered the rope and the spear that held up the tent. M. Lepidus, and Munat. Plancus being consuls, as also in the consulship of L. Paulus, and C. Metellus, swarms of Bees flying to Rome (as the augurs very well conjectured) did foretell the near approach of the enemy. Pompey likewise making war against Cæsar, when he had called his allies together, he set his army in order as he went out of Dyrrachium, Bees met him and sate so thick upon his ensigns that they could not be seen what they were. Philistus and Ælian relate, that while Dionysius the tyrant did in vain spur his horse that stuck in the mire, and there at length left him, the horse quitting himself by his own strength, did follow after his master the same way he went with a swarm of Bees sticking on his mane; intimating by that prodigy that tyrannical government which Dionysius affected over the Galeotæ. In the Helvetian History we read, that in the year 1385, when Leopoldus of Austria began to march towards Sempachum with his army, a swarm of Bees flew to the town and there sate upon the tyles; whereby the common people rightly foretold that some forain force was marching towards them. So Virgil, in 7 Æneid:

> The Bees flew buzzing through the liquid air:
> And pitcht upon the top o' th' laurel tree;
> When the Soothsayers saw this sight full rare,
> They did foretell th' approach of th' enemie.

That which Herodotus, Pausanias, Dio Cassius, Plutarch, Julius Cæsar, Julius Capitolinus, and other historians with greater observation then reason have confirmed. Saon Acrephniensis, when he could by no means finde the oracle Trophonius; Pausanias in his Bœticks saith he was lead thither by a swarm of Bees. Moreover, Plutarch, Pausanias, Ælian, Alex. Alexandrinus, Theocritus and Textor are authors that Jupiter Melitæus, Hiero of Syracuse, Plato,

Pindar, Apius Comatus, Xenophon, and last of all Ambrose, when their nurses were absent, had honey dropt into their mouths by Bees, and so were preserved."[1]

In East Norfolk, England, if Bees swarm on rotten wood, it is considered portentous of a death in the family.[2] This superstition is as old at least as the time of Gay, for, among the signs that foreshadowed the death of Blonzelind, it is mentioned:

> Swarmed on a rotten stick the Bees I spy'd
> Which erst I saw when Goody Dobson dy'd.[3]

In Ireland, the mere swarming of Bees is looked upon as prognosticating a death in the family of the owner.

In parts of England it is believed, that if a swarm of Bees come to a house, and are not claimed by their owner, there will be a death in the family that hives them.[4]

It is a very ancient superstition that Bees, by their acute sense of smell, quickly detect an unchaste woman, and strive to make her infamy known by stinging her immediately. In a pastoral of Theocritus, the shepherd in a pleasant mood tells Venus to go away to Anchises to be well stung by Bees for her lewd behavior.

> Now go thy way to Ida mount—
>   Go to Anchises now,
> Where mighty oaks, where banks along
>   Of square Cypirus grow,
> Where hives and hollow trunks of trees,
>   With honey sweet abound,
> Where all the place with humming noise
>   Of busie Bees resound.

Incontinence in men, as well as unchastity in women, was thought to be punished by these little insects. Thus in the lines of Pindarus:

> Thou painful Bee, thou pretty creature,
> Who honey-combs six angled, as the be,
> With feet doest frame, false Phœus and impure,
> With sting has prickt for his lewd villany.[5]

---

[1] *Theatr. Ins.*, p. 21-2. Topsel's *Hist. of Beasts and Serpents*, p. 645, 905.

[2] *N. and Q.*, vi. 480.

[3] Gay's *Pastorals*, v. 107–8.

[4] Chambers' *Book of Days*, i. 752.

[5] Plutarch, *Nat. Quest.*, 36. Holl. Trans., p. 831.

Pliny says: "Certain it is, that if a menstruous woman do no more but touch a Bee-hive, all the Bees will be gone and never more come to it again."[1]

In Western Pennsylvania, it is believed that Bees will invariably sting red-haired persons as soon as they approach the hives.

It is a common opinion that Bees in rough and boisterous weather, and particularly in a violent storm, carry a stone in their legs, in order to preserve themselves by its weight against the power of the wind. Its antiquity is also great, for in the writings of Plutarch we find an instance of this remarkable wisdom. "The Bees of Candi," says this philosopher, "being about to double a point or cape lying into the sea, which is much exposed to the winds, they ballase (ballast) themselves with small grit or petty stones, for to be able to endure the weather, and not be carried away against their wills with the winds through their lightness otherwise."[2]

Virgil, too, about a century earlier, mentions this curious notion in the following lines:

> And as when empty barks on billows float,
> With sandy ballast sailors trim the boat;
> So Bees bear gravel stones, whose poising weight
> Steers through the whistling winds their steady flight.[3]

Swammerdam, who has noticed this belief of the ancients, makes the following remarks: "But this, as Clutius justly observes, has not been hitherto remarked by any Bee-keeper, nor indeed have I myself ever seen it. Yet I should think that there may be some truth in this matter, and probably a certain observation, which I shall presently mention, has given rise to the story. There is a species of wild Bees not unlike the smallest kind of the Humble-Bee, which, as they are accustomed to build their nests near stone walls, and construct their habitations of stone and clay, sometimes carry such large stones that it is scarcely credible by what means so tender insects can sustain so great a load, and that even flying while they are obliged also to support their own body.

---

[1] *Nat. Hist.*, xxviii 7.   Holl. Trans., p. 308.
[2] Plutarch, *Land and Water Creatures Compared.*   Holl. Trans., p. 786.
[3] *Georg.* iv. 283-7.   Dryden's Trans.

Their nest by this means is often so heavy as to weigh one or two pounds."[1]

It was the general opinion of antiquity that Bees were produced from the putrid bodies of cattle. Varro says they are called βουγόναι by the Greeks, because they arise from petrified bullocks. In another place he mentions their rising from these putrid animals, and quotes the authority of Archelaus, who says Bees proceed from bullocks, and wasps from horses.[2] Virgil, however, is much more satisfactory, for he gives us the recipe in all its details for producing these insects :

> First, in a place, by nature close, they build
> A narrow flooring, gutter'd, wall'd, and til'd.
> In this. four windows are contriv'd, that strike
> To the four winds oppos'd, their beams oblique.
> A steer of two years old they take, whose head
> Now first with burnished horns begins to spread:
> They stop his nostrils, while he strives in vain
> To breathe free air, and struggles with his pain.
> Knock'd down, he dies : his bowels bruis'd within,
> Betray no wound on his unbroken skin.
> Extended thus, in his obscene abode,
> They leave the beast ; but first sweet flowers are strow'd
> Beneath his body, broken boughs and thyme,
> And pleasing Cassia, just renew'd in prime.
> This must be done, ere spring makes equal day,
> When western winds on curling waters play :
> Ere painted meads produce their flowery crops,
> Or swallows twitter on the chimney tops.
> The tainted blood, in this close prison pent,
> Begins to boil, and thro' the bones ferment.
> Then wond'rous to behold. new creatures rise,
> A moving mass at first, and short of thighs;
> Till shooting out with legs, and imp'd with wings,
> The grubs proceed to Bees with pointed stings:
> And more and more affecting air, they try
> Their tender pinions and begin to fly.[3]

This absurd notion was also promulgated by the great English chronicler, Hollingshed ; for, says this author, "Hornets, waspes, Bees, and such like, whereof we have

---

[1] Swam. *Hist. of Ins.*, Pt. I. p. 226.

[2] Martin's *Georg. of Virgil*, iv. 295, note.

[3] Dryden's *Virgil, Georg.* iv. 417–442. Democritus, said to have been contemporary with Socrates and Hippocrates, the learned Varro, Columella, and Florentinus, have severally given this same receipt. Vide Owen's *Geoponika*, ii. 199.

great store, and of which an opinion is conceived, that the first doo breed of the corruption of dead horses, the second of pears and apples corrupted, and the last of kine and oxen; which may be true, especiallie the first and latter in some parts of the beast, and not their whole substances, as also in the second, sith we never have waspes but when our fruit beginneth to wax ripe."[1]

To conclude the history of this belief, the following remarks of the learned Swammerdam will not be inappropriate. He says: "It is probable that the not rightly understanding Samson's adventure of the Lion, gave rise to the popular opinion of Bees springing from dead Lions, Oxen, and Horses; and this opinion may have been considerably strengthened, and indeed in a manner confirmed, by the great number of Worms that are often found during the summer months in the carcasses of such animals, especially as these Worms somewhat resemble those produced from the eggs of Bees. However ridiculous this opinion must appear, many great men have not been ashamed to adopt and defend it. The industrious Goedaert has ventured to ascribe the origin of Bees to certain dunghill Worms, and the learned de Mei joins with him in this opinion; though neither of them had any observation to ground their belief upon, but that of the external resemblance between the Bee and a certain kind of Fly produced from these Worms."[2]

The opinion that stolen Bees will not thrive, but pine away and die, is almost universal.[3] It is, too, of reverend antiquity, for Pliny mentions it: "It is a common received opinion, that Rue will grow the better if it be filched out of another man's garden; and it is as ordinarie a saying that stolen Bees will thrive worst."[4]

In South Northamptonshire, England, there is a superstition that Bees will not thrive in a quarrelsome family.[5] It might be well to promulgate this and the next preceding superstition. This prevails among us.

In Hampshire, England, it is a common saying that Bees are idle or unfortunate at their work whenever there are

---

[1] Hollings. *Chron.*, i. 384.
[2] Swam. *Hist. of Ins.*, Pt. I. p. 228.
[3] N. and Q., ii. 356.
[4] *Nat. Hist.*, xix. 7.　Holl. *Trans.*, p. 23. E.
[5] N. and Q., ii. 165.　Chamb. *Bk. of Days*, i. 752.

wars. A very curious observer and fancier says that this has been the case from the time of the movements in France, Prussia, and Hungary, up to the present time.[1]

In Bishopsbourne, England, there prevails the singular superstition of informing the Bees of any great public event that takes place, else they will not thrive so well.[2]

In Monmouthshire, England, the peasantry entertain so great a veneration for their Bees, that, says Bucke, some years since, they were accustomed to go to their hives on Christmas eve at twelve o'clock, in order to listen to their humming; which elicited, as they believed, a much more agreeable music than at any other period; since, at that time, they celebrated, in the best manner they could, the morning of Christ's nativity.[3]

Sampson, in his Statistical Survey of the County of Londonderry, 1802, p. 436, says that there "Bees must not be given away, but sold; otherwise neither the giver nor the taker will have *luck*."[4]

A clergyman in Devonshire, England, informs us that when any Devonian makes a purchase of Bees, the payment is never made in money, but in things (corn, for instance) to the value of the sum agreed upon; and the Bees are never removed but on a Good Friday.[5] In western Pennsylvania, it is thought by some of the old farmers that the vender of the Bees must be away from home when the hive is taken away, else the Bees will not thrive.

Another superstition is that if a swarm of Bees be met with in an open field away from any house, it is useless to hive them, for they will never do a bit of good.

In many parts of England, a popular opinion is that when Bees remove or go away from their hives, the owner of them will die soon after.[6]

It is commonly believed among us that if Bees come to a house, it forebodes good luck and prosperity; and, on the contrary, if they go away, bad luck.

A North German custom and superstition is, that if the master of the house dies, a person must go to the Beehive,

---

[1] *N. & Q.*, xii. 200.
[2] *Mag. of Nat. Hist.*, ii. 405.
[3] Bucke *on Nature*, i. 419.
[4] Brand's *Pop. Antiq.*, ii. 300.
[5] *Ibid.*   [6] *Ibid.*

knock, and repeat these words: "The master is dead, the master is dead," else the Bees will fly away.[1] This superstition prevails also in England, Lithuania, and in France.[2]

[Some years since, observes a correspondent of the Athenæum, quoted by Brande, a gentleman at a dinner table happened to mention that he was surprised, on the death of a relative, by his servant inquiring "whether his master would inform the Bees of the event, or whether *he* should do so." On asking the meaning of so strange a question, the servant assured him that Bees ought always to be informed of a death in a family, or they would resent the neglect by deserting the hive. This gentleman resides in the Isle of Ely, and the anecdote was told in Suffolk; and one of the party present, a few days afterward, took the opportunity of testing the prevalence of this strange notion by inquiring of a cottager who had lately lost a relative, and happened to complain of the loss of her Bees, "whether she had told them all she ought to do?" She immediately replied, "Oh, yes; when my aunt died I told every skep (*i.e.* hive) myself, and put them . . . .

"Into mourning." I have since ascertained the existence of the same superstition in Cornwall, Devonshire (where I have seen black crape put round the hive, or on a small black stick by its side), and Yorkshire. It probably exists in every part of the kingdom. . . . . The mode of communicating is by whispering the fact to each hive separately. . . . . In Oxford I was told that if a man and wife quarreled, the Bees would leave them.][3]

"In some parts of Suffolk," says Bucke, "the peasants believe, when any member of their family dies, that, unless the Bees are put in mourning by placing a piece of black cloth, cotton or silk, on the top of the hives, the Bees will either die or fly away.

"In Lithuania, when the master or mistress dies, one of the first duties performed is that of giving notice to the Bees, by rattling the keys of the house at the doors of their hives. Unless this be done, the Lithuanians imagine the

---

[1] Thorpe's *North. Mythol.*, iii. 161.
[2] Vide *N. and Q.* in Devon, v. 148; Essex, v. 437; Lincolnshire, iv. 270; Surrey, iv. 291; a Cornish superstition, too, xii. 88; in Buckinghamshire, Sussex, Lithuania, and France, iv. 308.
[3] Brande's *Pop. Antiq.*, ii. 300.

cattle will die; the Bees themselves perish, and the trees wither."[1]

At Bradfield, if Bees are not invited to funerals, it is believed they will die.[2]

In the Living Librarie, Englished by John Molle, 1621, p. 283, we read: "Who would beleeve without superstition (if experience did not make it credible), that most commonly all the Bees die in their hives, if the master or mistress of the house chance to die, except the hives be presently removed into some other place? And yet I know this hath hapned to folke no way stained with superstition."[3]

A similar superstition is, that Beehives belonging to deceased persons should be turned over the moment when the corpse is taken out of the house.[4] No consequence is given for the non-performance of this rite.

The following item is clipped from the Argus, a London newspaper, printed Sept. 13, 1790: "A superstitious custom prevails at every funeral in Devonshire, of turning round the Bee-hives that belonged to the deceased, if he had any, and that at the moment the corpse is carrying out of the house. At a funeral some time since, at Columpton, of a rich old farmer, a laughable circumstance of this sort occurred: for, just as the corpse was placed in the hearse, and the horsemen, to a large number, were drawn in order for the procession of the funeral, a person called out, 'Turn the Bees,' when a servant who had no knowledge of such a custom, instead of turning the hives about, lifted them up, and then laid them down on their sides. The Bees, thus hastily invaded, instantly attacked and fastened on the horses and their riders. It was in vain they galloped off, the Bees as precipitately followed, and left their stings as marks of indignation. A general confusion took place, attended with loss of hats, wigs, etc., and the corpse during the conflict was left unattended; nor was it till after a considerable time that the funeral attendants could be rallied, in order to proceed to the interment of their deceased friend."[5]

After the death of a member of a family, it has fre-

---

[1] Bucke *on Nature*, i. 413, note.
[2] *N. and Q.*, iv. 309.
[3] Brand's *Pop. Antiq.*, ii. 300.
[4] Fosbr. *Encycl. of Antiq.*, ii. 738.
[5] Brand's *Pop. Antiq.*, ii. 300.

quently been asserted that the Bees sometimes take their
loss so much to heart as to alight upon the coffin whenever
it is exposed. A clergyman told Langstroth, that he attended a funeral, where, as soon as the coffin was brought
from the house, the Bees gathered upon it so as to excite
much alarm. Some years after this occurrence, being engaged in varnishing a table, the Bees alighted upon it in
such numbers as to convince the reverend gentleman that
love of varnish, rather than sorrow or respect for the dead,
was the occasion of their conduct at the funeral.[1]

The following is an extract from a *Tour through Brit-
tany*, published in the Cambrian *Quarterly Magazine*, vol. ii.
p. 215: "If there are Bees kept at the house where a marriage feast is celebrated, care is always taken to dress up
their hives in red, which is done by placing upon them
pieces of scarlet cloth, or one of some such bright color;
the Bretons imagining that the Bees would forsake their
dwellings if they were not made to participate in the re-
joicings of their owners: in like manner they are all put
into mourning when a death occurs in a family."[2]

In the Magazine of Natural History we find the following
instance of singing psalms to Bees to make them thrive:
"When in Bedfordshire lately, we were informed of an old
man who sang a psalm last year in front of some hives
which were not doing well, but which, he said, would thrive
in consequence of that ceremony. Our informant could not
state whether this was a local or individual superstition."[3]

It is commonly said that if you sing to your Bees before
they swarm, it will prevent their leaving your premises
when they do swarm.

Peter Rotharmel, a western Pennsylvanian, had a singular notion that no man could have at one time a hundred
hives of Bees. He declared he had often as many as ninety-
nine, but could never add another to them.[4] I have since

---

[1] Langstroth *on Honey-Bee*, p. 80.
[2] *Mag. of Nat. Hist.*, iii. 211, note.
[3] *Ibid.*, i. 303. London, 1829.
[4] Peter Rotharmel had three specialties: Bees, Wheat, and Bonaparte. Concerning Bees, he had many strange notions, but the
above recorded is the only one of which I have any positive information. Concerning wheat, at one time in his life he purchased
an almanac, which indicated, among other things, the high and low
tides, and, from studying this, he got it into his head that the fluctua-

learned that this is not an individual superstition, but one that pretty generally prevails.

The Apiarians of Bedfordshire, England, have a custom of, as they call it, ringing their swarms with the door-key and the frying-pan; and if a swarm settles on another's premises, it is irrecoverable by the owner, unless he can prove the ringing, but it becomes the property of that person upon whose premises it settles.[1]

The practice of beating pans, and making a great noise to induce a swarm of Bees to settle, is, at least, as old as the time of Virgil.   He thus mentions it:

> But when thou seest a swarming cloud arise,
> That sweeps aloft, and darkens all the skies:
> The motions of their hasty flight attend;
> And know to floods, or woods, their airy march they bend.
> Then melfoil beat, and honey-suckles pound,
> With these alluring savors strew the ground,
> And mix with tinkling brass the cymbal's drowning sound.[2]

But concerning this practice, Langstroth says: "It is probably not a whit more efficacious than the hideous noises of some savage tribes, who, imagining that the sun, in an eclipse, has been swallowed by an enormous dragon, resort to such means to compel his snakeship to disgorge their favorite luminary."[3]

Dr. Toner, the author of that very interesting little work, "Maternal Instinct or Love," informs me that when a boy he witnessed a mode of alluring a swarm of Bees to settle, performed by a German man and his wife, which struck him at the time as being remarkable, and which was as follows: Having first put some pig-manure upon the hive into which

---

tions in the price of wheat were intimately connected with the rise and fall of the tides. So impressed was he with this idea, that he ever afterward yearly bought that particular almanac, and prophesied from it to his neighbors the probable value of their coming crops of wheat. On Sunday, he would walk fifteen and twenty miles through the country, to examine the different wheat-fields, and to afford him a topic of conversation for the ensuing week. But Napoleon was his principal study and his greatest mania. On him he would talk for hours, on the slightest provocation. The history of Bonaparte and his campaigns, which he only read, was an old German one.

[1] *Mag. of Nat. Hist.*, ii. 209.
[2] *Geog.*, Dryden's *Trans.*, iv. 82–9.
[3] *On the Honey-Bee*, p. 113.

they wished the Bees to go, they ran to and fro under the swarm, singing a monotonous German hymn; and this they continued till the Bees were settled and hived.

Another strange mode of alluring Bees into a new hive is practiced near Gloucester, England, but only when all the usual ways of preparing hives fail; it is this: When a swarm is to be hived, instead of moistening the inside of the hive with honey, or sugar and water, the Bee-master throws into it, inverted, about a pint of beans, which he causes a sow to devour, and immediately then, it is said, will the Bees take to it.[1]

Pliny, as follows, incidentally mentions another curious mode of preparing the hives to best suit the Bees: "Touching Baulme, which the Greeks call Melittis or Melissophyllon: if Bee-hives be rubbed all over and besmeared with the juice thereof, the Bees will never go away; for there is not a flower whereof they be more desirous and faine than of it."[2]

Borlase, in his Antiquities of Cornwall, p. 168, tells us of another strange practice in the hiving of Bees. He says: "The Cornish, to this day, invoke the spirit of Browny, when their Bees swarm; and they think that their crying Browny, Browny, will prevent them from returning into the former hive, and make them pitch and form a new colony."[3]

The Rev. Thomas P. Hunt, of Wyoming, Pa., has devised an amusing plan, by which he says he can, at all times, prevent a swarm of Bees from leaving his premises. Before his stocks swarm, he collects a number of dead Bees, and, stringing them with a needle and thread, as worms are strung for catching eels, makes of them a ball about the size of an egg, leaving a few strands loose. By carrying—fastened to a pole—this "*Bee-bob*" about his Apiary, when the Bees are swarming, or by placing it in some central position, he invariably secures every swarm.[4]

The barbarous practice of killing Bees for their honey, not yet entirely abolished, did not exist in the time of Aristotle, Varro, Columella, and Pliny. The old cultivators

---

[1] *N. and Q.*, 2d Ser., ix. 443.
[2] *Nat. Hist.*, xxi. 20, Holl *Trans.*, p. 106. K.
[3] Quot. in Brand's *Pop. Antiq.*, iii 225.
[4] *Langstroth on the Honey-Bee*, p. 132.

took only what their Bees could spare, killing no stocks except such as were feeble or diseased. The following epitaph, taken from a German work, might well be placed over every pit of these brimstoned insects:

HERE RESTS,

CUT OFF FROM USEFUL LABOR,

A COLONY OF

INDUSTRIOUS BEES,

BASELY MURDERED

BY ITS

UNGRATEFUL AND IGNORANT

OWNER.

To the epitaph also may be appended Thomson's verses:

Ah, see, where robbed and murdered in that pit,
Lies the still heaving hive! at evening snatched,
Beneath the cloud of guilt-concealing night,
And fixed o'er sulphur! while, not dreaming ill,
The happy people, in their waxen cells,
Sat tending public cares.
Sudden, the dark, oppressive steam ascends,
And, used to milder scents, the tender race,
By thousands, tumble from their honied dome
Into a gulf of blue sulphureous flame![1]

It is considered very cruel in Africa, as Campbell observes, to kill Bees in order to obtain their honey, especially as from flowers being there at all seasons, and most in winter, they can live comfortably all the year round. A Hottentot, who was accustomed to kill the Bees, was often reasoned with by the humane to give up so cruel a practice, yet he persisted in it till a circumstance occurred which determined him to relinquish it. He had a water-mill for grinding his corn, which went very slowly, from the smallness of the stream which turned it; consequently the flour dropped very gently. For some time much less than usual came into the sack, the cause of which he could not discover. At length he found that a great part of his flour, as it was ground, was carried off by the Bees to their hives: on examining this, he

___

[1] Quot. by Langstroth *on the Honey-Bee.* p. 231.

found it contained only his flour, and no honey. This robbery made him resolve to destroy no more Bees when their honey was taken, considering their conduct in robbing him of his property as a just punishment to him for his cruelty. The gentleman who related this story, Mr. Campbell says, was a witness to the Bees robbing the mill.[1]

An old English proverb, relative to the swarming of Bees, is,—

> A swarm of Bees in May,
> Is worth a load of hay;
> A swarm of Bees in June,
> Is worth a silver spoon;
> A swarm of Bees in July,
> Is not worth a fly.[2]

In Tusser's Five Hundred Points of Husbandry, under the month of May, are these lines:

> Take heed to thy Bees, that are ready to swarme,
> The losse thereof now is a crown's worth of harme.

On which is the following observation in Tusser Redivivus, 1744, p. 62: "The tinkling after them with a warming-pan, frying-pan, kettle, is of good use to let the neighbors know you have a swarm in the air, which you claim wherever it lights; but I believe of very little purpose to the reclaiming of the Bees, who are thought to delight in no noise but their own."

Ill fortune attends the killing of Bees,—a common saying. This, doubtless, arose from the thrift and usefulness of these insects.

That swarms of Bees, or fields, houses, stalls of cattle, or workshops, may not be affected by enchantment, Leontinus says: "Dig in the hoof of the right side of a sable ass under the threshold of the door, and pour on some liquid pitchy resin, salt, Heracleotic origanum, cardamonium, cumin, some fine bread, squills, a chaplet of white or of crimson wool, the chaste tree, vervain, sulphur, pitchy torches; and lay on some amaranthus every month, and lay on the mould; and, having scattered seeds of different kinds, let them remain."[3]

---

[1] Campbell's *Travels in S. Africa*, p. 339.
[2] *Percy Soc. Public.*, iv. 99.
[3] Owen's *Geoponika*, ii. 109–10.

To cure the stings of Bees, we have the following remedies :
"Rue," says Pliny, "is an hearbe as medicinable as the best
. . . and is available against the stings of Bees, Hornets,
and Wasps, and against the poison of the Cantharides and
Salamanders.[1]

"Yea, and it is an excellent thing for them that be stung,
to take the very Bees in drinke; for it is an approved
cure.[2]

"Baulme is a most present remedy not only against their
stings, but also of Wespes, Spiders, and Scorpions.[3]

"The Laurell, both leafe, barke, and berrie, is by nature
hot; and applied as a liniment, be singular good for the
pricke or sting of Wasps, Hornets, and Bees.[4]

"For the sting of Bees, Wasps, and Hornets, the Howlat
(owlet) is counted a soveraigne thing, by a certaine antipa-
thie in nature.[5]

"Moreover, as many as have about them the bill of a
Woodspeck (Woodpecker) when they come to take honey
out of the hive, shall not be stung by Bees."[6]

It is said that if a man suffers himself to be stung by Bees,
he will find that the poison will produce less and less effect
upon his system, till, finally, like Mithridates of old, he will
appear to almost thrive upon poison itself. When Langstroth' first became interested in Bees, according to his statement, a sting was quite a formidable thing, the pain being
often intense, and the wound swelling so as sometimes to
obstruct his sight. But, at length, however, the pain was
usually slight, and, if the sting was quickly extracted, no
unpleasant consequences ensued, even if no remedies were
used. Huish speaks of seeing the bald head of Bonner, a
celebrated practical Apiarian, covered with stings, which
seemed to produce upon him no unpleasant effects. The
Rev. Mr. Kleine advises beginners to suffer themselves to be
stung frequently, assuring them that, in two seasons, their
systems will become accustomed to the poison. An old
English Apiarian advises a person who has been stung, to

---

[1] *Nat. Hist.*, xx. 13.   Holl., p. 56. M.
[2] *Ibid.*, Holl., p. 95. A.
[3] *Ibid.*, xxi. 20.   Holl., p. 106. K.
[4] *Ibid.*, xxiii. 18.   Holl., p. 173. A.
[5] *Ibid.*, xxix. 4.   Holl., p. 361. D.
[6] *Ibid.*, xxx. 16.   Holl., p. 399. F.

catch as speedily as possible another Bee, and make it sting
on the same spot.[1]

It is generally believed among our boys that if the part
stung by a Bee be rubbed with the leaves of three different
plants at the same time, the pain will be relieved.

Willsford, in his Nature's Secrets, p. 134, says: "Bees,
in fair weather, not wandering far from their hives, presage
the approach of some stormy weather. . . . Wasps, Hornets,
and Gnats, biting more eagerly than they used to do, is a
sign of rainy weather."[2]

The prognostication drawn from a flight of Bees, in which
there is doubtless much truth, appears from the following
lines to have been known to Virgil:

> Nor dare they stay,
> When rain is promised, or a stormy day:
> But near the city walls their watering take,
> Nor forage far, but short excursions make.[3]

Bees were employed as the symbol of Epeses; they are
common also on coins of Elyrus, Julis, and Præsus.[4]

One of the most remarkable facts in the history of Bees
is that passage in the Bible[5] about the swarm of these insects and honey in the carcass of the lion slain by Samson.
Some look upon it as a paradox, others as altogether incredible; but it admits of easy explanation. The lion had
been dead some little time before the Bees had taken up
their abode in the carcass, for it is expressly stated that
"after a time," Samson returned and saw the Bees and
honey in the carcass, so that "if," as Oedman has well observed, "any one here represents to himself a corrupt and
putrid carcass, the occurrence ceases to have any true simili-
tude, for it is well known in these countries, at certain seasons of the year, the heat will in twenty-four hours so completely dry up the moisture of dead animals, and that without their undergoing decomposition, that their bodies long
remain, like mummies, unaltered, and entirely free from
offensive odor." To the foregoing quotation we may add
that very probably the larvæ of flies, ants, and other insects,

---

[1] Langstroth *on the Honey-Bee*, p. 316, note.
[2] Brand's *Pop. Antiq.*, iii. 225.
[3] *Georg.*, iv. 280-4; Dryden's *Trans.*
[4] Fosb. *Encycl. of Antiq.*, ii. 738.
[5] *Judg.* xiv. 8.

which at the time when Bees swarm, are to be found in great numbers, would help to consume the carcass, and leave perhaps in a short time little else than a skeleton.[1]

An instance of Bees tenanting a dead body is found in the following passage from the writings of Herodotus: "Now the Amathusians, having cut off the head of Onesilus, because he had besieged them, took it to Amatheus, and suspended it over the gates; and when the head was suspended, and had become hollow, a swarm of Bees entered it, and filled it with honey-comb. When this happened, the Amathusians consulted the oracle respecting it, and an answer was given them, 'that they should take down the head and bury it, and sacrifice annually to Onesilus, as to a hero; and if they did so, it would turn out better for them.' The Amathusians did accordingly, and continued to do so until my time."[2]

Another singular instance is mentioned by Napier in his Excursions on the shores of the Mediterranean: "Among this pretty collection of natural curiosities (in the cemetery of Algesiras), one in particular attracted our attention; this was the contents of a small uncovered coffin in which lay a child, the cavity of the chest exposed and tenanted by an industrious colony of Bees. The comb was rapidly progressing, and I suppose, according to the adage of the poet, they were adding sweets to the sweet, if not perfume to the violet."[3]

Butler, in his Feminine Monarchie, narrates the following curious story: "*Paulus Jovius* affirmeth that in *Muscovia*, there are found in the woods & wildernesses great lakes of honey, which the Bees have forsaken, in the hollow truncks of marvelous huge trees. In so much that hony & waxe are the most certaine commodities of that countrie. Where, by that occasion, he setteth down the storie reported by *Demetrius* a *Muscovite* ambassador sent to Rome. A neighbor of mine (saith he) searching in the woods for hony slipt downe into a great hollow tree, and there sunk into a lake of hony vp to his brest: where when he had stucke faste two daies calling and crying out in vaine for helpe, be-

---

[1] Cf. Swammerdam, *Hist. of Ins.*, Pt. I. p. 227, and Smith's *Dict. of the Bible.*

[2] Herod., v. 114–5.

[3] *Excursions*, i. 127.

cause no bodie in the meane while came nigh that solitarie place ; at length when he was out of all hope of life, hee was strangely delivered by the means of a great beare : which coming thither about the same businesse that he did, and smelling the hony stirred with his striving, clambered vp to the top of the tree, & thence began to let himselfe downe backward into it. The man bethinking himself, and knowing the worst was but death, which in that place he was sure of, beclipt the beare fast with both his hands aboit the loines, and withall made an outcry as lowd as he could. The beare being thus sodainely affrighted, what with the handling, & what with the noise, made vp againe withal speed possible : the man held, & the beare pulled, vntil with main force he had drawne *Dun out of the mire :* & then being let go, away he trots *more afeard than hurt*, leaving the smeered swaine in a joyful feare."[1]

By the Chinese writers, the composition of the characters for the Bee, Ant, and Mosquito, respectively, denote the *awl* insect, the *righteous* insect, and the *lettered* insect ; referring thereby to the sting of the first, the orderly marching and subordination of the second, and the letter-like markings on the wings of the last.[2]

In May, 1653, the remains of Childeric, King of the Franks, who died A.D. 481, and was buried at Tournay, were discovered ; and among the medals, coins, and books, which were found in his tomb, were also found above three hundred figures of, as Chiflet says, Bees, all of gold. Some of these figures were toads, crescents, lilies, spear-heads, and such like, but Chiflet, after much labor and research, was fully convinced they were Bees ; and, more than that, determines them to be the source whence the *Fleur de lis* in the Arms of France were afterward derived. Montfaucon, however, did not hesitate to say they were nothing more than ornaments of the horse-furniture.[3]

Napoleon I. and II. are said to have had their imperial robes embroidered with golden Bees, as claiming official descent from Carolus Magnus, who is said to have worn them on his coat of arms.[4]

---

[1] *Fem. Monarchie*, c. vi. 49.

[2] Williams' *Chinese Empire*, i. 275.

[3] Chiflet, 164–181 ; Montf. *Monarch. Franc.*, i. 12 ; Gough's *Sepul. Mon.*, vol. i. p. lxii.

[4] Cf. *N. & Q.*, vii. 478, 553 ; viii. 30.

On a Continental forty-five dollar bill. issued on the 14th of January, 1779, is represented an Apiary in which two Beehives are visible, and Bees are seen swarming about. The motto is "Sic floret Respublica—Thus flourishes the Republic." It conveys the simple lesson that by industry and frugality the Republic would prosper.[1]

Bees in the heroic ages it appears were not confined in hives; for, whenever Homer describes them, it is either where they are streaming forth from a rock,[2] or settling in bands and clusters on the spring flowers. Hesiod, however, soon after makes mention of a hive where he is uncourteously comparing women to drones :

> As when within their well-roof'd hives the Bees
> Maintain the mischief-working drones at ease,
> Their task pursuing till the golden sun
> Down to the western wave his course hath run,
> Filling their shining combs, while snug within
> Their fragrant cells, the drones, with idle din
> As princes revel o'er their unpaid bowls,
> On others' labors cheer their worthless souls.[3]

It may be surprising to many to know that Bees were not originally natives of this country. But such is the case; the first planters never saw any. The English first introduced them into Boston, and in 1670, they were carried over the Alleghany Mountains by a hurricane.[4] Since that time, it has been remarked they betray an invariable tendency for migrating southward.[5]

Bees for a long time were known to our Indians by the name of "English Flies;"[6] and they consider them, says Irving, as the harbinger of the white man, as the buffalo is of the red man, and say that in proportion as the Bee advances, the Indian and the buffalo retire.[7]

Longfellow, in his Song of Hiawatha, in describing the advent of the European to the New World, makes his Indian warrior say of the Bee and the white clover :

---

[1] Harper's *New Monthly Mag.*, xxvi. 441.
[2] *Il. ζ.* 87 ; *μ.* 67 ; *Odyss.*, v. 106.
[3] Hesiod, Theog., 594, seq.
[4] Bucke *on Nature*, ii. 75.
[5] Cf. Kalm, ii. 427 ; Schneider, Observ. sur Ulloa, ii. 198.
[6] *Ibid.*
[7] *Tour in the Prairies*, ch. ix.

> Wheresoe'er they move, before them
> Swarms the stinging fly, the Ahmo,
> Swarms the Bee, the honey-maker:
> Wheresoe'er they tread, beneath them
> Springs a flower unknown among us,
> Springs the White Man's Foot in blossom.

Many Apiarians contend that newly-settled countries are most favorable to the Bee; and an old German adage runs thus:

> Bells' ding dong,   But hoot of owl,
> And choral song,   And "wolf's long howl"
> Deter the bee   Incite to moil
> From industry: And steady toil.[1]

Hector St. John, in his Letters, gives the following curious account of the method which he employed in discovering Bees in our woods in early times: Provided with a blanket, some provisions, wax, vermilion, honey, and a small pocket compass, he proceeded to such woods as were at a considerable distance from the settlements. Then examining if they abounded with large trees, he kindled a small fire on some flat stones, close by which putting some wax, and, on another stone near by, dropping distinct drops of honey, which he encircled with the vermilion. He then retired to carefully watch if any Bees appeared. The smell of the burnt wax, if there were any Bees in the neighborhood, would unavoidably attract them; and, finding the honey, would necessarily become tinged with the vermilion, in attempting to get at it. Next, fixing his compass, he found out the direction of the hives by the flight of the loaded Bees, which is invariably straight when they are returning home. Then timing with his watch the absence of the Bee till it would come back for a second load, and recognizing it by the vermilion, he could generally guess pretty closely to the distance traversed by it in the given time. Knowing then the direction and the probable distance, he seldom failed in going directly to the right tree. In this way he sometimes found as many as eleven swarms in one season.[2]

The shepherds of the Alps, as we learn from Sausure quoted in the Insect Miscellanies, as soon as the snows are melted on the sides of the mountains, transfer their flocks from the

---

[1] Langstroth *on the Honey-Bee*, p. 236.
[2] *Letters.*

valleys below to the fresh pasture revived by the summer sun, in the natural parterres and patches of meadow-land formed at the foot of crumbling rocks, and sheltered by them from mountain storms; and so difficult sometimes is this transfer to be accomplished, that the sheep have to be slung by means of ropes from one cliff to another before they can be stationed on the little grass-plot above.[1] A similar artificial migration (if we may use the term), continues the author of the Miscellanies, is effected in some countries by the proprietors of Beehives, who remove them from one district to another, that they may find abundance of flowers, and by this means prolong the summer. Sometimes this transfer is performed by persons forming an ambulatory establishment, like that of a gipsy horde, and encamping wherever flowers are found plentiful. Bee-caravans of this kind are reported to be not uncommon in some districts of Germany;[2] and in parts of Greece,[3] Italy, and France,[4] the transportation of Bees was practiced from very early times. But a more singular practice in such transportation was to set the Beehives afloat in a canal or river, and we are informed that, in France, one Bee-barge was built of capacity enough for from sixty to one hundred hives, and by floating gently down the river, the Bees had an opportunity of gathering honey from the flowers along the banks.[5]

An instance of Bees being kept in this singular manner is found in the following quotation from the London Times, 1830: "As a small vessel was proceeding up the Channel from the coast of Cornwall, and running near the land, some of the sailors observed a swarm of Bees on an island; they steered for it, landed, and took the Bees on board; succeeded in hiving them immediately, and proceeded on their voyage;

---

[1] *Voyages dans les Alpes. Ins. Misc.*, p. 262.

[2] Brookes mentions the Duchy of Juliers, a district of Westphalia, Germany.—*Nat. Hist. of Ins.*, p. 160.

[3] Columella says the Greeks were accustomed, every year, to remove the hives from Achaia into Attica.—*Ibid.*

[4] One person in particular, in the territory called Gatonois, has been at the pains of removing his hives, after the harvest of Sainfoin, into the plains of Beauce, where the melilot abounds, and thence into Sologne, where it is well known the Bees may enjoy the advantage of buckwheat, till toward the end of September, for so long that plant retains its flowers.—*Ibid.*

[5] *Ins. Misc.*, p. 262.

as they sailed along shore, the Bees constantly flew from the vessel to the land, to collect honey, and returned again to their moving hive; and this was continued all the way up the Channel."[1]

In Lower Egypt, observes M. Maillet in his Description of Egypt, where the blossoming of flowers is about six weeks later than in the upper districts, the practice of transporting Beehives is much followed. The hives are collected from different villages along the banks, each being marked and numbered by individual proprietors, to prevent future mistakes. They are then arranged in pyramidal piles upon the boats prepared to receive them, which, floating gradually down the river, and stopping at certain stages of their passage, remain there a longer or a shorter time, according to the produce afforded by the surrounding country within two or three leagues. In this manner the Bee-boats sail for three months; the Bees, having culled the honey of the orange-flowers in the Said, and of the Arabian jasmine and other flowers in the more northern parts, are brought back to the places whence they had been carried. This procures for the Egyptians delicious honey and abundance of wax. The proprietors in return pay the boatmen a recompense proportioned to the number of hives which have been thus carried about from one extremity of Egypt to the other.[2] The celebrated traveler Niebuhr saw upon the Nile, between Cairo and Damietta, a convoy of 4000 hives in their transit from Upper Egypt to the coast of the Delta.[3]

In the Bienenzeitung for 1854, p. 83, appears the following statements: "Mr. Kaden, of Mayence, thinks that the range of the Bee's flight does not usually extend more than three miles in all directions. Several years ago, a vessel, laden with sugar, anchored off Mayence, and was soon visited by the Bees of the neighborhood, which continued to pass to and from the vessel from dawn to dark. One morning, when the Bees were in full flight, the vessel sailed up the river. For a short time, the Bees continued to fly as numerously as before; but gradually the number diminished, and, in course of half an hour, all had ceased to follow the vessel, which had, meanwhile, sailed more than four miles."[4]

---

[1] *Mag. of Nat. Hist*, iii 652.
[2] Wood's *Zoog.*, ii. 429.
[3] *Ins. Misc.*, p. 263.
[4] Quot. by Langstroth—*On Honey-Bee*, p. 305, note.

Aristomachus of Soli, says Pliny, made Bees his exclusive study for a period of fifty-eight years; and Philiocus, the Thracian, surnamed Agrius—"Wildman"—passed his life in desert spots tending swarms of Bees.[1]

Schomburgk says he saw, in his journey to the sources of the Takutu, an Indian, who was the conjuror or piaiman of his tribe, merely approach a nest of the wild Wampang-bees (*Wampisiana camniba*), and knocking with his fingers against it, drive out all the Bees without a single one injuring him. The piaiman, Schomburgk remarks, drew his fingers under the pits of his arms before he knocked against the hive.[2]

Brue, in his first voyage to Siratic, in Africa, met with what he called a "phenomenon" in a person entitling himself the "King of the Bees." His majesty accordingly came to the boat of the traveler entirely covered with these insects, and followed by thousands, over which he appeared to exercise the most absolute authority. These Bees were never known to injure either himself or those whom he took under his protection.[3]

Mr. Wildman, the most celebrated Bee-tamer, frequently asserted that armed with his friendly Bees he was defensible against the fiercest mastiffs; and, it is said, he actually did, at Salisbury, encounter three yard-dogs one after the other. The conditions of the engagement were, that he should have notice of the dog being set at him. Accordingly the first mastiff was set loose; and as he approached the man, two Bees were detached, which immediately stung him, the one on the nose, the other on the flank; upon receiving the wounds, the dog retired very much daunted. After this, the second dog entered the lists, and was foiled with the same expedition as the first. The third dog was at last brought against the champion, but the animal observing the ill success of his brethren, would not attempt to sustain a combat; so, in a cowardly manner, he retired with his tail between his legs.

Many other remarkable anecdotes are told of this gentleman, illustrating his wonderful control over Bees. He could also, indeed, tame wasps and hornets, with almost the same ease as he could Bees, and an instance is mentioned of his

---

[1] *Nat. Hist.*, x. 9.
[2] *Journ. of Geog. Soc.*, 1843, xiii. 40.
[3] Murray's *Africa*, i. 168.

hiving a nest of hornets which hung at the top of the inside of a high barn.   He, however, was stung twice in this undertaking.

Mr. Wildman frequently exhibited himself with his head and face almost covered with Bees, and with such a swarm of them hanging down from his chin as to resemble a venerable beard.   In this extraordinary dress he was once brought through the City of London sitting in a chair.   Before Earl Spencer, Mr. Wildman also made many wonderful performances.[1]

Says Dr. Evans:

> Such was the spell, which round a Wildman's arm
> Twined in dark wreaths the fascinated swarm,
> Bright o'er his breast the glittering legions led,
> Or with a living garland bound his head.
> His dexterous hand, with firm but hurtless hold,
> Could seize the chief, known by her scales of gold,
> Prune, 'mid the wondering throng, her filmy wing,
> Or o'er her folds the silken fetter fling.[2]

"Long experience has taught me," says Mr. Wildman himself, "that as soon as I turn up a hive, and give some taps on the sides and bottom, the queen immediately appears. Being accustomed to see her, I readily perceive her at the first glance; and long practice has enabled me to seize her instantly, with a tenderness that does not in the least endanger her person.   Being possessed of her, I can, without exciting any resentment, slip her into my other hand, and returning the hive to its place, hold her, till the Bees, missing her, are all on the wing and in the utmost confusion."   It was then, by placing the queen in view, he could make them light wherever he pleased, from their great attachment to her, and sometimes using a word of command to mystify the spectators, he would cause them to settle on his head, and to hang to his chin like a beard, from which he would order them to his hand, or to an adjacent window.   But, however easy such feats may appear in theory, Mr. Wildman cautions (probably with a view to deter rivals) those who are inexperienced not to put themselves in danger of attempting to imitate him.   A liberated Roman slave, C. F. Cnesinus, being accused before the tribunals of witchcraft, because his

---

[1] Scot's *Mag.*, Nov. 1766.   Chamb. *Journ.*, 1st S. xi. 184.
[2] *The Bees.*

crops were more abundant than those of his neighbors, produced as his witnesses some superior implements of husbandry, and well fed oxen. and pointing to them said: "These, Romans! are my instruments of witchcraft; but I cannot show you my toil, my perseverance, and my anxious cares." "So," says Wildman, "may I say, These, Britons! are my instruments of witchcraft; but I cannot show you my hours of attention to this subject, my anxiety and care for these useful insects; nor can I communicate to you my experience acquired during a course of years."[1]

Butler mentions two instances where the stings of Bees have been fatal to "cattaile":

"A horse," he informs us. "in the heate of the day look-ing over a hedge, on the other side whereof was a staule of Bees, while hee stood nodding with his head, as his manner is, because of the flies, the Bees fell vpon him and killed him. Likewise I heard of a teeme that stretching against a hedge overthrew a staule on the other side, and so two of the horses were stung to death."[2]

Mungo Park and his party were twice seriously attacked by large swarms of Bees. The first attack is mentioned in the account of his first journey; the second in the account of his second. The latter singular accident befell them in 1805, and is thus narrated in his journal: The coffle had halted at a creek, and the asses had just been unloaded, when some of his guide Isaaca's people, being in search of honey, unfortunately disturbed a large swarm of Bees near their resting-place. The Bees came out in immense num-bers, and attacked men and beasts at the same time. Luck-ily, most of the asses were loose, and galloped up the val-ley; but the horses and people were very much stung, and obliged to scamper off in all directions. The fire which had been kindled for cooking, being deserted, spread, and set fire to the bamboos, and the baggage had like to have been burned. In fact, for half an hour the Bees seemed to have completely put an end to the journey. In the evening when they became less troublesome, and the cattle could be col-lected, it was found that many of them were very much stung, and swollen about the head. Three asses were miss-ing; one died in the course of the evening, and one next

---

[1] *Treatise on Bees.* 1769. *Ins. Misc.*, p. 320–1.
[2] *Fem. Monarchie,* ch i. 39.

morning, and they were forced to leave one behind the next day. Altogether six were lost, besides which, the guide lost his horse, and many of the people were much stung about the face and hands.[1]

But in the Treasvrie of Avncient and Moderne Times, we find the following: "Anthenor, writing of the Isle of Crete (with whom also ioyneth Ælianus) saith, that a great multitude of Bees chased al the dwellers out of a City, and vsed their Houses instead of Hives."[2]

Montaigne mentions the following singular assistance rendered by Bees to the inhabitants of Tamly: The Portuguese having besieged the City of Tamly, in the territory of Xiatine, the inhabitants of the place brought a great many hives, of which there are great plenty in that place, upon the wall; and with fire drove the Bees so furiously upon the enemy that they gave over the enterprise, not being able to stand their attacks and endure their stings: and so the citizens, by this new sort of relief, gained liberty and the victory with so wonderful a fortune, that at the return of their defenders from the battle they found they had not lost so much as one.[3]

Lesser tells us that in 1525, during the confusion occasioned by a time of war, a mob assembling in Hohnstein (in Thuringia) attempted to plunder the house of the minister of Elende; who having spoken to them with no effect, as a last resort ordered his domestics to bring his Beehives, and throw them in the midst of the furious mob. The desired effect was instantaneous, for the mob dispersed immediately.[4]

Bees have also been employed as an article of food. Knox tells us that the natives of Ceylon, when they meet with a swarm of Bees hanging on a tree, hold burning torches under them to make them drop; and so catch and carry them home, where they boil and eat them, in their estimation, as excellent food.[5]

Peter Martyr, speaking of the Caribbean Islands, says:

---

1 *Travels*, p. 178, Harper's ed.
2 B. VII. c. xvi. p. 667. Printed, 1613.
3 Montaigne's *Works*, p. 243.
4 Lesser, ii. 171. K. & S. *Introd.*, ii. 247.
5 Knox, Pt. I. c. vi. p. 48.

"The Inhabitantes willingly eate the young Bees, rawe, roasted, and sometimes sodden."[1]

Bancroft tells us that when the negroes of Guiana are stung by Bees, they in revenge eat as many as they can catch.[2]

The following account of the Bee-eater of Selborne, England, is by the Reverend, and very accurate naturalist, Gilbert White: "We had in this village," says he, "more than twenty years ago (about 1765), an idiot boy, whom I well remember, who, from a child, showed a strong propensity to Bees: they were his food, his amusement, his sole object; and as people of this cast have seldom more than one point in view, so this lad exerted all his few faculties on this one pursuit. In the winter he dozed away his time, within his father's house, by the fireside, in a kind of torpid state, seldom departing from the chimney corner; but in the summer he was all alert, and in quest of his game in the fields and on sunny banks. Honey-bees, Humble-bees, and Wasps were his prey, wherever he found them: he had no apprehensions from their stings, but would seize *nudis manibus*, and at once disarm them of their weapons, and search their bodies for the sake of their honey-bags. Sometimes he would fill his bosom between his shirt and his skin with a number of these captives; and sometimes would confine them in bottles. He was a very *Merops apiaster*, or Bee-bird, and very injurious to men that kept Bees; for he would slide into their Bee-gardens, and, sitting down before the stools, would rap with his finger on the hives, and so take the Bees as they came out. He has been known to overturn hives for the sake of honey, of which he was passionately fond. Where metheglin was making, he would linger round the tubs and vessels, begging a draught of what he called *Bee-wine*. As he ran about he used to make a humming noise with his lips, resembling the buzzing of Bees. This lad was lean and sallow, and of a cadaverous complexion; and, except in his favorite pursuit, in which he was wonderfully adroit, discovered no manner of understanding."[3]

There is a peculiar substance formed by a species of Bee

---

[1] Martyr, p. 274.
[2] Banc. *Guiana*, p. 230.
[3] *Nat. Hist. of Selborne*, p. 293.

in the Orinoco country, which, says Captain Stedman, the
roosting tribes burn incessantly in their habitations, and
which effectually protects them from all winged insects.
They call it *Comejou;* Gumilla says it is neither earth nor
wax.[1]

Concerning the medicinal virtues of Bees, Dr. James
says: "Their salts are very volatile, and highly exalted;
for this reason, when dry'd, powder'd, and taken internally,
they are diuretic and diaphoretic. If this powder is mixed
in unguents, with which the head is anointed, it is said to
cure the Alopecia, and to contribute to the growth of hair
upon bald places."[2]

Another, an old writer, says: "If Bees, when dead, are
dried to powder, and given to either man or beast, this
medicine will often give immediate ease in the most excruciating pain, and remove a stoppage in the body when all
other means have failed." A tea made by pouring boiling
water upon Bees has recently been prescribed, by high medical authority, for violent strangury; while the poison of the
Bee, under the name of *apis,* is a great homœopathic
remedy.[3]

Concerning wax, Dr. James says: "All wax is heating,
mollifying, and moderately incarning. It is mixed in sorbile
liquors as a remedy for dysentery; and ten bits, of the size
of a grain of millet, swallowed, prevent the curdling of milk
in the breast of nurses."[4]

[If we might credit the history of former times, says Jamieson, in his Scottish Dictionary, sub. *Walx,* iv. 642–3, there
must have been a considerable demand for this article (wax)
for the purpose of witchcraft. It was generally found necessary, it would seem, as the medium of inflicting pain on the
bodies of men.

"To some others at these times he teacheth, how to make
pictures of waxe or clay, that by the wasting thereof, the
persons that they beare the name of, may be continually
melted or dried away by continuall sickenesse." K. James's
Dæmonologie, B. II. c. 5.

In order to cause acute pain in the patient, pins, we are

---

[1] *Trav.,* i. 9.
[2] *Med. Dict.*
[3] Langstroth *on Honey-Bee,* p. 315, note.
[4] *Med. Dict.*

told, were stuck in that part of the body of the image, in which they wished the person to suffer.

The same plan was adopted for inspiring another with the ardor of love.

> Then mould her form of fairest *wax*,
>   With adder's eyes and feet of horn ;
> Place this small scroll within its breast,
>   Which I, your friend, have hither borne.
>
> Then make a blaze of alder wood,
>   Before your fire make this to stand ;
> And the last night of every moon
>   The bonny May's at your command.
> *Hogg's Mountain Bard*, p. 35.

Then it follows :

> With fire and steel to urge her weel,
>   See that you neither stint nor spare ;
> For if the cock be heard to crow,
>   The charm will vanish into air.

The wounds given to the image were supposed to be productive of similar *stounds* of love in the tender heart of the maiden whom it represented.

> A female form, of melting *wax*,
>   Mess John surveyed with steady eye,
> Which ever an anon he *pierced*,
>   And forced the lady loud to cry.—P. 84.

The same horrid rites were observed on the continent. For Grilland (de Sortilegiis) says : Quidam solent apponere *imaginem cerae* juxta ignem ardentem, completis sacrificiis, de quibus supra, & adhibere quasdam preces nefarias, & turpia verba, ut quemadmodum imago illa igne consumitur & liquescit, eodem modo cor mulieris amoris calore talis viri feruenter ardeat, etc.   Malleus Malefic. T. H., p. 232.

It cannot be doubted that these rites have been transmitted from heathenism.   Theocritus mentions them as practiced by the Greeks in his time.   For he introduces Samoetha as using similar enchantments, partly for punishing, and partly for regaining her faithless lover.

> But strew the *salt*, and say in angry tones,
> "I scatter Delphid's, perjured Delphid's bones.",
> —First Delphid injured me, he raised my flame,
> And now I burn this bough in Delphid's name ;

> As this doth blaze, and break away in fume,
> How soon it takes, let Delphid's flesh consume,
> lynx, restore my false, my perjured swain,
> And force him back into my arms again.—
> As this devoted *wax* melts o'er the fire,
> Let Mindian Delphy melt in warm desire!
>
> *Idylliums*, p. 12, 13.

Samoetha burns the bough in the name of her false lover, and terms the wax *devoted*. With this the more modern ritual of witchcraft corresponded. The name of the person, represented by the image, was invoked. For according to the narrative given concerning the witches of Pollock-shaws, having bound the image on a spit, they "turned it before the fire,—saying, as they turned it, *Sir George Maxwell, Sir George Maxwell;* and that this was expressed by all of them." Glanvil's Sadducismus, p. 391.

According to Grilland, the image was baptized in the name of Beelzebub. Malleus, ut. sup., p. 229.

There is nothing analogous to the Grecian rite, mentioned by Theocritus, of strewing *salt*. For Grilland asserts that, in the festivals of the witches, salt was never presented. Ibid., p. 215. It was perhaps excluded from their infernal rites as having been so much used as a sacred symbol.]

The following are among the twenty-eight "singular vertues" attributed by Butler to Honey : ".... It breedeth good blood, it prolongeth old age .... yea the bodies of the dead being embalmed with honey have been thereby preserved from putrefaction. And *Athenæus* doth witness it to be as effectual for the living, writing out of Lycus, that the Cyrneans, or inhabitants of Corsica, were therefore long-lived, because they did dailie vse to feed on honey, whereof they had abundance : and no marvaile : seeing it is so sove-raigne a thing, and so many waies available for man's health, as well being outwardly as inwardly applied. It is drunke against the bite of a serpent or mad dogs : and it is good for them having eaten mushrooms, or drunke popy, etc.'"[1]

In the Treasvrie of Avncient and Moderne Times,[2] there are two chapters devoted to the "Vertues of Honey."

---

[1] *Fem. Monarchie*, c. x. 1.

[2] B. 3, c. xv. xvi. p. 274-9. See also extract from Works of Sir J. More, London, 1707, given by Langstroth—*on the Honey-Bee*, p. 287, note.

There is a story, that a man once came to Mohammed, and told him that his brother was afflicted with a violent pain in his belly; upon which the prophet bade him give him some honey. The fellow took his advice; but soon after coming again, told him that the medicine had done his brother no manner of service: Mohammed answered, "Go and give him more honey, for God speaks truth, and thy brother's belly lies." And the dose being repeated, the man, by God's mercy, was immediately cured.[1]

In the sixteenth chapter of the Koran, Mohammed has likewise mentioned honey as a medicine for men.[2]

Athenæus tells us that Democritus, the philosopher of Abdera, after he had determined to rid himself of life on account of his extreme old age, and when he had begun to diminish his food day by day, when the day of the Thesmophonian festival came round, and the women of his household besought him not to die during the festival, in order that they might not be debarred from their share of the festivities, was persuaded and ordered a vessel full of honey to be set near him: and in this way he lived many days with no other support than honey; and then some days after, when the honey had been taken away, he died. But Democritus, Athenæus adds, had always been fond of honey; and he once answered a man, who had asked him how he could live in the enjoyment of the best health, that he might do so if he constantly moistened his inward parts with honey and his outward man with oil. Bread and honey was the chief food of the Pythagoreans, according to the statement of Aristoxenus, who says that those who ate this for breakfast were free from disease all their lives.[3]

"The gall of a vulture," says Moufet, quoting Galen, in Euporist, "mingled with the juice of horehound (twice as much in weight as the gall is) and two parts of honey cures the suffusion of the eyes. Otherwise he mingles one part of the gall of the sea-tortoise, and four times as much honey, and anoints the eyes with it. Serenus prescribes such a receipt to cause one to be quick-sighted:

---

[1] *The Koran*, p. 219, note, Sale's.
[2] *Ibid.*, p. 219.
[3] Athen. *Deipn.*, B. 2, c. 26.

> Mingle Hyblæan honey with the gall
> Of Goats, 'tis good to make one see withall."[1]

We are told in the German Ephemerides, that a young country girl, having eaten a great deal of honey, became so inebriated with it, that she slept the whole day, and talked foolishly the day following.[2]

Bevan, in his work on the Honey-Bee, mentions the following instances of a curious use to which propolis is sometimes put by the Bees: A snail, says he, having crept into one of Mr. Reaumur's hives early in the morning, after crawling about for some time, adhered, by means of its own slime, to one of the glass panes. The Bees, having discovered the snail, surrounded it, and formed a border of propolis round the verge of its shell, and fastened it so securely to the glass that it became immovable.

> Forever closed the impenetrable door;
> It naught avails that in its torpid veins
> Year after year, life's loitering spark remains.
>
> EVANS.

Maraldi, another eminent Apiarian, states that a snail without a shell having entered one of his hives, the Bees, as soon as they observed it, stung it to death; after which, being unable to dislodge it, they covered it all over with an impervious coat of propolis.

> For soon in fearless ire, their wonder lost,
> Spring fiercely from the comb the indignant host,
> Lay the pierced monster breathless on the ground,
> And clap in joy their victor pinions round:
> While all in vain concurrent numbers strive
> To heave the slime-girt giant from the hive—
> Sure not alone by force instinctive swayed,
> But blest with reason's soul-directing aid,
> Alike in man or bee, they haste to pour,
> Thick, hard'ning as it falls, the flaky shower;
> Embalmed in shroud of glue the mummy lies,
> No worms invade, no foul miasmas rise.
>
> EVANS.[3]

Xenophon tells us that all the soldiers, who ate of the honey-combs, found in the villages on the mountains of the

---

[1] Moufet, *Theatr. Ins.*, p. 29.   Topsel's *Trans.*, p. 911.
[2] Brooke's *Nat. Hist. of Ins.*, p. 168.
[3] Quot. by Langstroth *on the Honey-Bee*, p. 78-9.

Colchians, lost their senses, and were seized with such violent vomiting and purging, that none of them were able to stand upon their legs: that those who ate but little, were like men very drunk, and those who ate much, like madmen, and some like dying persons. In this condition, this writer adds, great numbers lay upon the ground, as if there had been a defeat, and a general sorrow prevailed. The next day, they all recovered their senses, about the same hour they were seized; and, on the third and fourth days, they got up as if they had taken physic.[1]

Pliny accounts for this accident by saying there is found in that country a kind of honey, called from its effects, Thænomenon, that is, that those who eat it are seized with madness. He adds, that the common opinion is that this honey is gathered from the flowers of a plant called *Rhododendros*, which is very common in those parts. Tournefort thinks the modern *Laurocerasus* is the Rhododendros of Pliny, from the fact that the people of that country, at the present day, believe the honey that is gathered from its flowers will produce the effects described by Xenophon.[2]

The missionary Moffat in South Africa found some poisonous honey, which he unknowingly ate, but with no serious consequences. It was several days, however, before he got rid of a most unpleasant sensation in his head and throat. The plant from which the honey had been gathered was an Euphorbia.[3]

"In Podolia," says the chronicler Hollingshed, "which is now subject to the King of Poland, their hives (of Bees) and combes are so abundant, that huge bores, overturning and falling into them, are drowned in the honie, before they can recover & find the meanes to come out."[4]

Honey was offered up to the Sun by the ancient Peruvians.[5]

Dr. Sparrman has described a Hottentot dance, which he calls the Bee-dance. It is in imitation of a swarm of Bees; every performer as he jumps around making a buzzing noise.[6]

---

[1] *Anab.*, B. 4.
[2] Pliny, *Nat. Hist.*, xxi. 13. Tournefort, *Letters*, 17.
[3] *Mission. Lab.*, p. 121.
[4] Hollingsh. *Chron.*, i. 384.
[5] Hawk's *Peruvian Antiq.*, p. 198.
[6] *Voyage to C. of G. Hope*, i. 255.

"To have a Bee in one's bonnet" is a Scottish proverbial phrase about equivalent to the English, "To have a maggot in one's head"—to be hair-brained. Kelly gives this with an additional word: "There's a Bee in your bonnet-*case*." In Scotland, too, it is said of a confused or stupefied man, that his "head is in the Bees."[1] These proverbial expressions were also in vogue in England.[2]

The following beautiful epigram, on a Bee inclosed in amber, is from the pen of Martial: "The Bee is inclosed, and shines preserved, in a tear of the sisters of Phaëton, so that it seems enshrined in its own nectar. It has obtained a worthy reward for its great toils; we may suppose that the Bee itself would have desired such a death.

> The Bee inclosed, and through the amber shown,
> Seems buried in the juice that was her own.
> So honor'd was a life in labor spent:
> Such might she wish to have her monument."[3]

The Septuagint has the following eulogium on the Bee in Prov. vi. 8, which is not found in the Hebrew Scriptures: "Go to the Bee, and learn how diligent she is, and what a noble work she produces, whose labors kings and private men use for their health; she is desired and honored by all, and though weak in strength, yet since she values wisdom, she prevails."[4]

In Spain Bees are in great estimation; and this is evinced by the ancient proverb:

> Abeja y oveja,
> Y piedra que traveja,
> Y pendola trans oreja,
> Y parte en la Igreja,
> Desea a su hija, la vieja——

The best wishes of a Spanish mother to her son are, Bees, sheep, millstones, a pen behind the ear, and a place in the church.[5]

The following anecdote in the history of the Humble-bee

[1] Jamieson's *Scot. Dict.*
[2] Wright's *Prov. Dict.*
[3] *Epigrams,* B. iv. epigr. 32.
[4] Smith's *Dict. of the Bible.*
[5] Osbeck's *Travels,* i. 32–3.

(*Bombus*) is from the account of Josselyn of his voyages to
New England, printed in 1674: "Near upon twenty years
since there lived an old planter near *Blackpoint*, who on a
Sunshine day about one of the clock lying upon a green bank
not far from his house, charged his Son, a lad of 12 years of
age, to awaken him when he had slept two hours; the old
man falls asleep, and lying upon his back gaped with his
mouth wide open enough for a Hawke to —— into it; after
a little while the lad sitting by spied a Humble-bee creeping
out of his Father's mouth, which taking wing flew quite out
of sight, the hour as the lad guest being come to awaken his
Father, he jagged him and called aloud Father, Father, it is
two o'clock, but all would not rouse him, at last he sees the
Humble-bee returning, who lighted upon the sleeper's lip
and walked down as the lad conceived into his belly, and
presently he awaked."[1]

The following, on the different species of Humble-bees, is
one of the popular rhymes of Scotland :

> The todler-tyke has a very guide byke,
> And sae has the gairy Bee ;
> But weel's me on the little red-doup,
> The best o' a' the three.[2]

When the Archbishop of St. Andrews was cruelly murdered in 1679, "upon the opening of his tobacco box a living
humming bee flew out," which was explained to be a familiar
or devil.   A Scottish woman declared that a child was poisoned by its grandmother, who, together with herself, were
"in the shape of bume-bees," that the former carried the
poison "in her cleugh, wings, and mouth."   A great Bee
constantly resorted to another after receiving the Satanic
mark, and rested on it.[3]

An anecdote is related by M. Reaumur respecting the
thimble-shaped nest, formed of leaves, of the Carpenter-bee
(*Apis centuncularis?*), which is a striking instance of the
ridiculous superstition which prevails among the unedu-
cated, and which even sometimes has no slight influence on
those of better understandings.   "In the beginning of July,
1736, the learned Abbé Nollet, then at Paris, was surprised

---

[1] Josselyn's *Voy.*, p. 121.
[2] Chambers' *Pop. Rhymes of Scot.*, p. 292.   Edit. of 1841, p. 172.
[3] Dalyell's *Superst. of Scotland*, p. 563.

by a visit from an auditor of the chamber of accounts, whose estate lay at a distant village on the borders of the Seine, a few leagues from Rouen. This gentleman came accompanied, among other domestics, by a gardener, whose face had an air of much concern. He had come to Paris in consequence of having found in his master's ground many rows of leaves, unaccountably disposed in a mystical manner, and which he could not but believe were there placed by witchcraft, for the secret destruction of his lord and family. He had, after recovering from his first consternation, shown them to the curate of the parish, who was inclined to be of a similar opinion, and advised him without delay to take a journey to Paris, and make his lord acquainted with the circumstance. This gentleman, though not quite so much alarmed as the honest gardener, could not feel himself at perfect ease, and therefore thought it advisable to consult his surgeon upon the business, who, though a man eminent in his profession, declared himself utterly unacquainted with the nature of what was shown him, but took the liberty of advising that the Abbé Nollet, as a philosopher, should be consulted, whose well-known researches in natural knowledge might perhaps enable him to elucidate the matter. It was in consequence of this advice that the Abbé received the visit above mentioned, and had the satisfaction of relieving all parties from their embarrassment, by showing them several nests formed on a similar plan by other insects, and assuring them that those in their possession were the work of insects also."[1]

In an English paper, the Observer, of July 25, 1813, there is an account of a "swarm of Bees resting themselves on the inside of a lady's parasol." They were hived without any serious injury to the lady.

In the Annual Register, 1767, p. 117, there was published by M. Lippi, Licentiate in Physic of the army of Paris, an account of a petrified Beehive, discovered on the mountains of Siout, in Upper Egypt. Broken open it disclosed the larvæ of Bees in the cells, hard and solid, and Bees themselves dried up like mummies. Honey was also found in the cells![2] The account is curious, but not entitled to much credit.

---

[1] Shaw's *Zool.*, vi. 346–7. Wood's *Zoog.*, ii. 436–7.
[2] Kirby's *Wonderful Museum*, v. 390–1, given at length.

In the Liverpool Advertiser, and Times, of Nov. 24, 1817, there is a lengthy account of three Bees being found in a state of animation in a huge solid rock from the Western Point Quarry. Scientific attention was attracted, and as appears from the above-mentioned papers of Dec. 5, 1817, the mystery was cleared up by discovering in the rock "a sand hole" through which the insects had made their way.[1]

---

[1] Kirby's *Wond. Museum*, vi. 260–2, at length.

ORDER VI.

LEPIDOPTERA.

Papilionidæ—Butterflies.

THE lepidopterous insects in general, soon after they emerge from the pupa state, and commonly during their first flight, discharge some drops of a red-colored fluid, more or less intense in different species, which, in some instances, where their numbers have been considerable, have produced the appearance of a "shower of blood," as this natural phenomenon is commonly called.

Showers of blood have been recorded by historians and poets as preternatural—have been considered in the light of prodigies, and regarded where they have happened as fearful prognostics of impending evils.

There are two passages in Homer, which, however poetical, are applicable to a rain of this kind ; and among the prodigies which took place after the death of the great dictator, Ovid particularly mentions a shower of blood :

Sæpe faces visæ mediis ardere sub astris,
Sæpe inter nimbos guttæ cecidere cruentæ.

With threatening signs the lowering skies were fill'd,
And sanguine drops from murky clouds distilled.

Among the numerous prodigies reported by Livy to have happened in the year 214 B.C., it is instanced that, at Mantua, a stagnating piece of water, caused by the overflowing of the River Mincius, appeared as of blood ; and, in the cattle-market at Rome, a shower of blood fell in the Istrian Street. After mentioning several other remarkable phenomena that happened during that year, Livy concludes by saying that these prodigies were expiated, conformably to the answers of the Aruspices, by victims of the greater kinds, and supplication was ordered to be performed to all the deities who had shrines at Rome.[1] Again it is stated by Livy, that many alarming prodigies were seen at Rome in the year 181 B.C., and others reported from abroad; among which was a shower of blood, which fell in the courts of the temples of Vulcan and Concord. After mentioning that the image of Juno Sospita shed tears, and that a pestilence broke out in the country, this writer adds, that these prodigies, and the mortality which prevailed, alarmed the Senate so much, that they ordered the consuls to sacrifice to such gods as their judgment should direct, victims of the larger kinds, and that the Decemvirs should consult their books. Pursuant to their direction, a supplication for one day was proclaimed to be performed at every shrine at Rome; and they advised, besides, and the Senate voted, and the consul proclaimed, that there should be a supplication and public worship for three days throughout all Italy.[2] In the year 169 B.C., Livy also mentions that a shower of blood fell in the middle of the day. The Decemvirs were again called upon to consult their books, and again were sacrifices offered to the deities.[3] The account, also, of Livy, of the bloody sweat, on some of the statues of the gods, must be referred to the same phenomenon; as the predilection of those ages to marvel, says Thomas Brown, and the want of accurate investigation in the cases recorded, as well as the rare occurrence of these atmospherical depositions in our own times, inclines us to include them among the blood-red drops deposited by insects.[4]

In Stow's Annales of England, we have two accounts of showers of blood; and from an edition printed in London in 1592, we make our quotations: "Rivallus, sonne of Cunedagius, succeeded his father, in whose time (in the year 766 B.C.) it rained bloud 3 dayes: after which tempest ensued a great multitude of venemous flies, which slew much people, and then a great mortalitie throughout this lande, caused almost desolation of the same."[5] The second account is as follows: "In the time of Brithricus (A.D. 786) it rayned blood, which falling on men's clothes, appeared like crosses."[6]

---

[1] Livy, B. 34, c. 10. [2] Ibid., B. 40, c. 19. [3] Ibid., B. 43, c. 13.
[4] Brown's *Book of Butterflies*, i. 126.
[5] *Annales*, p. 15. [6] *Ibid.*

Hollingshed, Grafton, and Fabyan have also recorded these instances in their respective chronicles of England.[1]

A remarkable instance of bloody rain is introduced into the very interesting Icelandic ghost story of Thorgunna. It appears that in the year of our Lord 1009, a woman called Thorgunna came from the Hebrides to Iceland, where she stayed at the house of Thorodd: and during the hay season, a shower of blood fell, but only, singularly, on that portion of the hay she had not piled up as her share, which so appalled her that she betook herself to her bed, and soon afterward died. She left, to finish the story, a remarkable will, which, from not being executed, was the cause of several violent deaths, the appearance of ghosts, and, finally, a legal action of ejectment against the ghosts, which, it need hardly be said, drove them effectually away.[2]

In 1017, a shower of blood fell in Aquitaine;[3] and Sleidan relates that in the year 1553 a vast multitude of Butterflies swarmed through a great part of Germany, and sprinkled plants, leaves, buildings, clothes, and men with bloody drops, as if it had rained blood.[4] We learn also from Bateman's Doome, that these "drops of bloude upon hearbes and trees," in 1553, were deemed among the forewarnings of the deaths of Charles and Philip, dukes of Brunswick.[5]

In Frankfort, in the year 1296, among other prodigies, some spots of blood led to a massacre of the Jews, in which ten thousand of these unhappy descendants of Abraham lost their lives.[6]

In the beginning of July, 1608, an extensive shower of blood took place at Aix, in France, which threw the people of that place into the utmost consternation, and, which is a much more important fact, led to the first satisfactory and philosophical explanation of this phenomenon, but too late, alas! to save the Jews of Frankfort. This explanation was given by M. Peiresc, a celebrated philosopher of that place, and is thus referred to by his biographer, Gassendi: "Nothing in the whole year 1608 did more please him than that he observed and philosophized about, the *bloody rain*, which

---

[1] Holling., i. 449. Graft., i. 37. Fabyan, p. 17.
[2] Howitt's *North. Literat.*, i. 187.
[3] Bucke *on Nature*, i. 277.
[4] Moufet, p. 107.
[5] Hone's *Ev. Day Book*, p. 1127.
[6] Chambers' *Domest. Annals of Scotland*, ii. 489.

was commonly reported to have fallen about the beginning of July; great drops thereof were plainly to be seen, both in the city itself, upon the walls of the church-yard of the church, which is near the city wall, and upon the city walls themselves; also upon the walls of villages, hamlets, and towns, for some miles round about; for in the first place, he went himself to see those wherewith the stones were colored, and did what he could to come to speak with those husbandmen, who, beyond Lambesk, were reported to have been affrighted at the falling of said rain, that they left their work, and ran as fast as their legs could carry them into the adjacent houses. Whereupon, he found that it was a fable that was reported, touching those husbandmen. Nor was he pleased that naturalists should refer this kind of rain to vapours drawn up out of red earth aloft in the air, which congealing afterwards into liquor, fall down in this form; because such vapours as are drawne aloft by heat, ascend without color, as we may know by the alone example of red roses, out of which the vapours that arise by heat are congealed into transparent water. He was less pleased with the common people, and some divines, who judged that it was the work of the devils and witches who had killed innocent young children; for this he counted a mere conjecture, possibly also injurious to the goodness and providence of God.

"In the mean while an accident happened, out of which he conceived he had collected the true cause thereof. For, some months before, he shut up in a box a certain palmer-worm which he had found, rare for its bigness and form; which, when he had forgotten, he heard a buzzing in the box, and when he opened it, found the palmer-worm, having cast its coat, to be turned into a beautiful Butterfly, which presently flew away, leaving in the bottom of the box a red drop as broad as an ordinary sous or shilling; and because this happened about the beginning of the same month, and about the same time an incredible multitude of Butterflies were observed flying in the air, he was therefore of opinion that such kind of Butterflies resting on the walls had there shed such like drops, and of the same bigness. · Whereupon, he went the second time, and found, by experience, that those drops were not to be found on the house-tops, nor upon the round sides of the stones which stuck out, as it would have happened, if blood had fallen from the sky, but rather where the stones were somewhat hollowed, and in holes, where such small creatures might shroud and nestle themselves. Moreover, the walls which were so spotted, were not in the middle of towns, but they were such as bordered upon the fields, nor were they on the highest parts, but only so moderately high as Butterflies are commonly wont to fly.

"Thus, therefore, he interpreted that which Gregory of Tours relates, touching a bloody rain seen at Paris in divers places, in the days of Childebert, and on a certain house in the territory of Senlis; also that which is storied, touching raining of blood about the end of June, in the days of King Robert; so that the blood which fell upon flesh, garments, or stones could not be washed out, but that which fell on wood might; for it was the same season of Butterflies, and experience hath taught us, that no water will wash these spots out of the stones, while they are fresh and new. When he had said these and such like things to various, a great company of auditors being present, it was agreed that they should go together and search out the matter, and as they went up and down, here and there, through the fields, they found many drops upon stones and rocks; but they were only on the hollow and under parts of the stones, but not upon those which lay most open to the skies."[1]

This memorable shower of blood was produced by the *Vanessa urticæ*, or *V. polychloros*, most probably, since these species of Butterflies are said to have been uncommonly plentiful at the time when, and in the particular district where, the phenomenon was observed.[2][3]

---

[1] Gassendi's *Life of Peireskius*, p. 123–5; and Reaumur, i. 638, 667.

[2] Shaw, *Zool.*, vi. 206.

[3] The origin of red snow has likewise been a puzzle and query for ages, and many theories have been advanced by philosophers and naturalists to account for it. To those interested in the solution of this phenomenon, the following extract from the *Mag. of Nat. Hist.*, vol. ii. p. 322, may be curious, if not satisfactory. Mr. Thomas Nicholson, accompanied with two other gentlemen, made an excursion the 24th July, 1821, to Sowallick Point, near Bushman's Island, in Prince Regent's Bay, in quest of meteoric iron. "The summit of the hill," he says, "forming the point, is covered with huge masses of granite, whilst the side, which forms a gentle declivity to the bay, was covered with crimson snow. It was evident, at first view, that this colour was imparted to the snow by a substance lying on the surface. This substance lay scattered here and there in small masses, bearing some resemblance to powdered cochineal, surrounded by a lighter shade, which was produced by the colouring matter being partly dissolved and diffused by the deli-

Nicoll, in his Diary, p. 8, informs us that on the 28th of May, 1650, "there rained blood the space of three miles in the Earl of Buccleuch's bounds (Scotland), near the English border, which was verefied in presence of the Committee of State."[1]

We learn from Fountainhall that on Sunday, May 1st, 1687, a young woman of noted piety, Janet Fraser by name, the daughter of a weaver in the parish of Closeburn, Dumfriesshire, went out to the fields with a young female companion, and sat down to read the Bible not far from her father's house. Feeling thirsty, she went to the river-side (the Nith) to get a drink, leaving her Bible open at the place where she had been reading, which presented the verses of the 34th chapter of Isaiah, beginning—"My sword shall be bathed in heaven: behold, it shall come down upon Idumea, and upon the people of my curse, to judgment," etc. On returning, she found a patch of something like blood covering this very text. In great surprise, she carried the book home, where a young man tasted the substance with his tongue, and found it of a saltless or insipid flavor. On the two succeeding Sundays, while the same girl was reading the Bible in the open air, similar blotches of matter, like blood, fell upon the leaves. She did not perceive it in the act of falling till it was about an inch from the book. "It is not blood," our informant adds, "for it is as tough as

---

quescent snow. During this examination our hats and upper garments were observed to be daubed with a substance of a similar red colour, and a moment's reflection convinced us that this was the excrement of the little Auk (*Uria alle*, Temmink), myriads of which were continually flying over our heads, having their nests among the loose masses of granite. A ready explanation of the origin of the red snow was now presented to us, and not a doubt remained in the mind of any that this was the correct one. The snow on the mountains of higher elevation than the nests of these birds was perfectly white, and a ravine at a short distance, which was filled with snow from top to bottom, but which afforded no hiding-place for these birds to form their nests, presented an appearance uniformly white."

This testimony seems to be as clear and indisputable as the explanation given by Peiresc of the ejecta of the Butterflies at Aix. But though it will account, perhaps, for the red snow of the polar regions, it will not explain that of the Alps, the Apennines, and the Pyrenees, which are not, so far as is known, visited by the little Auk.—Vide *Ins. Transf.*, p. 352–5.

[1] Chamb. *Domes. Annals of Scotl.*, ii. 190.

glue, and will not be scraped off by a knife, as blood will; but it is so like blood, as none can discern any difference by the colour."[1]

On Tuesday, Oct. 9th, 1764, "a kind of rain of a red color, resembling blood, fell in many parts of the Duchy of Cleves, which caused great consternation. M. Bouman sent a bottle of it to Dr. Schutte, to know if it contained anything pernicious to health. Something of the like kind fell a'so at Rhenen, in the Province of Utrecht."[2]

Dr. Schutte, to whom was submitted a bottle of this red rain, gave it as his opinion that it was caused by particles of red matter, which had been raised into the atmosphere by a strong wind, and that it was in no way hurtful to mankind or beasts![3]

In 1819, a red shower fell in Carniola, which, being analyzed, says Bucke, was found to be impregnated with silex, alumine, and oxide of iron. Red rain fell also at Dixmude, in Flanders, November 2d, 1829; and on the following day at Schenevingen, the acid obtained from which was chloric acid, and the metal cobalt.[4]

In the year 1780, Rombeag noticed a shower of blood that had excited universal attention, and which he could satisfactorily show to be produced by the flying forth and casting of bees, as the phenomenon in the place around the beehives themselves was remarkably striking. From this fact it is evident that the appearance is attributable to other insects as well as the lepidoptera.[5]

Bloody rain has also been attributed, with much apparent reason, to other causes still, as the following accounts from reliable authorities show:

In 1848, Dr. Eckhard, of Berlin, when attending a case of cholera, found potatoes and bread within the house spotted with a red coloring matter, which, being forwarded to Ehrenberg, was found by him to be due to the presence of an animalcule, to which he gave the name of the *Monas prodigiosa*. It was found that other pieces of bread could be inoculated with this matter.[6]

---

[1] Chamb. *Domes. Annals of Scotl.*, ii. 417–8.
[2] *Gent. Mag.*, xxxiv. 496.
[3] *Ibid.*, xxxiv. 542.
[4] Bucke *on Nature*, i. 277.
[5] Brown's *Bk. of Butterflies*, i. 129.
[6] Chamb. *Domes. Annals of Scotl.*, ii. 448.

Swammerdam relates that, one morning in 1670, great excitement was created in the Hague by a report that the lakes and ditches about Leyden were turned to blood. Florence Schuyl, the celebrated professor of physic in the University of Leyden, went down to the canals, and taking home a quantity of this blood-colored matter examined it with a microscope, and found that the water was water still, and had not at all changed its color; but that it was full of small red animals, all alive and very nimble in their motions, the color and prodigious numbers of which gave a reddish tinge to the whole body of the water in which they lived. The animals which thus color the water of lakes and ponds are the *Pulices arborescentes* of Swammerdam, or the water fleas with branched horns. These creatures are of a reddish yellow or flame color. They live about the sides of ditches, under weeds, and among the mud; and are therefore the less visible, except at a certain time, which is in the month of June. It is at this time these little animals leave their recesses to float about the water, and meet for the propagation of their species; and by this means they become visible in the color which they give to the water. The color in question is visible, more or less, in one part or other of almost all standing waters at this season; and it is always at the same season that the bloody waters have alarmed the ignorant.[1]

The prodigy, mentioned by Livy, of a stagnating piece of water at Mantua appearing as of blood, was no doubt owing to the appearance of great numbers of the *Pulices arborescentes* in it.[2]

Concerning the origin of bloody rain, Swammerdam entertained the same idea as Peiresc; but he does not appear

---

[1] Swam. *Hist. of Ins.*, Pt. I. p. 40.

[2] Cf. the following verses from Ex. vii. 19: "And the LORD spake unto Moses, Say unto Aaron, Take thy rod, and stretch out thine hand upon the waters of Egypt, upon their streams, upon their rivers, and upon their ponds, and upon all their pools of water, that they may become blood; and that there may be blood throughout all the land of Egypt, both in vessels of wood and in vessels of stone.

"20. And Moses and Aaron did so, as the LORD commanded; and he lifted up the rod, and smote the waters that were in the river in the sight of Pharaoh, and in the sight of his servants; and all the waters that were in the river were turned to blood."

to have verified it from his own observation. He makes the following remarks: "Is it not possible that such red drops might issue from insects, at the time they come fresh from the nymphs, which distil a bloody fluid? This seems to happen especially when such insects are more than ordinarily multiplied in any particular year, as we often experience in the butterflies, flies, gnats, and others."[1]

Dust is commonly attributed as the cause of this phenomenon, but will satisfactorily explain only a few instances. A writer for Chambers' Journal, in an article on showers of red dust, bloody rain, etc., says: "In October, 1846, a fearful and furious hurricane visited Lyon, and the district between that city and Grenoble, during which occurred a fall of blood-rain. A number of drops were caught and preserved, and when the moisture was evaporated, there was seen the same kind of dust (as fell in showers in Genoa in 1846) of a yellowish brown or red color. When placed under the microscope, it exhibited a great proportion of fresh water and marine formations. Phytolytharia were numerous, as also 'neatly-lobed vegetable scales;' which, as Ehrenberg observes, is sufficient to disprove the assertion that the substance is found in the atmosphere itself, and is not of European origin. For the first time, a living organism was met with, the '*Eunota amphyoxis*, with its ovaries green, and therefore capable of life.' Here was a solution of the mystery: the dust, mingling with the drops of water falling from the clouds, produced the red rain. Its appearance is that of reddened water, and it cannot be called blood-like without exaggeration."[2]

To conclude the history of bloody rain, the following is most appropriate: In 1841, some negroes, in Wilson County, Tennessee, reported that it had rained blood in the tobacco field where they had been at work; that near noon there was a rattling noise like rain or hail, and drops of blood, as they supposed, fell from a red cloud that was flying over. Prof. Troost, of Nashville, was called upon to explain the phenomenon; and, after citing many instances of red rain, red snow, and so called showers of blood, he concluded his learned article with this opinion: "A wind might have

---

[1] Swam. *Hist. of Ins.*, Pt. I. p. 40.
[2] Chamb. *Journ.*, 2d S. xvii. 231.

taken up part of an animal, which was in a state of decomposition, and have brought it in contact with an electric cloud, in which it was kept in a state of partial fluidity or vicosity. In this case, the cloud which was seen by the negroes, as the state in which the materials were, is accounted for."

Prof. Troost published this profound solution in the forty-first volume of Silliman's Journal; but in the forty-fourth of the same magazine a much more satisfactory one is given, for it is there stated "that the whole affair was a hoax devised by the negroes, who pretended to have seen the shower for the sake of practicing on the credulity of their masters. They had scattered the decaying flesh of a dead hog over the tobacco leaves."[1]

Another phenomenon to be particularly noticed in the history of the Butterflies, is their appearance at certain times in countless numbers migrating from place to place. H. Kapp, a writer in the *Naturforsch*, observed on a calm sunny day a prodigious flight of the Cabbage-Butterfly, *Pontia brassicæ*, which passed from northeast to southwest, and lasted two hours.[2] Kalm, the Swedish traveler, saw these last insects midway in the British Channel.[3] Lindley tells us that in Brazil, in the beginning of March, 1803, for many days successively there was an immense flight of white and yellow Butterflies, probably of the same tribe as the *Pontia brassicæ*. They were observed never to settle, but proceeded in a direction from northwest to southeast. No buildings seemed to stop them from steadily pursuing their course; which being to the ocean, at only a small distance, they must all have inevitably perished. It is to be remarked that at this time no other kind of Butterfly was to be seen, though the country usually abounds in such a variety.[4]

A somewhat similar migration of Butterflies was observed in Switzerland on the 8th or 10th of June, 1828. The facts are as follows: Madame de Meuron Wolff and her family, established during the summer in the district of

---

[1] Sil. *Journ.*, xli. 403–4, and xliv. 216.

[2] *Naturforsch*, xi. 94.

[3] *Travels*, i. 13.

[4] *Royal Milit. Chron.* for March, 1815, p. 452. K. and S. *Introd.*, ii. 11.

Grandson, Canton de Vaud, perceived with surprise an immense flight of Butterflies traversing the garden with great rapidity. They were all of the species called *Belle Dame* by the French, and by the English the Painted Lady (*Vanessa cardui*, Stephens). They were all flying close together in the same direction, from south to north, and were so little afraid when any one approached, that they turned not to the right or to the left. The flight continued for two hours without interruption, and the column was about ten or fifteen feet broad. They did not stop to alight on flowers; but flew onward, low and equally. This fact is the more singular, when it is considered that the larvæ of the *Vanessa cardui* are not gregarious, but are solitary from the moment they are hatched; nor are the Butterflies themselves usually found together in numbers. Professor Bonelli, of Turin, however, observed a similar flight of the same species of Butterflies in the end of March preceding their appearance at Grandson, when it may be presumed they had just emerged from the pupa state. Their flight, as at Grandson, was from south to north, and their numbers were so immense, that at night the flowers were literally covered with them. As the spring advanced, their numbers diminished; but even in June a few still continued. A similar flight of Butterflies is recorded about the end of the last century by M. Loche, in the Memoirs of the Turin Academy. During the whole season, these Butterflies, as well as their larvæ, were very abundant, and more beautiful than usual.[1]

Pallas once saw such vast flights of the orange-tipped Butterfly, *Pontia cardamines*, in the vicinity of Winofka, that he at first mistook them for flakes of snow.[2] At Barbados, some days previous to the hurricane in 1780, the trees and shrubs were entirely covered with a species of Butterfly of the most beautiful colors, so as to screen from the sight the branches, and even the trunks of the trees. In the afternoon before the gale came on, and when it was quite still, they all suddenly disappeared. The gale came on soon after.[3] Darwin tells us that several times, when the "Beagle"

---

[1] *Mag. of Nat. Hist.*, i. 387, and *Mem. de la Soc. de Phys. et d'Hist. Nat. de Genève.*

[2] *Penny Mag.*, 1844, p. 3.

[3] *Gent. Mag.*, liv. 744.

had been some miles off the mouth of the Plata, and at other times when off the shores of Northern Patagonia, the air was filled with insects : that one evening, when the ship was about ten miles from the Bay of San Blas, vast numbers of Butterflies, in bands or flocks of countless myriads, extended as far as the eye could range. The seamen cried out "It was raining Butterflies," and such in fact, continues Darwin, was the appearance. Several species were in this flock, but they were chiefly of a kind very similar to, but not identical with, the common English *Colias edusa.* Some moths and hymenopterous insects accompanied the Butterflies; and a fine beetle (*Calosoma*) flew on board.[1] Captain Adams mentions an extraordinary flight of small Butterflies, with spotted wings, which took place at Annamaboo, on the Guinea coast, after a tornado. The wind veered to the northward, and blew fresh from the land, with thick mist, which brought off from the shore so many of these insects, that for one hour the atmosphere was so filled with them as to represent a snow-storm driving past the vessel at a rapid rate, which was lying at anchor about two miles from the shore.[2]

Mr. Charles J. Anderson encountered, in South-western Africa, for two consecutive days, such immense myriads of lemon-colored Butterflies that the sound caused by their wings was such as to resemble "the distant murmuring of waves on the sea-shore." They always passed in the same direction as the wind blew, and, as numbers were constantly alighting on the flowers, their appearance at such times was not unlike "the falling of leaves before a gentle autumnal breeze."[3]

In Bermuda, October 10, 1847, the Butterfly, *Terias lisa* of Boisduval, suddenly appeared in great abundance, hundreds being seen in every direction. Previous to that occasion, Mr. Hurdis, the observer of this flight, had never met with this Butterfly. In the course of a few days, they had all disappeared.[4]

In Ceylon, in the months of April and May, migrations of Butterflies (mostly the *Callidryas hilariæ, C. alcmeone,* and *C. pyranthe,* with straggling individuals of the genus

[1] *Researches,* ch. viii. p. 158.
[2] Brown's *Bk. of Butterf.,* p. 101.
[3] *Lake Nyami,* p. 267.
[4] *Naturalist in Bermuda,* p. 120.

Euplœa, E. coras, and *E. prothoe*) are quite frequent. Their passage is generally in a northeasterly direction. The flights of these delicate insects appear to the eye of a white or pale yellow hue, and apparently to extend miles in breadth, and of such prodigious length as to occupy hours, and even days, in their uninterrupted passage. A friend of Tennent, traveling from Kandy to Kornegalle, drove for *nine miles* through such a cloud of white Butterflies, which was passing *across* the road by which he went. Whence these immense numbers of Butterflies come no one knows, and whither going no one can tell. But the natives have a superstitious belief that their flight is ultimately directed to Adam's Peak, and that their pilgrimage ends on reaching that sacred mountain.[1]

Monfet says: "Wert thou as strong as Milo or Hercules, and wert fenced or guarded about with an host of giants for force and valour, remember that such an army was put to the worst by an army of Butterflies flying in troops in the air, in the year 1104, and they hid the light of the sun like a cloud. Licosthenes relates, that on the third day of August, 1543, that no hearb was left by reason of their multitudes, and they had devoured all the sweet dew and natural moisture, and they had burned up the very grasse that was consumed with their dry dung."[2]

The most beautiful as well as pleasing emblem among the Egyptians was exhibited under the character of Psyche—the Soul. This was originally no other than a Butterfly: but it afterwards was represented as a lovely female child with the beautiful wings of that insect. The Butterfly, after its first and second stages as an egg and larva, lies for a season in a manner dead; and is inclosed in a sort of coffin. In this state it remains a shorter or longer period; but at last bursting its bonds, it comes out with new life, and in the most beautiful attire. The Egyptians thought this a very proper picture of the soul of man, and of the immortality, to which it aspired. But they made it more particularly an emblem of Osiris; who having been confined in an oak or coffin, and in a state of death, at last quitted his prison, and enjoyed a renewal of life.[3] This symbol passed

---

[1] Tennent's *Nat. Hist. of Ceylon,* ch. xii. p. 407.
[2] *Theatr. Ins.,* p. 107. Topsel's *Hist. of Beasts,* p. 974.
[3] Bryant's *Anct. Mythol.,* ii. 386.

over to the Greeks and Romans, who also considered the Butterfly as the symbol of Zephyr.[1]

Among the coats of arms of several of our most celebrated tribes of Indians, Baron Lahontan mentions one, that of the "Illinese," which bore a beech-leaf with a Butterfly argent.[2]

The sight of a trio of Butterflies is considered an omen of death.[3] An English superstition.

If a Butterfly enters a house, a death is sure to follow shortly in the family occupying it; if it enters through the window, the death will be that of an infant or very young person. As far as I know this superstition is peculiar to Maryland.

If a Butterfly alights upon your head, it foretells good news from a distance. This superstition obtains in Pennsylvania and Maryland.

The first Butterfly seen in the summer brings good luck to him who catches it. This notion prevails in New York.

In Western Pennsylvania, it is believed that if the chrysalides of Butterflies be found suspended mostly on the under sides of rails, limbs, etc., as it were to protect them from rain, that there will soon be much rain, or, as it is termed, a "rainy spell"; but, on the contrary, if they are found on twigs and slender branches, that the weather will be dry and clear.

Du Halde and Grosier tell us that the Butterflies of the mountain of Lo-few-shan, in the province of Quang-tong, China, are so much esteemed for their size and beauty, that they are sent to court, where they become a part of certain ornaments in the palaces. The wings of these Butterflies are very large, and their colors surprisingly diversified and lively.[4] Dionysius Kao, a native of China, also remarks, in his Geographical Description of that Empire, that the Butterflies of Quang-tong are generally sent to the emperor, as they form a part of the furniture of the imperial cabinets.[5]

Osbeck says the Chinese put up insects in boxes made of coarse wood, without covering, and lined with paper,

---

[1] Fosbroke, *Encycl. of Antiq.*, ii. 738.

[2] *Travels.* He doubtless refers to an Indian *totem.*

[3] *N. and Q.*, iii. 4.

[4] Du Halde, *China*, p. 21-2; Grosier's *China*, i. 570; Williams' *Mid. Kingd.*, i. 273; Astley's *Col. of Voy. and Trav.*, iv. 512.

[5] Harris's *Col. of Voy. and Trav.*, ii. 987.

which they carry round to sell; each box bringing half a piastre. Of the Butterflies, which were the principal insects thus sold, he enumerates twenty-one species.[1]

The Chinese children make Butterflies of paper, with which "they play after night by sending them, like kites, into the air."[2]

We learn from Captain Stedman, that even in the forests of Guiana, some people make Butterfly-catching their business, and obtain much money by it. They collect and arrange them in paper boxes, and send them off to the different cabinets of Europe.[3]

Butterflies are now extensively worn by French and American ladies on their head-dresses.

From the relations of Sir Anthony Shirley, quoted in Burton's Anatomy of Melancholy,[4] we learn that the kings of Persia were wont to hawk after Butterflies with sparrows and stares, or starlings, trained for the purpose; and we are also told that M. de Luisnes (afterward Prime Minister of France), in the nonage of Louis XIII., gained much upon him by making hawks catch little birds, and by making some of those little birds again catch Butterflies.[5]

In the Zoological Journal, No. 13, it is recorded that at a meeting of the Linnæan Society, March 11, 1832, Mr. Stevens exhibited a remarkable freak of nature in a specimen of *Vanessa urtica*, which possessed five wings, the additional one being formed by a second, but smaller, hinder wing on one side.[6]

J. A. de Mandelsloe, who made a voyage to the East Indies in 1639, tells us that not far from the Fort of Ternate grows a certain shrub, called by the Indians *Catopa*, from which falls a leaf, which, by degrees, is supposed to be metamorphosed into a Butterfly.[7]

De Pauw tells us that, not long before his time, the French peasants entertained a kind of worship for the chrysalis of the caterpillar found on the great nettle (the pupa of *Vanessa cardui?*), because they fancied that it revealed evident traces of Divinity; and quotes M. Des Landes in

[1] Osbeck, *Travels*, i. 331.   [2] *Ibid.*, i. 324.
[3] Stedman, *Surinam*, i. 279. Cf. Bancroft, *Guiana*, p. 229.
[4] *Anat. of Melanch.*, 1651, p. 268.
[5] *Life of Lord Herbert of Cherbury*, p. 134.
[6] *The Mirror*, xxv. 160.
[7] Harris's *Col. of Voy. and Trav.*, i. 790.

saying that the curates had even ornamented the altars with these pupæ.[1]

The Butterfly (Aug. Sax. *Buttor-fleoge*, or *Buter-flege*) is so named from the common yellow species, or from its appearing in the butter season. Its German names are *Schmetterling*, from *schmetten*, cream; and *Molkendieb*, the Whey-thief. The association with milk in its three forms, in butter, cream, and whey, is remarkable.

The African Bushmen eat the caterpillars of Butterflies; and the Natives of New Holland eat the caterpillars of a species of Moth, and also a kind of Butterfly, which they call *Bugong*, which congregates in certain districts, at particular seasons, in countless myriads. On these occasions the native blacks assemble from far and near to collect them; and after removing the wings and down by stirring them on the ground, previously heated by a large fire, winnowing them, eat the bodies, or store them up for use, by pounding and smoking them. The bodies of these Butterflies abound in oil, and taste like nuts. When first eaten, they produce violent vomitings and other debilitating effects; but these go off after a few days, and the natives then thrive and fatten exceedingly on this diet, for which they have to contend with a black crow, which is also attracted by the Butterflies, and which they dispatch with their clubs and use also as food.

Another practice in Australia is to follow up the flight of the Butterflies, and to light fires at nightfall beneath the trees in which they have settled. The smoke brings the insects down, when their bodies are collected and pounded together into a sort of fleshy loaf.[2]

Bennet tells us the larva of a Lepidopterous insect (the *Bugong?*) that destroys the green-wattle (*Acacia decurrens*) is much sought after, and considered a delicacy, by the blacks of Australia. These people eat also the pink grubs found in the wattle-trees, either roasted or uncooked. Europeans, who have tasted of this dish, say it is not disagreeable.[3]

Swammerdam, treating of the metamorphoses of larvæ into pupæ and thence into perfect insects, makes the following

---

[1] *Egypt. and Chinese*, ii. 106.
[2] Simmond's *Curios. of Food*, p. 312.
[3] *Gatherings of a Nat. in Austral.*, p. 288.

curious comparison: "The worms, after the manner of the brides in Holland, shut themselves up for a time, as it were to prepare, and render themselves more amiable, when they are to meet the other sex in the field of Hymen."[1]

Sphingidæ—Hawk-moths.

To the superstitious imaginations of the Europeans, the conspicuous markings on the back of a large evening moth, the *Sphinx Atropos*, represent the human skull, with the thigh-bones crossed beneath; hence is it called the *Death's-head Moth*, the *Death's-head Phantom*, the *Wandering Death-bird*, etc. Its cry,[2] which closely resembles the noise caused by the creaking of cork, or the plaintive squeaking of a mouse, certainly more than enough to frighten the ignorant and superstitious, is considered the voice of anguish, the moaning of a child, the signal of grief; and it is regarded "not as the creation of a benevolent being, but as the device

[1] *Hist. of Ins.*, p. 3.

[2] Reaumur considers this cry to be produced by the friction of the palpi against the proboscis (*Memoires*, ii. 293). Huber, but without mentioning the particulars, says he has ascertained that Reaumur was quite mistaken (*On Bees*, p. 313, note). Schroeter ascribes the sound to the rubbing of the tongue against the head; and Rösel to the friction of the chest upon the abdomen. M. de Johet thinks it is produced by the air being suddenly propelled against these scales by the action of the wings. M. Lorry states that the sound arises from the air escaping rapidly through peculiar cavities communicating with the spiracles, and furnished with a fine tuft of hairs on the sides of the abdomen (Cuv. *An. Kingd.—Ins.*, ii. 678). Mr. E. L. Layard seems to be of the same opinion (Tennent's *Nat. Hist. of Ceylon*, p. 427). But M. Passerini, curator of the Museum of Nat. Hist. at Florence, has lately investigated the subject more minutely. He traced the origin of the sound to the interior of the head, in which he discovered a cavity at the passage where muscles are placed for impelling and expelling the air. M. Dumeril has since discovered a sort of membrane stretched over this cavity, like, as he says, to the head of a drum. M. Duponchel has also confirmed by experiment the opinions of Passerini and Dumeril, and confutes Lorry, whose notion was generally adopted, by stating that the noise is produced from the head when the body of the insect is removed (*Annales des Sci. Nat.*, Mars., 1828).

of evil spirits "—spirits, enemies to man, conceived and fabricated in the dark; and the very shining of its eyes is supposed to represent the fiery element whence it is thought to have proceeded. Flying into their apartments in the evening, it at times extinguishes the light, foretelling war, pestilence, famine, and death to man. The sudden appearance of these insects, we are informed by Latrielle, during a season while the people were suffering from an epidemic disease, tended much to confirm the notions of the superstitious in that district, and the disease was attributed by them entirely to their visitation.[1] Jaeger says, at a very recent day, that this large Moth first attracted his "attention during the prevalence of a severe and fatal epidemic, and of course nothing more was necessary than its appearance at such a time to induce an ignorant people to believe it the veritable prophet and forerunner of death. A curate in Bretagne, France," continues this author, "made a most horrible and fear-exciting description of this animal, describing the very loud and dreadful sound which it emitted as a sort of lamentation for the awful calamity which was coming on the earth."[2] Reaumur informs us that all the members of a female convent in France were thrown into the greatest consternation at the appearance of one of these insects, which happened to fly in during the evening at one of the windows of the dormitory.[3]

In the Isle of France, the natives believe that the dust (scales) cast from the wings of the Death's-head Moth, in flying through an apartment, is productive of blindness to the visual organs on which it falls.[4]

There is a quaint superstition in England that the Death's-head Moth has been very common in Whitehall ever since the martyrdom of Charles I.[5]

Illustrative of the tough texture of the skin with which many soft larvæ are provided for protection, the following may be instanced : Bonnet squeezed under water the caterpillar of the privet Hawk-moth, *Sphinx ligustris*, till it was as flat and empty as the finger of a glove, yet within an

---

[1] Cf. *Penny Encycl.*, *sub.* Sphinx, and *The Mirror*, **xix.** 212.
[2] *Hist of Ins.*, p. 191.
[3] Reaumur, ii. 289. Shaw, *Zool.*, vi. 217.
[4] *Saturday Mag.*, xix. 102.
[5] *Notes and Queries*, xii. 200.

hour it became plump and lively as if nothing had happened.[1]

The name Sphinx is applied to this genus of insects from a fancied resemblance between the attitude assumed by the larvæ of several of the larger species, when disturbed, and that of the Egyptian Sphinx.

Bombicidæ—Silk-worm Moths.

The notices of the cultivation of the mulberry and the rearing of Silk-worms, found in Chinese works, have been industriously collected and published by M. Julien, by order of the French government. From his work it appears that credible notices of the culture of the tree and the manufacture of silk are found as far back as B.C. 780; and in referring its invention to the Empress Siling, or Yuenfi, wife of the Emperor Hwangti, B.C. 2602 (Du Halde says 2698), the Chinese have shown their belief of its still higher antiquity. The Shi King contains this distich:

The legitimate wife of Hwangti, named Siling Shi, began to rear Silk-worms:
At this period Hwangti invented the art of making clothing.

Du Halde says this invention raised the Empress to the rank of a divinity, under the title of Spirit of the Silk-worm, and of the Mulberry-tree.[2]

The Book of Rites contains a notice of the festival held in honor of this art, which corresponds to that of plowing by the emperor. "In the last month of spring, the young empress purified herself and offered sacrifice to the goddess of Silk-worms. She went into the eastern fields and collected mulberry-leaves. She forbade noble dames and the ladies of statesmen adorning themselves, and excused her attendants from their sewing and embroidery, in order that they might give all their care to the rearing of Silk-worms."[3]

---

[1] Bonnet, *Œuvres*, ii. 124.
[2] *China*, p. 253.   Astley's *Col. of Voy. and Trav.*, iv. 138.
[3] Williams' *Middle Kingdom*, ii. 121-2.

The manufacture of silk has been known in India from time immemorial, it being mentioned in the oldest Sanscrit books.[1] It is the opinion of modern writers, however, that the culture of the Silk-worm passed from China into India, thence through Persia, and then, after the lapse of several centuries, into Europe. But long before this, wrought silk had been introduced into Greece from Persia. This was effected by the army of Alexander the Great, about the year 323 before Christ.

The Greeks fabled silk to have first been woven in the Island of Cos by Pamphila, the daughter of Plateos.[2] Of its true origin they were, in a great measure, ignorant, but seem to have been positive that it was the work of an insect. Pausanias thus describes the animal and its culture: "But the thread, from which the Ceres (an Ethiopian race) make garments, is not produced from a tree, but is procured by the following method: A worm is found in their country which the Greeks call *Seer*, but the Ceres themselves, by a different name. This worm is twice as large as a beetle, and, in other respects, resembles spiders which weave under trees. It has, likewise, eight feet as well as the spider. The Ceres rear these insects in houses adapted for this purpose both to summer and winter. What these insects produce is a slender thread, which is rolled round their feet. They feed them for four years on oatmeal; and on the fifth (for they do not live beyond five years) they give them a green reed to feed on: for this is the sweetest of all food to this insect. It feeds, therefore, on this till it bursts through fullness, and dies: after which, they draw from its bowels a great quantity of thread."[3]

Aristotle seems to have had a much clearer idea of the origin of silk, for he says it was unwound from the *pupa* (he does not expressly say the *pupa*, but this we must suppose) of a large horned caterpillar.[4] The *larva* he means could not, however, be the common Silk-worm, since it is rather small and without horns.

Pliny, who, most probably, obtained the most of his ideas from Pausanias and Aristotle, was of opinion that silk was

[1] Colebrook, *Asiat. Research.*, v. 61.
[2] Aristotle, v. 17–9. Pliny, ix. 20.
[3] Paus. *Hist. of Greece*, B. 6, c. 26.
[4] Aristot. *Hist. An.*, v. 19.

the produce of a worm which built clay-nests and collected wax. At first these worms, he says, assume the appearance of small butterflies with naked bodies, but soon after, being unable to endure the cold, they throw out bristly hairs, which assume quite a thick coat against the winter, by rubbing off the down that covers the leaves, by the aid of the roughness of their feet. This they compress into balls by carding it with their claws, and then draw it out and hang it between the branches of the trees, making it fine by combing it out, as it were : last of all, they take and roll it round their body, thus forming a nest in which they are enveloped. It is in this state that they are taken ; after which they are placed in earthen vessels in a warm place, and fed upon bran. A peculiar sort of down soon shoots forth upon the body, on being clothed with which they are sent to work upon another task.[1]

The first kinds of silk dresses worn by the Roman ladies were from the Island of Cos, and, as Pliny says, were known by the name of *Coæ vestes*.[2] These dresses, of which Pliny says in such high praise, "that while they cover a woman, they at the same time reveal her charms," were indeed so fine as to be transparent, and were sometimes dyed purple, and enriched with stripes of gold. They had their name from the early reputation which Cos acquired by its manufacture of silk. But silk was a very scarce article among the Romans for many ages, and so highly prized as to be valued at its weight in gold. Vospicius informs us that the Emperor Aurelian, who died A.D. 125, refused his empress a robe of silk, which she earnestly solicited, merely on account of its dearness. Galen, who lived about A.D. 173, speaks of the rarity of silk, being nowhere then but at Rome, and there only among the rich. Heliogabalus is said to have been the first Roman that wore a garment entirely of silk.

We learn from Tacitus, that early in the reign of Tiberius, about A.D. 17, the Senate enacted "that men should not defile themselves by wearing garments of silk."[3] Pliny says, however, that in his time men had become so degenerate as

---

[1] Pliny, *Nat. Hist.*, xi. 23.
[2] *Ibid.*, xi. 22.
[3] Tacitus, *Ann.*, B. 2, c. 33.

to not even feel ashamed to wear garments of this material.[1]

The mode of producing and manufacturing silk was not known to Europe until long after the Christian era, being first learned about the year 555 by two Persian monks, who, under the encouragement of the Emperor Justinian, procured in India the eggs of the Silk-worm Moth, with which, concealing them in hollow canes, they hastened to Constantinople. They also brought with them instructions for hatching the eggs, rearing and feeding the worms, and drawing, spinning, and working the silk.[2]

From Constantinople, the culture of the Silk-worm spread over Greece, so that in less than five centuries that portion of this country, hitherto called the Peloponnesus, changed its denomination into that of Morea, from the immense plantations of the *Morus alba*, or white mulberry.[3] Large manufactories were set up at Athens, Thebes, and Corinth. The Venetians, soon after this, commencing a commerce with the Grecians, supplied all the western parts of Europe with silks for many centuries. Several kinds of modern silk manufactures, such as damasks, velvets, satins, etc., were as yet unknown.

About the year 1130, Roger II., King of Sicily, having conquered the Peloponnesus, transported the Silk-worms and such as cultivated them to Palermo and to Calabria. Such was the success of the speculation in Calabria, that it is doubtful whether, even at the present moment, it does not produce more silk than the whole of the rest of Italy.[4]

By degrees the rest of Italy, as well as Spain, learned from the Sicilians and Calabrians the management of Silk-worms and the working of silk; and at length, during the wars of Charles VIII., in 1499, the French acquired it, by right of neighborhood, and soon large plantations of the mulberry were raised in Provence. Henry I. is reported

---

[1] *Nat. Hist.*, xi. 22.

[2] Cf. Gibbon's *Decl. and Fall of Rom. Em.*, c. 40.

[3] Some authors, however, assert that the name was suggested by the resemblance of the Morea to the shape of the mulberry-leaf, a less plausible opinion by far than the former.

[4] Thuanus, in contradiction to most other writers, makes the manufacture of silk to be introduced into Sicily two hundred years later, by Robert the Wise, King of Sicily and Count of Provence.

to have been the first French king who wore silk stockings. The invention, however, originally came from Spain, whence silk stockings were brought over to England to Henry VIII. and Edward VI.

It is stated, that at the celebration of the marriage between Margaret, daughter of Henry III., and Alexander III. of Scotland, in the year 1251, a most extravagant display of magnificence was made by one thousand English knights appearing in suits of silk. It appears also by the 33d of Henry VI., cap. 5, that there was a company of silk-women in England as early as the year 1455; but these were probably employed rather in embroidering and making small haberdasheries, than in the broad manufacture, which was not introduced till the year 1620.

Sir Thomas Gresham, in a letter to Sir William Cecil, Elizabeth's great minister, dated Antwerp, April 30th, 1560, says: "I have written into Spain for silk hose both for you and my lady, your wife, to whom, it may please you, I may be remembered." These silk hose, of a black color, were accordingly soon after sent by Gresham to Cecil.[1]

Hose were, in England, up to the time of Henry VIII., made out of ordinary cloth: the King's own were formed of yard-wide taffata. It was only by chance that he might obtain a pair of silk hose from Spain. His son, Edward VI., received as a present from Sir Thomas Gresham—Stow speaks of it as a great present—"a pair of long Spanish silk stockings." For some years longer, silk stockings continued to be a great rarity. "In the second year of Queen Elizabeth," says Stow, "her silk-woman, Mistress Montague, presented her Majesty with a pair of black knit-silk stockings for a New-Year's gift; the which, after a few days' wearing, pleased her Highness so well, that she sent for Mistress Montague, and asked her where she had them, and if she could help her to any more; who answered, saying, 'I made them very carefully, of purpose only for your Majesty, and, seeing these please you so well, I will presently set more in hand.' 'Do so,' quoth the Queen, 'for indeed I like silk stockings so well, because they are pleasant, fine, and delicate, that henceforth I will wear no more

---

[1] Burgon's *Life of Sir Thomas Gresham.* 1839. i. 110, 302.

cloth stockings.' And from that time to her death the Queen never wore cloth hose, but only silk stockings."[1]

James I., while King of Scotland, is said to have once written to the Earl of Mar, one of his friends, to borrow a pair of silk stockings, in order to appear with becoming dignity before the English Ambassador; concluding his letter with these words: "For ye would not, sure, that your King should appear like a scrub before strangers." This shows the great rarity of silk articles at that period in Scotland.

In 1629, the manufacture of silk was become so considerable in London, that the silk trowsters of the city and parts adjacent were incorporated; and in 1661, this company employed above forty thousand persons. The revocation of the Edict of Nantes, in 1685, contributed in a great degree to promote the manufacture of this article; and the invention of the silk-throwing machine at Derby, in 1719, added so much to the reputation of English manufactures, that even in Italy, according to Keysler, the English silks bore a higher price than the Italian.[2]

Rev. Stephen Olin tells us that the Mohammedans of Arabia will not allow strangers to look into their cocooneries, on account of their superstitious fear of the evil eye, of the influence of which the Silk-worms are thought to be peculiarly susceptible.[3]

The silk of the nests of the social caterpillar of the *Bombyx Madrona*, was an object of commerce in Mexico in the time of Montecusuma; and the ancient Mexicans pasted together the interior layers, which may be written upon without preparation, to form a white, glossy pasteboard. Handkerchiefs are still manufactured of it in the Intendency of Oaxaca.[4]

A complete nest of these Silk-worms, called in Brazil *sustillo*, was sent by the Academy of Sciences and Natural History to the King of Spain. The naturalist, Don Antonio Pineda, sent also a piece of this natural silk paper, measuring a yard and a half, of an elliptical shape, which, however, is peculiar to them all.[5]

---

[1] Stow's *Chronicle*, edit. 1631, p. 887.

[2] Keysler, *Trav.*, i. 289.

[3] Olin, *Travels*.

[4] *Polit. Essay on N. Spain*, iii. 59.

[5] Skinner's *Pres. State of Peru*, p. 346, note. Southey's *Hist. of Brazil*, iii. 644. Calancha's *Augustine Hist. of Peru*, i. 66.

The Chinese fix on rings with threads the females of two species of wild *Bombyx*, whose caterpillars produce silk, and place these insects on a tree, or on some body situated in the open air, to allow the males, guided by their scent, to visit them.[1]

"'The manner of the Chinese is," we read in Purchas's Pilgrims, "in the Spring time to revive the Silke-worms (that lye dead all the Winter) by laying them in the warme sunne, and (to hasten their quickening, that they may sooner goe to worke) to put them into bagges, and so hang them under their childrens armes.'"[2]

In China, the pupæ of the Silk-worms after the silk is wound off, and the larvæ of a species of Sphinx-moth, furnish articles for the table, and are considered delicacies.[3] The natives of Madagascar, who eat all kinds of insects, consider also Silk-worms a great luxury.[4]

Aldrovandus states that the German soldiers sometimes fry and eat Silk-worms.[5]

Dr. James says: "Silk-worms dried, and reduced to a powder, are, by some, applied to the crown of the head for removing vertigos and convulsions. The silk, and case or coat, are of a due temperament between heat and cold, and corroborate and recruit the vital, natural, and animal spirits."[6] The cocoons are also the basis of Goddard's *Drops*, and enter into several other compositions, such as the *Confectio de Hyacintho*, when made in the best manner.[7]

With respect to the coloring of silk, we find in "Tseën Tse Wan," or thousand character classic, a work that has been a school-book in China for the last 1200 years, that an ancient sage by the name of Mih, seeing the white silk colored, wept on account of its original purity being destroyed.[8]

Some of the eggs of a wild species of Silk-worm being sent overland from China to Paris, proved a source of considerable anxiety to different parties who received them during the transit, the instructions on the box, instead of

---

[1] Cuvier, *An. King.*—*Ins.*, ii. 634.
[2] *Pilgrims*, iii. 442.
[3] Darwin, *Phytolog.*, p. 364.   Donovan's *Ins. of China*, p. 6.
[4] Hollman, *Travels*, p. 473.
[5] Donovan's *Ins. of China*, p. 6.
[6] *Med. Dict.*
[7] Geoffroy, *Treat. on Subst. used in Physic*, p. 383.
[8] *Twelve Years in China*, p. 14.

simply stating that it contained the eggs of the *wild* Silk-worm Moth, was couched in the following manner by the French savant who forwarded them : "Must be kept far from the engines; this box contains *savage* worms."[1]

About twenty-five years ago, during a mania for rearing Silk-worms, to meet the demand for the eggs of these insects, fish-spawn was distributed throughout the country. The humbug was quite as successful as it was curious.

It has been said that the search after the "Golden Fleece" may be ascribed to the desire to obtain silk.[2]

As a protection against rifle-balls, the Chinese, who were engaged in the rebellion of 1853, state that they wore dresses thickly padded with floss silk; they said that while the ball had a twist in it, revolving in its course, it caught up the silk and fastened itself in the garment. One man declared that he took out six so caught, in one day, after a severe fight. They said the dress was of more use within a hundred yards than at long range, when the ball had lost its revolving motion.[3]

Vaucanson, the inventor of the famous "automaton duck," to revenge himself upon the silk-weavers of Lyons, who had stoned him because he attempted to simplify the ordinary loom, is said to have invented a loom on which a donkey worked silken cloth.[4]

The following curious Welsh epigram on the Silk-worm is composed entirely of vowels, and can be recited without closing or moving lips or teeth :

> O'i wiw wy i ê â, a'i weuaw
> O'i wyau y weua;
> E' weua ei wî aia',
> A'i weuau yw ieuau iâ.

> I perish by my art ; dig mine own grave;
> I spin the thread of life ; my death I weave.[5]

---

[1] *Twelve Years in China*, p. 14. [2] *Ibid.*
[3] *Ibid.*, p. 194.
[4] *Memoires of Robt. Houdin*, p. 161.
[5] *Mag. of Nat. Hist.*, vi. 9.

Arctiidæ—Wooly-bear Moths.

In 1783, the larvæ of the Moth, *Arctia chrysorrhœa*, were so destructive in the neighborhood of London that subscriptions were opened to employ the poor in cutting off and collecting the webs; and it is asserted that not less than eighty bushels were collected and burnt in one day in the parish of Clapham. And even in some places prayers were offered up in the churches to avert the calamities of which they were supposed by the ignorant to be the forerunner.[1]

If a caterpillar spins its cocoon in a house, it foretells its desolation by death; if in your clothes, it warns you you will wear a shroud before the year is out. This superstition obtains in the Middle States, Virginia, and Maryland.

If Moths, flying in a candle, put it out, it forebodes a calamity amounting to almost death. This superstition is pretty general.

Why Moths fly in a candle: Kempfer tells us, there is found in Japan an insect, which, by reason of its incomparable beauty, is kept by the Japanese ladies among the curiosities of their toilets. He calls it a Night-fly, and describes it as being "about a finger long, slender, round-bodied, with four wings, two of which are transparent and hid under a pair of others, which are shining as it were polished, and most curiously adorned with blue and golden lines and spots." The following little fable, which accounts so beautifully for the flying of Moths in a candle, owes its origin to the unparalleled beauty of this insect, and is well worthy of being preserved: The Japanese say that all other Night-flies (Moths, etc.) fall in love with this particular one, who, to get rid of their importunities, maliciously bids them, under the pretense of trying their constancy, to go and bring to her fire. And the blind lovers, scrupling not to obey her command, fly to the nearest fire or candle, in which they never fail to burn themselves to death.[2]

The following verses, embodying the above fable (except in several minor particulars) are from the pen of Mrs. A. L. Ruter Dufour:

---

[1] Baird's *Encycl. of Nat. Sci.* Shaw's *Zool.*, vi. 229.
[2] Pinkerton's *Col. of Voy. and Trav.*, vii. 705.

One summer night, says a legend old,
  A Moth a Firefly sought to woo:
"Oh, wed me, I pray, thou bright star-child,
  To win thee there's nothing I'd dare not do."

"If thou art sincere," the Firefly cried,
  "Go—bring me a light that will equal my own;
Not until then will I deign be thy bride;"—
  Undaunted the Moth heard her mocking tone.

Afar he beheld a brilliant torch,
  Forward he dashed, on rapid wing,
Into the light to bear it hence;—
  When he fell a scorched and blighted thing.—

Still ever the Moths in hope to win,
  Unheeding the lesson, the gay Firefly,
Dash, reckless, the dazzling torch within,
  And, vainly striving, fall and die!

WASHINGTON, D. C., June 24, 1864.

Moufet says: "Our North, as well as our West countrymen, call it (the Moth, *Phalaina*) *Saule, i. e. Psychen, Animam*, the soul; because some silly people in old time did fancy that the souls of the dead did fly about in the night seeking light."[1] "Pliny commends a goat's liver to drive them away, yet he shews not the means to use it."[2]

One of the most highly prized curiosities in the collection of Horace Walpole, was the silver bell with which the popes used to curse the caterpillars. This bell was the work of Benvenuto Cellini, one of the most extraordinary men of his extraordinary age, and the relievos on it representing caterpillars, butterflies, and other insects, are said to have been wonderfully executed.[3]

In Purchas's Pilgrims, we read of worms being sprinkled with holy water to kill them.[4]

Apuleius says, that if you take the caterpillars from another garden, and boil them in water with anethum, and let them cool, and besprinkle the herbs, you will destroy the existing caterpillars.[5]

---

[1] *Theatr. Ins.*, p. 88.   Topsel's *Hist. of Beasts*, p. 958.
[2] Moufet, p. 108.   Topsel, p. 975.
[3] *Monthly Mag.*, 7 (Pt. I.) xxxix. 1799.
[4] *Pilgrims*, ii. 1034.
[5] Owen's *Geoponika*, ii. 99.

Pliny says, that "if a woman having a catamenia strips herself naked, and walks round a field of wheat, the caterpillars, worms, beetles, and other vermin, will fall off the ears of the grain!" This important discovery, according to Metrodurus of Scepsos, was first made in Cappadocia; where, in consequence of such multitudes of " Cantharides" being found to breed there, it was the practice for women to walk through the middle of the fields with their garments tucked up above the thighs.[1] Columella[2] has described this practice in verse, and Ælian[3] also mentions it. Pliny says further that in other places, again, it is the usage of women to go barefoot, with the hair disheveled and the girdle loose : due precaution, however, he seriously observes, must be taken that this is not done at sunrise, for if so the crop will wither and dry up.[4] Apuleius,[5] Columella,[6] and Palladius[7] relate the same story. Constantinus, likewise, whose verses, as translated in Moufet's Theater of Insects, are as follows :

> But if against this plague no art prevail,
> The Trojan arts will do't, when others fail.
> A woman barefoot with her hair untied,
> And naked breasts must walk as if she cried,
> And after Venus' sports she must surround
> Ten times, the garden beds and orchard ground.
> When she hath done, 'tis wonderful to see,
> The caterpillars fall off from the tree,
> As fast as drops of rain, when with a crook,
> For acorns or apples the tree is shook.[8]

This remarkable superstitious remedy for destroying caterpillars was frequently practiced by the Indians of America. Schoolcraft, treating of the peculiar superstitions connected with the menstrual lodge of these people, says :

"This superstition does not alone exert a malign influence, or spell, on the human species. Its ominous power,

---

[1] Pliny, *Nat. Hist.*, xxviii. 7 (23).
[2] Col. B. x.
[3] Ælian, B. xi. c. 3.
[4] Pliny, *Nat. Hist.*, xxviii. 7 (23).
[5] Vide Owen's *Geoponika*, ii. 99.
[6] Col. *In Hort.*, v. 357.
[7] Pallad. B. i. c. 35.
[8] *Theatr. Ins.*, p. 193. Topsel's *Hist. of Beasts*, p. 1041 and 670.

or charm, is equally effective on the animate creation, at least on those species which are known to depredate on their little fields and gardens. To cast a protective spell around these, and secure the fields against vermin, insects, the sciurus, and other species, as well as to protect the crops against blight, the mother of the family chooses a suitable hour at night, when the children are at rest and the sky is overcast, and having completely divested herself of her garments, trails her *machecota* behind her, and performs the circuit of the little field."[1]

The fat of bears, says Topsel, "some use superstitiously beaten with oil, wherewith they anoynt their grape-sickles when they go to vintage, perswading themselves that if nobody know thereof, their tender vine-branches shall never be consumed by caterpillars. Others attribute this to the vertue of bears' blood."[2]

Nicander used "a caterpillar to procure sleep: for so he writes; and Hieremias Martius thus translates him:

> Stamp but with oyl those worms that eat the leaves,
>   Whose backs are painted with a greenish hue,
> Anoint your body with 't, and whilst that cleaves,
>   You shall with gentle sleep bid cares adieu."[3]

Of a caterpillar that feeds upon cabbage leaves, the *Eruca officinalis* of Schroder, Dr. James says : "Bruised, or a powder of them, raise a blister like cantharides, and take off the skin. Moufet says, they will cause the teeth to fall out of their sockets, and Hippocrates writes, that they are good for a Quinsey."[4]

Psychidæ—Wood-carrying Moth, etc.

The larvæ of the Wood-carrying Moth (of the genus *Oiketicus*, or *Eumeta*, Wlk.) of Ceylon, surround themselves with cases made of stems of leaves, and thorns or pieces of

---

[1] *Hist. of Indians of U. S.*, v. p. 70.

[2] *Hist. of Beasts*, p. 30.

[3] Moufet, *Theatr. Ins.*, p. 194. Topsel's *Hist. of Beasts*, pp. 670, 1041.

[4] *Med. Dict.*

twigs bound together by threads, till the whole resembles a miniature Roman fasces; in fact, an African species of these insects has obtained the name of "Lictor." The Germans have denominated the group *Sackträger*, and the Singhalese call them Darra-kattea or "billets of fire-wood," and regard the inmates, Tennent says, as human beings, who, as a punishment for stealing wood in some former state of existence, have been condemned to undergo a metempsychosis under the form of these insects.[1]

Noctuidæ—Antler-moth, Cut-worm, etc.

The Antler-moth, *Noctua graminis*, Linn., has been particularly observed in Sweden, Norway, Northern Germany, and even in Greenland, where it does great mischief to grass-plots and meadows. It is recorded to have done very great injury in the eastern mountains of Georgenthal, as well as at Töplitz in Bohemia, where larvæ were in such large numbers that in four days and a half 200 men found 23 bushels of them, or 4,500,000 in the 60 bushels of mould which they examined. In Germany it seems to be confined to high and dry districts, and it never appears there in wet meadows, but its devastations are sometimes most extensive, as happened in the Hartz territory in 1816 and '17, when whole hills that in the evening were clad in the finest green, were brown and bare the following morning; and such vast numbers of the caterpillars were there that the ruts of the roads leading to the hills were full of them, and the roads being covered with them were even rendered slippery and dirty by their being crushed in some places.[2]

The notorious astrologer, William Lilly, alluding to the comet which appeared in 1677, says: "All comets signify wars, terrors, and strange events in the world;" and gives the following curious explanation of the prophetic nature of these bodies: "The spirits, well knowing what accidents shall come to pass, do form a star or comet, and give it

---

[1] Tennent, *Nat. Hist. of Ceylon*, p. 431.
[2] Köllar's *Treat. on Ins.*, Lond. Trans., p. 105–36. Curtis's *Farm Insects*, p. 507.

what figure or shape they please, and cause its motion through the air, that people might behold it, and thence draw a signification of its events." Further, a comet appearing in the Taurus portends "mortality to the greater part of cattle, as horses, oxen, cows, etc.," and also "prodigious shipwrecks, damage in fisheries, monstrous floods, and destruction of fruit by caterpillars and other vermin."[1]

Josselyn, in the account of his voyage to New England, printed in London in 1674, has the following relation of an insect which is doubtless a species of *Agrotis*, probably the *Agrotis telifera:* "There is also (in New England) a dark dunnish Worm or Bug of the bigness of an Oaten-straw, and an inch long, that in the Spring lye at the Root of Corn and Garden plants all day, and in the night creep out and devour them; these in some years destroy abundance of *Indian* Corn and Garden plants, and they have but one way to be rid of them, which the *English* have learned of the Indians; And because it is somewhat strange, I shall tell you how it is, they go out into a field or garden with a Birchen-dish, and spudling the earth about the roots, for they lye not deep, they gather their dish full which may contain a quart or three pints, then they carrie the dish to the Sea-side when it is ebbing water and set it a swimming, the water carrieth the dish into the Sea, and within a day or two you go into your field you may look your eyes out sooner than find any of them."[2]

The Army-worm (larva of *Leucania unipunctata* of Haworth), during this our great rebellion, is thought, by many persons in Western Pennsylvania, to prognosticate the success or defeat of our armies by the direction it travels. If toward the North, the South will be victorious; and if toward the South, the North will conquer. An old gentleman, who believes that a frog's foot drawn in chalk above the door will keep away witches, tells me this worm invariably travels southward.

This larva was noticed but a few years before the war began, and then appearing, as it were, in armies, it was called the Army-worm. The superstitious omen from it has followed not preceded the name.

Lindenbrog, in his Codex Legum Antiquarum, cum

---

[1] Lilly's *Prophetical Merlin*, pub. in 1644.
[2] Josselyn's *Voy.*, p. 116.

Glossario, fol. Francof. 1613, mentions the following super-stition : " The peasants, in many places in Germany, at the feast of St. John, bind a rope around a stake drawn from a hedge, and drive it hither and thither, till it catches fire. This they carefully feed with stubble and dry wood heaped together, and they spread the collected ashes over their pot-herbs, confiding in vain superstition, that by this means they can drive away Canker-worms. They therefore call this Nodfeur, q. *necessary fire.*" •

These fires were condemned as sacrilegious, not as if it had been thought that there was anything unlawful in kindling a fire in this manner, but because it was kindled with a superstitious design. They are, however, Du Cange says, still kindled in France, on the eve of St. John's day.[1]

Geometridæ—Span-worms.

The Measuring-worm, crawling on your clothes, is thought to foretell a new suit; on your hands, a pair of gloves, etc.

Tineidæ—Clothes'-moth, Bee-moth, etc.

In Newton's Journal of the Arts for December, 1827, there is the following mention of a new kind of cloth fabri-cated by insects : The larvæ of the Moth, *Tinea punctata*, or *T. padilla*, have been directed by M. Habenstreet, of Munich, so as to work on a paper model suspended from a ceiling of a room. To this model he can give any form and dimensions, and he has thus been enabled to obtain square shawls, an air balloon four feet high, and a woman's com-plete robe, with the sleeves, but without seams. One or two larvæ can weave a square inch of cloth. A great num-ber are, of course, employed, and their motions are inter-dicted from the parts of the model not to be covered, by oiling them. The cloth exceeds in fineness the lightest

---

[1] Jamieson's *Scot. Dict.*, ii. 144.

gauze, and has been worn as a robe over her court dress by the Queen of Bavaria.[1]

Authors are of opinion that the ancients possessed some secret for preserving garments from the Moth, *Tinia tapetzella.* We are told the robes of Servius Tullius were found in perfect preservation at the death of Sejanus, an interval of more than five hundred years. Pliny gives as a precaution "to lay garments on a coffin;" others recommend "cantharides hung up in a house, or wrapping them in a lion's skin"—"the poor little insects," says Reaumur, "being probably placed in bodily fear of this terrible animal."[2]

Moufet says: "They that sell woollen clothes use to wrap up the skin of a bird called the king's-fisher among them, or else hang one in the shop, as a thing by a secret antipathy that Moths cannot endure."[3]

Among the various contrivances resorted to as a safe-guard against the Bee-moth, *Galleria cereana*, Fabricius, perhaps the most ingenious is that, mentioned by Langstroth, of "governing the entrances of all the hives by a long lever-like *hen-roost*, so that they may be regularly closed by the crowing and cackling tribe when they go to bed at night, and opened again when they fly from their perch to greet the merry morn."[4]

An intelligent man informed Langstroth that he paid ten dollars to a "Bee-quack" professing to have an infallible secret for protecting Bees against the Moth; and, after the quack had departed with his money, learned that the secret consisted in "always keeping strong stocks."[5]

---

[1] *Mag. of Nat. Hist.*, i. 66.
[2] Harper's *New Monthly Mag.*, xxii. 41.
[3] *Theatr. Ins.*, p. 274. Topsel's *Hist. of Beasts*, p. 1100.
[4] *On the Honey-Bee*, p. 248.
[5] *Ibid.*, p. 238, note.

ORDER VII.

HOMOPTERA.

Cicadidæ—Harvest-flies.

THE Cicadas, *C. plebeja*, Linn., called by the ancient Greeks, (by whom, as well as by the Chinese, they were kept in cages for the sake of their song,) *Tettix*, seem to have been the favorites of every Grecian bard, from Homer and Hesiod to Anacreon and Theocritus. Supposed to be perfectly harmless, and to live only upon dew, they were addressed by the most endearing epithets, and were regarded as almost divine. Thus sings the muse of Anacreon:

> Happy creature! what below
> Can more happy live than thou?
> Seated on thy leafy throne,
> Summer weaves thy verdant crown.
> Sipping o'er the pearly lawn,
> The fragrant nectar of the dawn,
> Little tales thou lov'st to sing,
> Tales of mirth—an insect king.
> Thine the treasures of the field,
> All thy own the seasons yield;
> Nature paints thee for the year,
> Songster to the shepherds dear;
> Innocent, of placid fame,
> What of man can boast the same?
> Thine the loudest voice of praise,
> Harbinger of fruitful days;
> Darling of the tuneful nine,
> Phœbus is thy sire divine;
> Phœbus to thy note has given
> Music from the spheres of heaven;
> Happy most as first of earth,
> All thy hours are peace and mirth;
> Cares nor pains to thee belong,
> Thou alone art ever young.
> Thine the pure immortal vein,
> Blood nor flesh thy life sustain;
> Rich in spirits—health thy feast,
> Thou art a demi-god at least.

(250)

But the old witticism, attributed to the incorrigible Rhodian sensualist, Xenarchus, gives quite a different reason to account for the supposed happiness of these insects:

> Happy the Cicadas' lives,
> Since they all have voiceless wives![1]

Plutarch, reasoning upon that singular Pythagorean precept which forbid the wife to admit swallows in the house, remarks: "Consider, and see whether the swallow be not odious and impious . . . . because she feedeth upon flesh, and, besides, killeth and devoureth especially grasshoppers (Cicadas), which are sacred and musical."[2]

The Athenians were so attached to the Cicadas, that their elders were accustomed to fasten golden images of them in their hair. Thucidides incidentally remarks that this custom ceased but a little before his time. He adds, also, that the fashion prevailed, too, for a long time with the elders of the Ionians, from their affinity to the Athenians.[3]

This singular form, for their ornamental combs, seems to have been adopted originally from the predilection of the Athenians for whatever bore any affinity to themselves, who boasted of being autochthones or aboriginal. It is sung of the Athenians:

> Blithe race! whose mantles were bedeck'd
> With golden grasshoppers, in sign that they
> Had sprung. like those bright creatures. from the soil
> Whereon their endless generations dwelt.

Mr. Michell supposes the Athenians to have imitated in this instance their prototypes, the Egyptians; for as they, he adds, wore their favorite symbol, the Scarabæus, in this manner, so Attic pride set up a rival in the head-dress thus introduced by Cecrops and his followers.[4]

From a very ancient writer,[5] we have similar ornaments

---

[1] It is a philosophical fact that the female Cicadas are not capable of making any noise—the above distich evinces its early discovery.

[2] *Symposiaques.* B. 8. Holl. *Trans.*, p. 630.

[3] Thuc. B. 1, vi. (Bohn's ed.).

[4] On Aristoph., *Vesp.* 230.

[5] Cited by Athen., 525.

ascribed to the Samians. They also most probably derived this fashion from the early Athenians.[1]

It seems, from the following lines of Asius,[2] that Cicadas were also worn as ornaments on dresses:

> Clad in magnificent robes, whose snow-white folds
> Reach'd to the ground of the extensive earth,
> And golden knobs on them like grasshoppers.

The sound of the Cicada and that of the harp were called by the Greeks by one and the same name; and a Cicada sitting upon a harp was the usual emblem of the science of music. This was accounted for by the following very pleasing and elegant tale: Two rival musicians, Eunomis of Locris and Aristo of Rhegium, when alternately playing upon the harp, the former was so unfortunate as to break a string of his instrument, and by which accident would certainly have lost the prize, when a Cicada, flying to him and sitting upon his harp, supplied the place of the broken string with its melodious voice, and so secured to him an easy victory over his antagonist.[3]

To excel the Cicada in singing was the highest commendation of a singer, and the music of Plato's eloquence was only comparable to the voice of this insect. Homer compared his good orators to the Cicadæ, "which, in the woods, sitting on a tree, send forth a delicate voice."[4] But Virgil speaks of them as insects of a disagreeable and stridulous tone, and accuses them of bursting the very shrubs with their noise,—

> Et cantu querulæ rumpent arbusta Cicadæ.[5]

Moufet says: "The Cicadæ, abounding in the end of spring,

---

[1] Cicada-combs are alluded to in Aristoph., Eq. 1331. Cf. also Philostr. *Imag.*, p. 837. Heracl. Pont., cited by Athen., p. 512. Bloomfield's *Thucid.*, i. 14.

[2] Cited by Athen., p. 842 (Bohn's ed.).

[3] Strabo, *Geog.* B. 6.

[4] *Iliad*, iii. 152. Buckley's translation, p. 53.

[5] *Georg.* iii. 328. Cf. Bucol. ii. Sir J. E. Smith, Tour., iii. 95, says also that the common Italian species makes a most disagreeable and dull chirping. The Cicadas of Africa, it is said, may be heard half a mile off; and the sound of one in a room will put a whole company to silence. Thunberg asserts that those of Java utter a sound as shrill and piercing as that of a trumpet. Captain Hancock informed Messrs. Kirby and Spence that the Brazilian Cicadas sing

do foretel a sickly year to come, not that they are the cause
of putrefaction in themselves, but only shew plenty of putrid
matter to be, when there is such store of them appear.
Oftentimes their coming and singing doth portend the
happy state of things: so also says Theocritus. Niphus
saith that what year but few of them are to be seen, they
presage dearness of victuals, and scarcity of all things
else . . . . .

"The Egyptians, by a Cicada painted, understood a
priest and an holy man; the latter makers of hieroglyphics
sometimes will have them to signifie musicians, sometimes
pratlers or talkative companions, but very fondly. How
ever the matter be, the Cicada hath sung very well of herself, in my judgement, in this following distich :

> Although I am an insect very small,
> Yet with great virtue am endow'd withall."[1]

Sir G. Staunton, in his account of China, remarks: "The
shops of Hai-tien, in addition to necessaries, abounded in
toys and trifles, calculated to amuse the rich and idle of
both sexes, even to cages containing insects, such as the
noisy Cicada, and a large species of the Gryllus."[2]
S. Wells Williams tells us that the Chinese boys often
capture the male Cicada of their country, and tie a straw
around the abdomen, so as to irritate the sounding apparatus, and carry it through the streets in this predicament,
to the great annoyance of every one, for the stridulous
sound of this insect is of deafening loudness.[3]

When in Quincy, Illinois, in the summer of 1864, I was
shown by a boy a toy, which he called a "Locust," with
which he imitated the loud rattling noise of the *Cicada
septemdecim* with great accuracy. It consisted of a horse-

---

as loud as to be heard at the distance of a mile. *Introd.*, ii. 400.
The sound of our American species, *C. septemdecim*, has been compared to the ringing of horse-bells. The tettix of the Greeks, says
Dr. Shaw, *Travels*, 2d edit., p. 186, must have had quite a different
voice, more soft surely and more melodious; otherwise the fine
orators of Homer, who are compared to it, can be looked upon as
no better than loud, loquacious scolds.

[1] *Theatr. Ins.*, p. 134. Topsel's *Hist. of Beasts*, p. 994.
Vide Pierius' *Hieroglyph.*, p. 270-1. Initiatus sacris; Dicacita-
tis castigatio; Vana garrulitas; Nobilitas generis; Musica.

[2] V. 2, c. 4, Donovan's *Ins. of China*, p. 32.

[3] *Middle Kingd.*

hair tied to the end of a short stick, and looped in a cap of stiff writing-paper placed over the hole of a spool. To make the sound, then, the toy was whirled rapidly through the air, when the stiff paper acted as a sounding-board to the vibrating hair.

At Surinam, Madame Merian tells us, the noise of the *Cicada tibicen* is still supposed to resemble the sound of a harp or lyre, and hence called the *Lierman*—the harper.[1] Another species, in Ceylon, which makes the forest re-echo with a long-sustained noise so curiously resembling that of a cutler's wheel, has acquired the highly appropriate name of the *Knife-grinder*.[2]

It is said of our *Cicada septemdecim*, the so-called, but very improperly, "Seventeen-year Locust," that, when they first leave the earth, when they are plump and full of juices, they have been made use of in the manufacture of soap.

The larva of a Chinese species of Cicada, the *Flata limbata*, which scarcely exceeds the domestic fly in size, forms a sort of grease, which adheres to the branches of trees and hardens into wax. In autumn the natives scrape this substance, which they call *Pela*, from off the trees, melt, purify, and form it into cakes. It is white and glossy in appearance, and, when mixed with oil, is used to make candles, and is said to be superior to the common wax for use. The physicians employ it in several diseases; and the Chinese, as we are informed by the Abbe Grosier, when they are about to speak in public, or when any occasion is likely to occur on which it may be necessary to have assurance and resolution, eat an ounce of it to prevent swoonings or palpitations of the heart.[3]

On the large cheese-like cakes of this wax, hanging in the grocers' and tallow-chandlers' shops at Hankow, are often seen the inscription written : "It mocks at the frost, and rivals the snow." The price, in 1858, was forty dollars a picul, or about fifteen pence a pound.[4]

The Greeks, notwithstanding their veneration for the Cicada, made these insects an article of food, and accounted them delicious. Aristotle says, the larva, when it is grown

---

[1] *Surinam*, 49.
[2] Tennent, *Nat. Hist. of Ceylon*, p. 432.
[3] *Desc. of China*, i. 442.
[4] Oliphant's *Lord Elgin's Miss. to China*, p. 565.

in the earth, and become a tettigometra (pupa), is the sweetest; when changed to the tettix, the males at first have the best flavor, but after impregnation the females are preferred, on account of their white ova.[1] Athenæus and Aristophanes also mention their being eaten; and Ælian is extremely angry with the men of his age that an animal sacred to the Muses should be strung, sold, and greedily devoured.[2] The *Cicada septemdecim*, Mr. Collinson in 1763 said, was eaten by the Indians of America, who plucked off the wings and boiled them.[3]

Osbeck tells us that the *Cicada chinensis*, along with .the *Buprestis maxima*, and several species of Butterflies, is made an article of commerce by the Chinese, being sold in their shops.[4]

Fulgoridæ—Lantern-flies.

The Lantern-fly, *Fulgora lanternaria* of Linnæus, found in many parts of South America, is supposed to emit a vivid light from the large hood, or lantern, which projects from its body, and to be frequently serviceable to benighted travelers; hence the specific name, *lanternaria*. This story originated about a century and a half ago, from the work of the celebrated Madame Merian, who lived several years in Surinam. Her account contains the following anecdote: "The Indians once brought me, before I knew that they shone by night, a number of these Lantern-flies, which I shut up in a wooden box. In the night they made such a noise that I awoke in a fright, and ordered a light to be brought; not knowing whence the noise proceeded. As soon as we found that it came from the box, we opened it; but were still much more alarmed, and let it fall to the ground, in a fright at seeing a flame of fire come out of it; and as so many animals as came out, so many flames of fire appeared. When we found this to

---

[1] *Hist. An.*, B. 5, c. 24, § 3, 4. Bohn's edit.
[2] Cf. Bochart, *Hieroz.*, ii. 491.
[3] *Phil. Trans.*, 1763, n. 10.
[4] *Travels*, i. 331.
Baird says, but on what authority he does not state, that Cicadas are frequently to be seen represented on the Egyptian monuments, and are said to be emblems of the ministers of religion.—*Encycl. of Nat. Sci.*

be the case, we recovered from our fright, and again collected the insects, highly admiring their splendid appearance."[1]

Dr. Darwin, in a note to some lines relative to luminous insects, in his poem, the Loves of the Plants, makes Madame Merian affirm that she drew and finished her figure of the insect by its own light. This story is without foundation.

The Indians of South America say and believe that the Lyerman, *Cicada tibicen*, is changed into the Lantern-fly; and that the latter emits a light similar to that of a lantern.[2]

This story of the Lantern-fly being luminous is the more remarkable since the veracity of its author is unimpeached. She doubtless has confounded it with the *Cucujus*, *Elater noctilucus*. Donovan, however, states that the Chinese Lantern-fly, *Fulgora candelaria*, has an illuminated appearance in the night.[3]

From the loud noise the Lantern-fly makes at night, which is said to be somewhat between the grating of a razor-grinder and the clang of cymbals, it is called by the Dutch, in Guiana, *Scare-sleep*.[4] Ligon, in his History of Barbados, printed in 1673, probably refers to this insect, when he says: "They lye all day in holes and hollow trees, and as soon as the Sun is down they begin their tunes, which are neither singing nor crying, but the shrillest voyces that ever I heard; nothing can be so nearly resembled to it, as the mouths of a pack of small beagles at a distance." This author, however, thought this sound by no means unpleasant. "So lively and chirping," he continues, "the noise is, as nothing can be more delightful to the ears, if there were not too much of it, for the musick hath no intermission till morning, and then all is husht."[5]

---

[1] *Insects of Surinam*, p. 49.
[2] Jaeger, *Life of N. A. Ins.*, p. 73.
[3] *Ins. of China*, p. 30. That the Lantern-fly emits no light, see *Dict. d'Hist. Nat.*; M. Richards' statement in *Encyclop.*, art. *Fulgora*; *Berlin Mag.*, i. 153; Kirby and Spence, *Introd.*, ii. 414, note; Jaeger, *qua supra*.
[4] Stedman, *Surinam*, ii. 37.
[5] *Hist. of Barbados*, p. 65.

Aphidæ—Plant-lice.

The Aphides are remarkable for secreting a sweet, viscid fluid, known by the name of Honey-dew, the origin of which has puzzled the world for ages. Pliny says "it is either a certaine sweat of the skie, or some unctuous gellie procceding from the starres, or rather a liquid purged from the aire when it purifyeth itself."[1]

Amyntas, in his Stations of Asia, quoted by Athenæus, gives a curious account of the manner of collecting this article, which was supposed to be superior to the nectar of the Bee, in various parts of the East, particularly in Syria. In some cases they gathered the leaves of trees, chiefly of the linden and oak, for on these the dew was most abundantly found,[2] and pressed them together. Others allowed it to drop from the leaves and harden into globules, which, when desirous of using, they broke, and, having poured water on them in wooden bowls, drank the mixture. In the neighborhood of Mount Lebanon, Honey-dew was collected plentifully several times in the year, being caught by spreading skins under the trees, and shaking into them the liquid from the leaves. The Dew was then poured into vessels, and stored away for future use. On these occasions the peasants used to exclaim, "Zeus has been raining honey!"[3]

In the Treasvrie of Avncient and Moderne Times, we read: "*Galen* saith, that there fell such great quantity of this Dew (in his time) in his Countrey of *Pergamus*, that the Countrey people (greatly delighted therein) gave thankes therefor to *Iupiter*. *Ælianus* writeth also that there fell such plenty thereof in *India*, in the Region which is called *Prasia*, and so moistened the Grasse, that the Sheepe, Kine, and Goates feeding thereon, yeelded Milke sweete like Hony, which was very pleasing to drinke. And when they used that Milke in any disease, they needed not to put any Hony therein, to the end it should not corrupt in the stomacke: as it is appointed in Hecticke Feauers, Consumption,

---

[1] Nat. Hist., xi. 12. Holl. *Trans.*, i. 315. E.

[2] Theoph. *Hist. Plant.*, iii. 7, 6. Cf. Hes. *Opp. et Dies*, 232, seq. and Bacon, *Syl. Sylvarum*, 496.

[3] St. John's *Anct. Greeks*, ii. 299.

Tisickes, and for others that are ulcered in the intestines, as is confirmed by the Histories of *Portugall*."[1]

The Aphides, like many other insects, sometimes migrate in clouds; and among other instances on record of these migrations, Mr. White informs us that about three o'clock in the afternoon of the first of August, 1785, the people of the village of Selborne were surprised by a shower of Aphides which fell in those parts. Persons who walked in the street at this time found themselves covered with them, and they settled in such numbers in the gardens and on the hedges as to blacken every leaf. Mr. White's annuals were thus all discolored with them, and the stalks of a bed of onions were quite coated over for six days afterward. These swarms, he remarks, were then no doubt in a state of emigration, and might have come from the great hop-plantations of Kent and Sussex, the wind being all that day in the east. They were observed at the same time in great clouds about Farnham, and all along the vale from Farnham to Alton.[2] A similar emigration of these insects Mr. Kirby once witnessed, to his great annoyance, when traveling later in the year in the Isle of Ely. The air was so full of them, that they were incessantly flying into his eyes and nostrils, and his clothes were covered by them; and in 1814, in the autumn, the Aphides were so abundant for a few days in the vicinity of Ipswich, as to be noticed with surprise by the most incurious observers.[3] Neither Mr. White nor Mr. Kirby informs us what particular species formed these immense flights, but it is most probable they belonged to the Hop-fly, *Aphis humuli*.

Reaumur tells us that in the Levant, Persia, and China, they use the galls of a particular species of *Aphis* for dyeing silk crimson.[4]

In England, the mischief caused by the Hop-fly, *Aphis humuli*, in some seasons, as in 1802, has brought the duty of hops down from £100,000 to £14,000.

A quite common, though erroneous, belief in England is, that Aphides are produced, or brought by, a northern or eastern wind. Thomson has fallen into the error; he has also confounded the mischief of caterpillars with that of the Aphis:

---

[1] B. 3, c. xvi. p. 278. Printed 1613.
[2] *Nat. Hist. of Selborne*, p. 366.
[3] K. and S. *Introd.*, ii. 9.
[4] Reaumur, iii. xxxi. Pref.

For oft, engendered by the hazy north,
Myriads on myriads insect armies warp,
Keen in the poison'd breeze, and wasteful eat
Through buds and bark into the blackened core
Their eager way.   A feeble race!   Yet oft
The sacred sons of vengeance, on whose course
Corrosive famine waits, and kills the year.

Coccidæ—Shield-lice.

The Kermes-dye, or scarlet, made from the *Coccus ilicis* of Linnæus, an insect found chiefly on a species of oak, the *Quercus ilex*, in the Levant, France, Spain, and other parts of the world, was known in the East in the earliest ages, even before the time of Moses, and was a discovery of the Phœnicians in Palestine, who also first employed the murex and buccinum for the purpose of dyeing.

Tola or *Thola* was the ancient Phœnician name for this insect and dye, which was used by the Hebrews, and even by the Syrians; for it is employed by the Syrian translator.[1] Among the Jews, after their captivity, the Aramæan *zehori* was more common.   This dye was known also to the Egyptians in the time of Moses; and it is most probable that the color mentioned in Exodus[2] as one of the three which were prescribed for the curtains of the tabernacle, and for the "holy garments" of Aaron, and which the English translators have rendered by the word *scarlet* (not the color now so called, which was not known in James the First's reign when the Bible was translated), was no other than the blood-red color dyed from the *Coccus ilicis.*

The Arabs received the name *Kermes* or *Alkermes* for the insect and dye, from Armenia and Persia, where the insect was indigenous, and had long been known; and that name banished the old name in the East, as the name scarlet has in the West.   For the first part of this assertion we must believe the Arabs.   The Kermes, however, were not indigenous to Arabia, as the Arabs appear to have no name for them.   To the Greeks this dye was known under the

---

[1] Isaiah, ch. i. v. 18.
[2] Ex. ch. xxvi. xxviii. xxix.

name of *Coccus*, as appears from Dioscorides, and other Greek writers.[1]

From the epithets *kermes* and *coccus*, and that of *vermiculus* or *vermiculum*, given to the Kermes in the middle ages, when they were ascertained to be insects, have sprung the Latin *coccineus*, the French *carmesin, carmine, cramoisi* and *vermeil*, the Italian *chermisi, cremisino*, and *chermesino*, and our *crimson* and *vermilion*.

The imperishable reds of the Brussels and other Flemish tapestries were derived from the Kermes; and, in short, previous to the discovery of cochineal, this was the material universally used for dyeing the most brilliant red then known. At the present time the Kermes are only gathered in Europe by the peasantry of the provinces in which they are found, but they still continue to be employed as of old in a great part of India and Persia.[2]

Brookes says the women gather the harvest of Kermes insects before sunrise, tearing them off with their nails; and, for fear there should be any loss from the hatching of the insects, they sprinkle them with vinegar. They then lay them in the sun to dry, where they acquire a red color.[3]

The scarlet grain of Poland, *Coccus polonicus*, found on the German knot-grass or perennial knawel (*Scleranthus perennis*), was at one time collected in large quantities in the Ukraine and other provinces of Poland (here under the name of *Czerwiec*), and also in the great duchy of Lithuania. But though much esteemed and still employed by the Turks and Armenians for dyeing wool, silk, and hair, as well as for staining the nails of women's fingers, it is now rarely used in Europe except by the Polish peasantry. A similar neglect has attended the Coccus found on the roots of the Burnet (*Poterium sanguisorba*, Linn.), which was used, particularly by the Moors, for dyeing wool and silk a rose color; and the *Coccus uvæ-ursi*, which with alum affords a crimson dye.[4]

Cochineal, the *Coccus cacti*, is doubtless the most valuable product for which the dyer is indebted to insects, and

---

[1] Diosc. iv. 48, p. 260. Pausan. B. x. p. 890.
[2] Beckman's *Hist. of Inventions*, ii. 163–195. Bancroft *on Perm. Colors*, i. 393–408.
[3] *Nat. Hist. of Ins.*, p. 77.
[4] Bancroft *on Permanent Colors*, i. 408–9.

with the exception perhaps of indigo, the most important of dyeing materials. It is found on a kind of fig, called in Mexico, where the insect is produced in any quantity, Nopal or Tuna, which generally has been supposed to be the *Cactus cochinilifer*, but according to Humboldt is unquestionably a distinct species, which bears fruit internally white.

Cochineal was discovered by the Spaniards, on their first arrival in Mexico, about the year 1518; but who first remarked this valuable production, and made it known in Europe, Mr. Beckman says, he has been unable to discover. Some assert that the native Mexicans, before the landing of Cortes, were acquainted with cochineal, which they employed in painting their houses and dyeing their clothes; but others maintain the contrary. Be that as it may, however, the Spanish ministry, as early as the year 1523, as Herrera informs us, ordered Cortes to take measures for multiplying this valuable commodity; and soon after it must have begun to be quite an object of commerce, for Guicciardini, who died in 1589, mentions it among the articles procured then by the merchants of Antwerp from Spain.

Professor Beckman, who has given the subject particular attention, thinks that with the first cochineal, a true account of the manner in which it was procured must have reached Europe, and become publicly known. Acosta in 1530, and Herrera in 1601, as well as Hernandez and others, gave so true and complete a description of it, that the Europeans could entertain no doubt respecting its origin. The information of these authors, however, continues this gentleman, was either overlooked or considered as false, and disputes arose whether cochineal was insects or worms, or the berries or seeds of certain plants. The Spanish name *grana*, confounded with *granum*, may have given rise to this contest.

Illustrative of this great difference of opinion, Mr. Beckman narrates the following anecdote: "A Dutchman, named Melchior de Ruusscher, affirmed in a society, from oral information he had received in Spain, that cochineal was small animals. Another person, whose name he has not made known, maintained the contrary with so much heat and violence, that the dispute at length ended in a bet. Ruusscher charged a Spaniard, one of his friends, who was going to Mexico, to procure,for him in that country authentic proofs of what he had asserted. These proofs, legally confirmed in October, 1725, by the court of justice in the city of Antiquera, in the valley of Oaxaca, arrived at Amsterdam in the autumn of the year 1726. I have been informed that Ruusscher upon this got possession of the sum betted, which amounted to the whole property of the loser; but that, after keeping it a certain time, he again returned it, deducting only the expenses he had been at in procuring the evidence, and in causing it to be published. It formed a small octavo volume, with the following title printed in red letters: *The History of Cochineal proved by Authentic documents.* These proofs sent from New-Spain are written in Dutch, French, and Spanish."[1]

Among the important discoveries made by accident, the following in the history of Cochineal may be instanced: "The well-known Cornelius Drebbel, who was born at Alemaar, and died at London in 1634, having placed in his window an extract of Cochineal, made with boiling water, for the purpose of filling a thermometer, some aqua-regia dropped into it from a phial, broken by accident, which stood above it, and converted the purple dye into a most beautiful dark red. After some conjectures and experiments, he discovered that the tin by which the window frame was divided into squares had been dissolved by the aqua-regia, and was the cause of this change. He communicated his observation to Kuffelar, an ingenious dyer at Leyden. The latter brought the discovery to perfection, and employed it some years alone in his dye-house, which gave rise to the name of Kuffelar's color."[2]

That innocent cosmetic, so much used by the ladies, and commonly known by the French term Rouge, is no other than a preparation of Cochineal.[3]

Kermes-berries, *Coccus ilicis,* and Cochineal, *C. cacti,* Geoffroy says, "are esteemed to be greatly cordial and sudorific, being very full of volatile salt. They are given also to prevent abortion from any strain or hurt."[4]

Lac is the produce of an insect supposed by Amatus

---

[1] *Hist. of Inventions,* ii. 184.
[2] *Ibid.,* 192.
[3] *Shaw's Zool.,* vi. 192.
[4] *Subst. used in Physic,* p. 370.

Lusitanus to be a kind of ant, and by others a bee, but now ascertained to be a species belonging to the Coccidæ—the *Coccus ficus* or *C. lacca*. It is collected from various trees in India, where it is found so abundantly, that, were the consumption ten times greater than it is, it could be readily supplied.

Lac is known in Europe by the different appellations of *stick-lac*, when in its natural state, adhering to, and often completely surrounding, for five or six inches, the twigs on which it is produced by the insects contained in its cells; *seed-lac*, when broken into small pieces, garbled, and the greater part of the coloring matter extracted by water; when it appears in a granulated form; *lump-lac*, when melted and made into cakes; and *shell-lac*, when strained and formed into transparent laminæ.

Lac, in its different forms, is made use of in the manufacture of varnishes, japanned ware, sealing-wax, beads, rings, arm-bracelets, necklaces, water-proof hats, etc., etc. Mixed with fine sand it forms grindstones; and added to lamp or ivory black, being first dissolved in water with the addition of a little borax, it composes an ink not easily acted upon by dampness or water. It has been applied also to a still more important purpose, originally suggested by Dr. Roxburgh about the year 1790—that of a substitute for Cochineal in dyeing scarlet.[1] From this suggestion, under the direction of Dr. Bancroft, large quantities of a substance termed *lac-lake*, consisting of the coloring matter of stick-lac precipitated from an alkaline lixivium by alum, were manufactured at Calcutta and sent to England, where at first the consumption was so great, that, according to the statement of Dr. Bancroft, in 1806, and the two following years, the sales of it at the India House equaled in point of coloring matter half a million of pounds' weight of Cochineal. Soon after this, a new preparation of lac color, under the name of *lac-dye*, was substituted for the lac-lake, and with such advantage, that in a few months £14,000 were saved by the East India Company in the purchase of scarlet cloths dyed with this color and Cochineal conjointly, and without any inferiority in the color obtained.[2]

---

[1] *Phil. Trans.* for 1791.

[2] Bancroft *on Permanent Colors*, ii. 1–59.

The Coccidæ, although they furnish an invaluable dye and many articles of commerce, are among the most hurtful of insects in gardens and hot-houses. In 1843, the orange-trees of the Azores or Western Islands were nearly entirely destroyed by the *Coccus Hesperidum:* and in Fayal, an island which had usually exported twelve thousand chests of oranges annually, not one was exported.[1]

---

[1] Baird's *Cyclop. of Nat. Sci.*

HETEROPTERA.

Cimicidæ—Bed-bugs.

"In the year 1503," says Moufet, "Dr. Penny was called in great haste to a little village, called Mortlake, near the Thames, to visit two noble ladies (*duas nobiles*), who were much frightened by the appearance of bug-bites (*ex cimicum vestigiis*), and were in fear of I know not what contagion; but when the matter was known, and the insects caught, he laughed them out of all fear."[1]

This fact disproves the statement of Southall, that the *Cimex lectularius* was not known in England before 1670, and that of Linnæus, and the generality of later writers, that this insect is not originally a native of Europe, but was introduced into England after the great fire of London in 1666, having been brought in timber from America.

The original English names of the *C. lectularius*, were *Chinche*, *Wall-louse*, and *Punaise* (from the French); and the term *Bug*, which is a Celtic word, signifying a ghost or goblin, was applied to them after the time of Ray,[2] most probably because they were considered as "terrors of the night."[3]

In the Nicholson's Journal[4] there is mention of a man who, far from disliking Bed-bugs, took them under his protecting care, and would never suffer them to be disturbed, or his bedsteads removed, till in the end they swarmed to an

---

[1] *Theatr. Ins.*, p. 270.

[2] Ray, *Hist. Ins.*, 7.

[3] Hence the English word *Bug-bear*. In Matthew's Bible, the passage of the Psalms (xci. 5), "Thou shalt not be afraid of *the terror by night*," is rendered, "Thou shalt not nede to be afraid of any *bugs by night*." *Bug* in this sense often occurs in Shakspeare. *Winter's Tale*, A. iii. Sc. 2, 3; *Henry VI.*, A. v. Sc. 2; *Hamlet*, A. v. Sc. 2.

[4] *Journal*, xvii. 40.

incredible degree, crawling up even the walls of his drawing-room; and after his death millions were found in his bed and chamber furniture.

Gemelli, in 1695, visited the Banian hospital at Surat, and says that what amazed him most, though he went there for that express purpose, was to see "a poor wretch, naked, bound down hands and feet, to feed the Bugs or Punaises, brought out of their stinking holes for that purpose."[1]

Mr. Forbes, speaking of this remarkable institution for animals, says: "At my visit, the hospital contained horses, mules, oxen, sheep, goats, monkeys, poultry, pigeons, and a variety of birds. The most extraordinary ward was that appropriated to rats and mice, Bugs, and other noxious vermin. The overseers of the hospital frequently hire beggars from the streets, for a stipulated sum, to pass a night among the Fleas, Lice, and Bugs, on the express condition of suffering them to enjoy their feast without molestation."[2]

Navarette says that a species of Bugs (most probably a *Cimex*), which swarm in some parts of China, are a source of great amusement to the natives; for they take particular delight in killing them with their fingers, and then clapping them to their noses.[3]

Democritus says that the feet of a hare, or of a stag, hung round the feet of the bed at the bottom of the couch, does not suffer Bugs to breed; but, in traveling, Didymus adds, if you fill a vessel with cold water and set it under the bed, they will not touch you when you are asleep.[4]

A superstition prevails among us that beds, in order to rid them effectually of Bugs, must be cleaned during the dark of the moon.

The medicinal virtues of the Cimex are given by Pliny (doubtless quoting Dioscorides, ii. 36) as follows: "The Bug is said to be a neutralizer of the venom of serpents, asps in particular, and to be a preservative against all kinds of poisons. As a proof of this, they tell us that the sting of an asp is never fatal to poultry, if they have eaten Bugs that day; and that, if such is the case, their flesh is remark-

---

[1] Churchill's *Col. of Voy. and Trav.*, iv. 190.
[2] *Oriental Memoirs*, i. 256.
[3] Astley's *Col. of Voy. and Trav.*, iv. 513. Churchill's *same*, i. 31.
[4] Owen's *Geoponika*, ii. 160.

ably beneficial to persons who have been stung by serpents. Of the various recipes given in reference to these insects, the least revolting are the application of them externally to the wound, with the blood of a tortoise; the employment of them as a fumigation to make leeches loose their hold; and the administering of them to animals in drink when a leech has been accidentally swallowed. Some persons, however, go so far as to crush Bugs with salt and woman's milk, and anoint the eyes with the mixture; in combination, too, with honey and oil of roses, they use them as an injection for the ears. Field-bugs, again, and those found upon the mallow (perhaps the *Cimex pratensis* is meant here; neither this nor the *Cimex juniperinus*, the *C. brassicæ*, or the *Lygæus hyoscami*, has the offensive smell of the *C. lectularius*) are burnt, and the ashes mixed with oil of roses as an injection for the ears.

"As to the other remedial virtues attributed to Bugs for the cure of vomiting, quartan fevers, and other diseases, although we find recommendations given to swallow them in an egg, some wax, or in a bean,[1] I look upon them as utterly unfounded, and not worthy of further notice. They are employed, however, for the treatment of lethargy, and with some fair reason, as they successfully neutralize the narcotic effects of the poison of the asp; for this purpose seven of them are administered in a cyathus of water; but in the case of children, only four. In cases, too, of strangury they have been injected into the urinary channel.[2] So true it is that nature, that universal parent, has engendered nothing without some powerful reason or other. In addition to these particulars, a couple of Bugs, it is said, attached to the left arm in some wool that has been stolen from the shepherds, will effectually cure nocturnal fevers; while those recurrent in the daytime may be treated with equal success by inclosing the Bugs in a piece of russet-colored cloth."[3]

---

[1] Dr. James says: "Given to the number of seven, as food with beans, they help those who are afflicted with a quartan ague, if they be eaten before the accession of the fit."—*Med. Dict.*

[2] An excellent method, Ajasson remarks, of adding to the tortures of the patient.

[3] Pliny, *Nat. Hist.*, xxix. 17. Bostock and Riley's *Trans.*, v. 393.

Guettard, a French commentator on Pliny, recommends Bugs to be taken internally for hysteria; and Dr. James says "the smell of them relieves under hysterical suffocations!"[1]

At the present time the Bed-bug is sometimes given by the country people of Ohio as a cure for the fever and ague.

Moufet says: "The verses of Quintus Serenus show that they are good for tertian agues:

> Shame not to drink three Wall-lice mixt with wine,
> And garlick bruised together at noon-day.
> Moreover a bruised Wall-louse with an egg, repine
> Not for to take, 'tis loathsome, yet full good I say.

"Gesner in his writings confirms this experiment, having made trial of it among the common and meaner sort of people in the country. The ancients gave seven to those that were taken with a lethargy, in a cup of water, and four to children. Pliny and Serenus consent to this in these verses:

> Some men prescribe seven Wall-lice for to drink,
> Mingled with water, and one cup they think
> Is better than with drowsy death to sink."[2]

Anatolius says that if an ox, or other quadruped, swallows a leech in drinking, having pounded some Bugs, let the animal smell them, and he immediately throws up the leech.[3]

Mr. Mayhew, in his work on the London poor and their labor, has an interesting chapter devoted to the Destroyers of Vermin, from which we have taken the liberty of quoting pretty largely in the course of this work. His statements can be relied on, and we give them as nearly in his own words as possible. Concerning Bugs and Fleas, and the trade carried on in the manufacture and vending of poisons to destroy these pests, we learn from him: The vending of bug-poison in the London streets is seldom followed as a regular source of living. He has met with persons who remembered to have seen men selling packets of vermin poison; but to find out the venders themselves was next to an

---

[1] *Med. Dict.*

[2] *Theatr. Ins.*, p. 270–1.	Topsel's *Hist. of Beasts*, p. 1098.

[3] Owen's *Geoponika*, ii. 157.

impossibility. The men seem to take merely to the business as a living when all other sources have failed. All, however, agree in acknowledging that there is such a street trade; but that the living it affords is so precarious that few men stop at it longer than two or three weeks.

The most eminent firm, perhaps, of the bug-destroyers in London now is that of Messrs. Tiffin and Son. They have pursued their calling in the streets, but now rejoice in the title of "Bug-Destroyers to Her Majesty and the Royal Family."

Mr. Tiffin, the senior party in this house, kindly obliged Mr. Mayhew with the following statement. It may be as well to say that Mr. Tiffin appears to have paid much attention to the subject of Bugs, and has studied with much earnestness the natural history of this vermin. He said:

"We can trace our business back as far as 1695, when one of our ancestors first turned his attention to the destruction of bugs. He was a lady's stay-maker—men used to make them in those days, though, as far back as that is concerned, it was a man that made my mother's dresses. This ancestor found some bugs in his house—a young colony of them, that had introduced themselves without his permission, and he didn't like their company, so he tried to turn them out of doors again, I have heard it said, in various ways. It is in history, and it has been handed down in my own family as well, that bugs were first introduced into England, after the fire in London, in the timber that was brought for the rebuilding of the city, thirty years after the fire, and it was about that time that my ancestor first discovered the colony of bugs in his house. I can't say whether he studied the subject of bug-destroying, or whether he found out his stuff by accident, but he certainly *did* invent a compound which completely destroyed the bugs, and, having been so successful in his own house, he named it to some of his customers who were similarly plagued, and that was the commencement of the present connection, which has continued up to this time.

"At the time of the illumination for the Peace, I thought I must have something over my shop, that would be both suitable for the event and to my business; so I had a transparency done, and stretched on a big frame, and lit up by gas, on which was written

MAY THE
DESTROYERS OF PEACE
BE DESTROYED BY US.

TIFFIN & SON,

BUG-DESTROYERS TO HER MAJESTY.

"Our business was formerly carried on in the Strand, where both my father and myself were born; in fact, I may say I was born to the bug business.

"I remember my father as well as possible; indeed, I worked with him for ten or eleven years. He used, when I was a boy, to go out to his work killing bugs at his customers' houses with a sword by his side and a cocked-hat and bag-wig on his head—in fact, dressed up like a regular dandy. I remember my grandmother, too, when she was in the business, going to the different houses, and seating herself in a chair, and telling the men what they were to do, to clean the furniture and wash the woodwork.

"I have customers in our books for whom our house has worked these 150 years; that is, my father and self have worked for them and their fathers. We do the work by contract, examining the house every year. It's a precaution to keep the place comfortable. You see, servants are apt to bring bugs in their boxes; and, though there may be only two or three bugs perhaps hidden in the woodwork and the clothes, yet they soon breed if let alone.

"We generally go in the spring, before the bugs lay their eggs; or, if that time passes, it ought to be done before June, before their eggs are hatched, though it's never too late to get rid of a nuisance.

"I mostly find the bugs in the bedsteads. But, if they are left unmolested, they get numerous and climb to the tops of the rooms, and about the corners of the ceilings. They colonize anywhere they can, though they're very high-minded and prefer lofty places. Where iron bedsteads are used, the bugs are more in the *rooms*, and that's why such things are bad. They don't keep a bug away from a person sleeping. Bugs'll come if they're thirty yards off.

"I knew a case of a bug who used to come every night about thirty or forty feet—it was an immense large room—

from the corner of the room to visit an old lady. There was only one bug, and he'd been there for a long time. I was sent for to find him out. It took me a long time to catch him. In that instance I had to examine every part of the room, and when I got him I gave him an extra nip to serve him out. The reason why I was so bothered was, the bug had hidden itself near the window, the last place I should have thought of looking for him, for a bug never, by choice, faces the light; but when I came to inquire about it, I found that this old lady never rose till three o'clock in the day, and the window-curtains were always drawn, so that there was no light like.

"Lord! yes, I am often sent for to catch a single bug. I've had to go many, many miles—even 100 or 200—into the country, and perhaps only catch half a dozen bugs after all; but then that's all that are there, so it answers our employer's purpose as well as if they were swarming.

"I work for the upper classes only; that is, for carriage-company and such like approaching it, you know. I have noblemen's names, the first in England, on my books.

"My work is more method; and I may call it a scientific treating of the bugs rather than wholesale murder. We don't care about the thousands, it's the last bug we look for, whilst your carpenters and upholsterers leave as many behind them, perhaps, as they manage to catch.

"The bite of the bug is very curious. They bite all persons the same (?); but the difference of effect lies in the constitutions of the parties. I've never noticed that a different kind of skin makes any difference in being bitten. Whether the skin is moist or dry, it don't matter. Wherever bugs are, the person sleeping in the bed is sure to be fed on, whether they are marked or not; and as a proof, when nobody has slept in the bed for some time, the bugs become quite flat; and, on the contrary, when the bed is always occupied, they are round as a lady-bird.

"The flat bug is more ravenous, though even he will allow you time to go to sleep before he begins with you; or at least till he thinks you ought to be asleep. When they find all quiet, not even a light in the room will prevent their biting; but they are seldom or never found under the bedclothes. They like a clear ground to get off, and generally bite round the edges of the nightcap or the nightdress. When they are found *in* the bed, it's because the parties have been tossing about, and have curled the sheets round the bugs.

"The finest and fattest bugs I ever saw were those I found in a black man's bed. He was the favorite servant of an Indian general. He didn't want his bed done by me; he didn't want it touched. His bed was full of 'em, no bee-hive was ever fuller. The walls and all were the same, there wasn't a patch that was not crammed with them. He must have taken them all over the house wherever he went.

"I've known persons to be laid up for months through bug-bites. There was a very handsome fair young lady I knew once, and she was much bitten about the arms, and neck, and face, so that her eyes were so swelled up she couldn't see. The spots rose up like blisters, the same as if stung with a nettle, only on a very large scale. The bites were very much inflamed, and after a time they had the appearance of boils.

"Some people fancy, and it is historically recorded, that the bug smells because it has no vent; but this is fabulous, for they *have* a vent. It is not the human blood neither that makes them smell, because a young bug who has never touched a drop will smell. They breathe, I believe, through their sides; but I can't answer for that, though it's not through the head. They haven't got a mouth, but they insert into the skin the point of a tube, which is quite as fine as a hair, through which they draw up the blood. I have many a time put a bug on the back of my hand, to see how they bite; though I never felt the bite but once, and then I sup-pose the bug had pitched upon a very tender part, for it was a sharp prick, something like that of a leech-bite.

"I once had a case of lice-killing, for my process will an-swer as well for them as for bugs, though it's a thing I never should follow by choice. Lice seem to harbor pretty much the same as bugs do. I find them in the furniture. It was a nurse that brought them into the house, though she was as nice and clean a looking woman as ever I saw. I should almost imagine the lice must have been in her, for they say there is a disease of that kind; and if the tics breed in sheep, why should not lice breed in us? for we're but live matter, too. I didn't like myself at all for two or three days after that lice-killing job, I can assure you; it's the only case of the kind I ever had, and I can promise you it shall be the last.

"I was once at work on the Princess Charlotte's own bedstead. I was in the room, and she asked me if I had found anything, and I told her no; but just at that minute I *did* happen to catch one, and upon that she sprang up on the bed, and put her head on my shoulder, to look at it. She had been tormented by the creature, because I was ordered to come directly, and that was the only one I found. When the Princess saw it, she said, 'Oh, the nasty thing! That's what tormented me last night; don't let him escape.' I think he looked all the better for having tasted royal blood.

"I also profess to kill beetles, though you never can destroy them so effectually as you can bugs; for, you see, beetles run from one house to another, and you can never perfectly get rid of them; you can only keep them under. Beetles will scrape their way and make their road round a fire-place, but how they go from one house to another I can't say, but they *do*.

"I never had patience enough to try and kill Fleas by my process; it would be too much of a chivey to please me.

"I never heard of any but one man who seriously went to work selling bug-poison in the streets. I was told by some persons that he was selling a first-rate thing, and I spent several days to find him out. But, after all, his secret proved to be nothing at all. It was train-oil, linseed and hempseed, crushed up all together, and the bugs were to eat it till they burst.

"After all, secrets for bug-poisons ain't worth much, for all depends upon the application of them. For instance, it is often the case that I am sent for to find out one bug in a room large enough for a school. I've discovered it when the creature had been three or four months there, as I could tell by his having changed his jacket so often, for bugs shed their skins, you know. No, there was no reason that he should have bred; it might have been a single gentleman or an old maid.

"A married couple of bugs will lay from forty to fifty eggs at one laying. The eggs are oval, and are each as large as the thirty-second part of an inch; and when together are in the shape of a caraway comfit, and of a bluish-white color. They'll lay this quantity of eggs three times in a season. The young ones are hatched direct from the egg, and, like young partridges, will often carry the broken eggs

about with them, clinging to their back. They get their
fore-quarters out, and then they run about before the other
legs are completely cleared.

"As soon as the bugs are born they are of a cream color,
and will take to blood directly ; indeed, if they don't get it
in two or three days, they die ; but after one feed they will
live a considerable time without a second meal. I have
known old bugs to be frozen over in a horse-pond—when
the furniture had been thrown in the water—and there they
have remained for a good three weeks ; still, after they have
got a little bit warm in the sun's rays, they have returned to
life again.

" I myself kept bugs for five years and a half without
food, and a housekeeper at Lord H———'s informed me
that an old bedstead that I was then moving from a store-
room was taken down forty-five years ago, and had not been
used since, but the bugs in it were still numerous, though as
thin as living skeletons. They couldn't have lived upon the
sap of the wood, it being worm-eaten and dry as a bone. A
bug will live for a number of years, and we find that when
bugs are put away in old furniture without food, they don't
increase in number ; so that, according to my belief, the bugs
I just mentioned must have existed forty-five years : besides,
they were large ones, and very dark colored, which is another proof of age.

" It is a dangerous thing for bugs when they are shedding
their skins, which they do about four times in the course of
a year ; when they throw off their hard shell and have a
soft coat, so that the least touch will kill them ; whereas at
other times they will take a strong pressure. I have plenty
of bug-skins, which I keep by me as curiosities, of all sizes
and colors, and sometimes I have found the young bugs
collected inside the old ones' skins for warmth, as if they had
put on their father's great-coat. There are white bugs—
albinoes you may call 'em—freaks of nature like."[1]

Notonectidæ—Water-boatmen.

Humboldt mentions that he saw insects' eggs sold in the markets of Mexico, which were collected on the surface of lakes. Under the name of *Axayacat*, these eggs, or those of some other species of fly, deposited on rush mats, are sold as a caviare in Mexico. Rev. Thomas Smith, who makes the same statement, also says the Mexicans likewise eat the flies themselves, ground and made up with saltpetre. Something similar to these eggs, found in the pools of the desert of Fezzan, serves the Arabs for food, having the taste of caviare.

In the Bulletin de la Société Impériale Zoologique d'Acclimation, M. Guerin Méneville has published a paper on a sort of bread which the Mexicans make of the eggs of three species of heteropterous insects.

According to M. Craveri, by whom some of the Mexican bread, and of the insects yielding it, were brought to Europe, these insects and their eggs are very common in the fresh waters of the lagunes of Mexico. The natives cultivate, in the lagune of Chalco, a sort of carex called touté, on which the insects readily deposit their eggs. Numerous bundles of these plants are made, which are taken to a lagune, the Texcuco, where they float in great numbers in the water. The insects soon come and deposit their eggs on the plants, and in about a month the bundles are removed from the water, dried, and then beaten over a large cloth to separate the myriad of eggs with which the insects have covered them. These eggs are then cleaned and sifted, put into sacks like flour, and sold to the people for making a sort of cake or biscuit called "hautlé," which forms a tolerably good food, but has a fishy taste, and is slightly acid. The bundles of carex are replaced in the lake, and afford a fresh supply of eggs, which process may be repeated for an indefinite number of times.

It appears that these insects have been used from an early period, for Thomas Gage, a religionist, who sailed to Mexico in 1625, says, in speaking of articles sold in the markets, that they had cakes made of a sort of scum collected from the lakes of Mexico, and that this was also sold in other towns.

Brantz Mayer, in his Mexico as it was and as it is, 1844,

says: "On the lake of Texcuco I saw men occupied in collecting the eggs of flies from the surface of plants, and cloths arranged in long rows as places of resort for the insects. These eggs, called *agayacath*, formed a favorite food of the Indians long before the conquest; and when made into cakes, resemble the roe of fish, having a similar taste and appearance. After the use of frogs in France, and birds'-nests in China, I think these eggs may be considered a delicacy, and I found that they are not rejected from the tables of the fashionable inhabitants of the capital."

The more recent observations of MM. Saussure, Sallé, Virlet d'Aoust, etc. have confirmed the facts already stated, at least in the most essential particulars.

"The insects which principally produce this animal farinha of Mexico," says a writer in the Journal de Pharmacie, "are two species of the genus *Corixa* of Geoffroy, hemipterous (heteropterous) insects of the family of water-bugs. One of the species has been described by M. Guerin Méneville as new, and has been named by him *Corixa femorata:* the other, identified in 1831 by Thomas Say as one of those sold in the market at Mexico, bears the name of *Corixa mercenaria.* The eggs of these two species are attached in innumerable quantities to the triangular leaves of the carex forming the bundles which are deposited in the waters. They are of an oval form with a protuberance at one end and a pedicle at the other extremity, by means of which they are fixed to a small round disk, which the mother cements to the leaf. Among these eggs, which are grouped closely together, there are found others, which are larger, of a long and cylindrical form, and which are fixed to the same leaves. These belong to another larger insect, a species of *Notonecta,* which M. Guerin Méneville has named *Notonecta unifasciata.*"

It appears from M. Virlet d'Aoust, that in October the lakes of Chalco and Texcuco, which border on the City of Mexico, are haunted by millions of "small flies," which, after dancing in the air, plunge down into the water, to the depth of several feet, and deposit their eggs at the bottom.

"The eggs of these insects are called hautle (haoutle) by the Mexican Indians, who collect them in great numbers, and with whom they appear to be a favorite article of food.

They are prepared in various ways, but usually made into cakes, which are eaten with a sauce flavored with chillies."[1]

Rev. Thomas Smith enumerates the following insects as eaten by the ancient Mexicans: The *Atelepitz*, "a marsh beetle, resembling in shape and size the flying beetles, having four (?) feet, and covered with a hard shell." The *Atopinan*, "a marsh grasshopper of a dark color and great size, being no less than six inches long and two broad."(!) The *Ahuihuitla*, "a worm which inhabits the Mexican lakes, four inches long, and of the thickness of a goose quill, of a tawny color on the upper part of the body, and white upon the under part; it stings with its tail, which is hard and poisonous." And the *Ocuiliztac*, "a black marsh worm, which becomes white on being roasted."[2]

---

[1] *Annals of Nat. Hist.* Simmond's *Curiosities of Food*, p. 308–311.
[2] *Nature and Art*, xii. 198.

DIPTERA.

Culicidæ—Gnats.[1]

CONCERNING the generation of Gnats, Moufet says : "Countrey people suppose them, and that not improbably, to be procreated from some corrupt moisture of the earth."[2]

A battle of Gnats (probably an appearance of Ephemera) is recorded in Stow's Chronicles of England, p. 509, to have been fought in the reign of King Richard II.: "A fighting among Gnats at the King's maner of *Shine*, where they were so thicke gathered, that the aire was darkened with them : they fought and made a great battaile. Two partes of them being slayne, fel downe to the grounde; the thirde parte hauing got the victorie, flew away, no man knew whither. The number of the deade was such that they might be swept uppe with besomes, and bushels filled weyth them."[3]

In the year 1736 the Gnats, *Culex pipiens*, were so numerous in England, that, as it is recorded, vast columns of them were seen to rise in the air from the steeple of the cathedral at Salisbury, which, at a little distance resembling columns of smoke, occasioned many people to think the edifice was on fire.[4] At Sagan, in Silesia, in July, 1812, a similar occurrence gave rise in like manner to an alarm that the church was on fire.[5] In May of the following year at Norwich, at about six o'clock in the evening, the inhabitants of that city were alarmed by the appearance of smoke issuing from the upper window of the spire of the cathedral,

---

[1] The numerous family of *Culicidæ* are confounded under the common names of Gnat and Mosquito; hence many mistakes will necessarily arise.

[2] *Theat. Ins.*, p. 81. Topsel's *Hist. of Beasts*, p. 952.

[3] Quot. in N. & Q., ix. 303

[4] *Phil. Trans.*, lvii. 113; Bingley's *Anim. Biog.*, iv. 205.

[5] Germar's *Mag. der Entomol.*, i. 137.

for which at the time no satisfactory account could be given, but which was most probably produced by the same cause.[1] And in the year 1766, in the month of August, they appeared in such incredible numbers at Oxford as to resemble a black cloud, darkening the air and almost intercepting the rays of the sun. Mr. John Swinton mentions, that in the evening of the 20th, about half an hour before sunset, he was in the garden of Wadham College, when he saw six columns of these insects ascending from the tops of six boughs of an apple-tree, two in a perpendicular, three in an oblique direction, and one in a pyramidal form, to the height of fifty or sixty feet. Their bite was so envenomed, that it was attended by violent and alarming inflammation; and one when killed usually contained as much blood as would cover three or four square inches of wall.[2] A similar column, of two or three feet in diameter and about twenty feet in height, was seen at eight o'clock in the evening of Sunday, July 14th, 1833, in Kensington Gardens. The upper portion of the column being curved to the east, the whole resembled the letter J inverted. The Gnats in every part of the column were in the liveliest motion.[3] The author of the "Faerie Queene" seems to have witnessed the like curious phenomenon, which furnished him with the following beautiful simile:

> As when a swarme of gnats at eventide
> Out of the fennes of Allan doe arise,
> Their murmuring small trumpets sownden wide,
> Whiles in the air their clust'ring army flies,
> That as a cloud doth seem to dim the skies;
> Ne man nor beast may rest or take repast,
> For their sharp wounds and noyous injuries,
> Till the fierce northern wind with blust'ring blast
> Doth blow them quite away, and in the ocean cast.

Ligon, in his History of Barbados, makes the following curious observation relative to a species of insects which he calls "Flyes," but which are more probably Gnats or Mosquitoes: "There is not only a race of all these kinds, that go in a generation, but upon new occasions, new kinds; as, after a great downfall of rain, when the ground has been

---

[1] K. & S. *Introd.*, i. 114.
[2] *Phil. Trans.*, lvii. 112–3.
[3] *Mag. of Nat. Hist.*, vi. 545.

extremely moistened, and softened with the water, I have walk'd out upon a dry walk (which I made my self) in an evening, and there came about me an army of such Flyes, as I had never seen before, nor after; and they rose, as I conceived, out of the earth: They were as big bodied as Bees, but far larger wings, harm they did us none, but only lighted on us; their colour between ash-colour and purple."[1]

If Gnats swarm in the summer in globular masses, it is supposed to prognosticate a storm. Moufet says: "If Gnats near sunset do play up and down in open air, they presage heat; if in the shade, warm and milde showers; but if they altogether sting those that passe by them, then expect cold weather and very much rain. . . . If any one would finde water either in a hill or valley, let him observe (saith Paxanus in Geoponika) the sun rising, and where the Gnats whirle round in form of an obelisk, underneath there is water to be found. Yea, if Apomasaris deceive us not, dreams of Gnats do foretell news of war or a disease, and that so much the more dangerous as it shall be apprehended to approach the more principall parts of the body."[2]

"On the 14th of December, 1830, at Oremburg, snow fell accompanied by a multitude of small black Gnats, whose motions were similar to those of a flea." This singular phenomenon was described at the session of the Academy of St. Petersburg, held February 21st, 1831.[3]

The pertinacity of the *Culicidæ* frequently renders them a most formidable pest. Humboldt tells us "that between the little harbor of Higuerote and the mouth of the Rio Unare, the wretched inhabitants are accustomed to stretch themselves on the ground, and pass the night buried in the sand three or four inches deep, exposing only the head, which they cover with a handkerchief."[4] As another proof of the terrible state to which man is sometimes reduced by Mosquitoes, Captain Stedman relates that in one of his

---

[1] *Hist. of Barbados*, p. 63.

[2] *Theatr. Ins.*, p. 86.   Topsel's *Hist. of Beasts*, p. 956.

[3] Silliman's *Journal*, xxii. 375.

[4] *Personal Narrative*, E. T. v. 87.   Humboldt has given a detailed account of these insect plagues, by which it appears that among them there are diurnal and crepuscular, as well as nocturnal species, or genera: the Mosquitoes, signifying *little flies* (*Simulia*), flying in the day; the *Temporaneros*, flying during twilight; and the Zancudos, meaning *long-legs* (*Culices*), in the night.

dreadful marches, the clouds of them were such, that the soldiers dug holes with their bayonets in the earth, into which they thrust their heads, stopping the entry and covering their necks with their hammocks, while they lay with their bellies on the ground: to sleep in any other position was absolutely impossible. He himself, by a negro's advice, climbed to the top of the highest tree he could find, and there slung his hammock among the boughs, and slept exalted nearly a hundred feet above his companions, "whom," says he, "I could not see for the myriads of mosquitoes below me, nor even hear, from the incessant buzzing of these troublesome insects."[1]

"The Gnats in America," says Moufet, "do so plash and cut, that they will pierce through very thick clothing; so that it is excellent sport to behold how ridiculously the barbarous people, when they are bitten, will skip and frisk, and slap with their hands their thighs, buttocks, shoulders, arms, and sides, even as a carter doth his horses."[2]   Isaac Weld tells us that "these insects were so powerful and bloodthirsty that they actually pierced through General Washington's boots."[3]   They probably crept within the boots, but the story is not incredible if we believe Moufet.   This naturalist says: "In Italy, near the Po, great store and very great ones are to be seen, terrible for biting, and venomous, piercing through a thrice-doubled stocking, and boots likewise (*morsu crudeles et venenati, triplices caligas, imo ocreas, item perforantes*), sometimes leaving behind them impoysoned, hard, blue tumours, sometimes painful bladders, sometimes itching pimples, such as Hippocrates hath observed in his Epidemics, in the body of one Cyrus, a fuller, being frantic."[4]

The poet Spenser, in his View of Ireland, says the Irish "goe all naked except a mantle, which is a fit house for an outlaw—a meet bed for a rebel—and an apt cloak for a thiefe. It coucheth him strongly against the Gnats, which, in that country, doe more to annoy the naked rebels, and

---

[1] Stedm. *Surinam*, ii. 93.
[2] *Ins. Theatr.*, p. 82.
[3] *Travels*, 8vo. edit. p. 205.
[4] *Ins. Theatr.*, p. 81.

doe more sharply wound them, than all their enemies' swords and speares, which can seldom come nigh them."

Stewart says that the negroes of Jamaica, who cannot afford mosquito-nets, get into a mechanical habit of driving away these troublesome nocturnal visitors, that even when apparently wrapt in profound sleep, there is a continual movement of the hands.[1]

Herodotus says : " The means devised by the Egyptians to avoid the Gnats, which swarm in prodigious numbers, are these : Those who reside at some elevation above the marshes, avail themselves of towers which they ascend to sleep ; for the Gnats, to avoid the winds, do not fly high. While those who dwell on the very margins of the marshes, instead of towers, practise another contrivance. Every man possesses a net, which, during the day, he employs in catching fish, and which at night he uses as his bed-chamber, where he places it over his couch, and so sleeps within it. For if any one," he concludes, " sleeps wrapped in a cloak or cloth, the Gnats will bite him through it ; but they never attempt to penetrate the net."[2]   With regard to the conclusion of Herodotus, that nets with meshes will effectually exclude Gnats, Tennent says he has " been satisfied by painful experience that (if the theory be not altogether fallacious) at least the modern mosquitoes of Ceylon are uninfluenced by the same considerations which restrained those of the Nile under the successors of Cambyses."[3]

Jackson complains that after a fifty-miles journey in Africa, the Gnats would not suffer him to rest, and that his hands and face appeared, from their bites, as if he was infected with the small-pox in its worst stage.[4]   Dr. Clarke relates that in the neighborhood of the Crimea, the Russian soldiers are obliged to sleep in sacks to defend themselves from the mosquitoes ; and even this, he adds, is not a sufficient security, for several of them die in consequence of mortification produced by these furious blood-suckers.[5]

When we consider these circumstances, it is not incredible that the army of Julian the Apostate should be so

---

[1] *View of Jamaica*, p 91.
[2] Herod.   Taylor's *Trans.*, p. 141.
[3] *Nat. Hist. of Ceylon*, p. 435.
[4] Jackson's *Morocco*, p. 57.
[5] *Travels*, i. 388.

fiercely attacked by these insects as to be driven back; or that the inhabitants of various cities, as Mouffet has collected from different authors,[1] should, by an extraordinary multiplication of this plague, have been compelled to desert them. Also the latter part of the following story, related by Theodoret, seems entitled to belief: When Sapor, King of Persia, says this historian, was besieging the Roman City of Nisibis in the year 360, James, Bishop of that city, ascended one of the towers, and "prayed that Flies and Gnats might be sent against the Persian hosts, that so they might learn from these small insects the great power of Him who protected the Romans." Scarcely had the Bishop concluded his prayer, continues Theodoret, when swarms of Flies and Gnats appeared like clouds, so that the trunks of the elephants were filled with them, as also were the ears and nostrils of the horses and of the other beasts of burden; and that, not being able to get rid of these insects, the elephants and horses threw their riders, broke the ranks, left the army, and fled away with the utmost speed; and this, he concludes, compelled the Persians to raise the siege.[2]

"As the Cossacks of the Black Sea are no agriculturists," says Jaeger, "but derive their subsistence from their numerous herds of horses, oxen, sheep, goats, and hogs, they suffer immensely, at times, from the ravages of the mosquitoes. Although they are fortunately not seen every year, these blood-suckers may be considered a real Egyptian plague among the herds of these Cossacks; for they soon transform the most delightful plains into a mournful, solitary desert, killing all the beasts, and completely stripping the fields of every animated creature. One thousand of these insatiate tormentors enter the nostrils, ears, eyes, and mouth of the cattle, who shortly after die in convulsions, or from secondary inflammation, or from absolute suffocation. In the small town of Elizabethpol alone, during the month of June, thirty horses, forty foals, seventy oxen, ninety calves, a hundred and fifty hogs, and four hundred sheep were killed by these flies."[3]

Ammianus Marcellinus, in his Roman History, treating

---

[1] *Ins. Theatr.*, p 85.
[2] Theod. *Eccles. Hist.*, B. ii. ch. xxx.
[3] *N. A. Ins.*, p. 317.

of the wild beasts in Mesopotamia, gives us the following curious zoological theory on the destruction of lions by mosquitoes:

"The lions wander in countless droves among the beds of rushes on the banks of the rivers of Mesopotamia, and in the jungles, and lie quiet all the winter, which is very mild in that country. But when the warm weather returns, as these regions are exposed to great heat, they are forced out by the vapours, and by the size of the Gnats, with swarms of which every part of that country is filled. And these winged insects attack the eyes, as being both moist and sparkling, sitting on and biting the eyelids; the lions, unable to bear the torture, are either drowned in the rivers, to which they flee for refuge, or else, by frequent scratchings, tear their eyes out themselves with their claws, and then become mad. And if this did not happen, the whole of the East would be overrun with beasts of this kind."[1]

I have never heard of mosquitoes being turned to any good account save in California; and there, it seems, according to Rev. Walter Colton, they are sometimes made the ministers of justice. A rogue had stolen a bag of gold from a digger in the mines, and hid it. · Neither threats nor persuasions could induce him to reveal the place of its concealment. He was at last sentenced to a hundred lashes, and then informed that he would be let off with thirty, provided he would tell what he had done with the gold; but he refused. The thirty lashes were administered, but he was still stubborn as a mule. He was then stripped naked, and tied to a tree. The mosquitoes with their long bills went at him, and in less than three hours he was covered with blood. Writhing and trembling from head to foot with exquisite torture, he exclaimed, "Untie me, untie me, and I will tell where it is." "Tell first," was the reply. So he told where it might be found. Then some of the party with wisps kept off the still hungry mosquitoes, while others went where the culprit directed, and recovered the bag of gold. He was then untied, washed with cold water, and helped to his clothes, while he muttered, as if talking to himself, "I couldn't stand that anyhow."[2]

The largest kind of mosquito in the valley of the lower

---

[1] *Roman History*, B. xviii. c. 7, § 5.
[2] *Three Years in California.* p. 250.

Mississippi is called the "Gallinipper." It is peculiarly described, by the boatmen, to be as large as a goose, and that it flies about at night with a brickbat under its wings with which it sharpens its "sting."

They tell a good story to show the superiority of the Gallinipper, over the ordinary Mosquito, in this wise. Some fellow made a bet that, for a certain length of time, he could stand the stings of the mosquitoes inflicted upon his bare back while he lay on his face. He stripped himself for the ordeal, and was bearing it manfully, when some mischievous spectator threw a live coal of fire on him. He winced, and, looking up by way of protest, exclaimed, "I bar (debar) the Gallinipper."

The Culicidæ, say Kirby and Spence, like other conquerors who have been the torment of the human race, have attained to fame, and have given their names to bays, towns, and even to considerable territories; and instance Mosquito Bay in St. Christopher's; Mosquito, a town in the Island of Cuba; and the Mosquito Shore of Central America.[1]

Democritus says: "Horse-hair, stretched through the door, and through the middle of the house, destroys Gnats."[2]

St. Macarius, Alban Butler says, was a confectioner of Alexandria, who, in the flower of his age, spent upwards of sixty years in the deserts in labor, penance, and contemplation. "Our Saint," continues Butler, "happened one day to inadvertently kill a Gnat that was biting him in his cell; reflecting that he had lost the opportunity of suffering that mortification, he hastened from the cell for the marshes of Scete, which abound with great flies, whose stings pierce even wild boars. There he continued six months, exposed to those ravaging insects; and to such a degree was his whole body disfigured by them with sores and swellings, that when he returned he was only to be known by his voice."[3]

In the old English translation of the Bible, the observation of our Saviour to the Pharisees, "Ye blind guides, which strain *at* a Gnat, and swallow a camel," is rendered

---

[1] *Introd.*, i. 119.
[2] Owen's *Geoponika*, ii. 150.
[3] *Lives of the Saints*, i. 50.

"which strain *out* a Gnat," and Bishop Pearce observes that this is conformable to the sense of the passage. An allusion is made to the custom which prevailed in Oriental countries of passing their wine and other liquors through a strainer, that no Gnats or flies might get into the cup. In the Fragments to Calmet, we are informed that there is a modern Arabic proverb to this effect, "He swallowed an elephant, but was strangled by a fly."[1]

Tipulidæ—Crane-flies.

The larvæ of a species of Agaric-Gnat (*Mycetophila*) live in society, and emigrate in files in a very soldier-like manner. First goes one, next follow two, then three, etc., so as to exhibit a singular serpentine appearance. The common people of Germany call this file *heerwurm*, and, it is said, view them with great dread, regarding them as ominous of war.[2]

Maupertuis, in describing his ascent to Mount Pulinga, in Lapland, says: "They had to fell a whole wood of large trees, and the Flies (most probably *Tipulidæ*) attack'd 'em with that fury, that the very soldiers, tho' harden'd to the greatest fatigues, were obliged to rap up their faces, or cover them with tar. These insects poison'd their victuals, for no sooner was a dish serv'd, but it was quite covered with them."[3] Maupertuis, in another place, says: "These Flies make Lapland less tolerable in the summer than the cold does in the winter."[4] The severity with which the Tipulidæ torment the Laplanders is attested also by Acerby,[5] Linnæus,[6] De Geer,[7] and Reaumur.[8]

[1] Lawson's *Bible Cyclop.*, ii. 558, 3 v. 8vo.
[2] Kirb. and Sp. *Introd.*, ii. 8.
[3] *Gent. Mag.*, 1738, viii. 577.   [4] *Ibid.*, xxiv. 274.
[5] *Travels*, ii. 5; 34–5; 51. Lond. 1802. 4to.
[6] *Lach. Lapp.*, ii. 108. *Flor. Lapp.*, 380.
[7] V. vi. p. 603–4.
[8] V. ix. p. 573.

Muscidæ—Flies.

Among the instances recorded of Flies appearing in immense numbers, the following are the most remarkable:

"When the Creole frigate was lying in the outer roads of Buenos Ayres, in 1819, at a distance of six miles from the land, her decks and rigging were suddenly covered with thousands of Flies and grains of sand. The sides of the vessel had just received a fresh coat of paint, to which the insects adhered in such numbers as to spot and disfigure the vessel, and to render it necessary partially to renew the paint. Capt. W. H. Smyth was obliged to repaint his vessel, the Adventure, in the Mediterranean, from the same cause. He was on his way from Malta to Tripoli, when a southern wind blowing from the coast of Africa, then one hundred miles distant, drove such myriads of Flies upon the fresh paint that not the smallest point was left unoccupied by the insects."[1]

"In May, 1699, at Kerton," records Mrs. Thoresby, p. 15, "in Lincolnshire, the sky seemed to darken north-westward at a little distance from the town, as though it had been a shower of hailstones or snow; but when it came near the town, it appeared to be a prodigious swarm of Flies, which went with such a force toward the south-east that persons were forced to turn their backs of them."[2]

On the morning of the 17th of September, 1831, a small dipterous insect, belonging to Meigen's genus *Chlorops*, and nearly allied to, if not identical with, his *C. læta*, appeared suddenly, and in such immense quantities, in one of the upper rooms of the Provost's Lodge, in King's College, Cambridge, that the greater part of the ceiling toward the window of the room was so thickly covered as not to be visible. They entered by a window looking due north, while the wind was blowing steadily N. N. W. So it appears they came from the direction of the River Cam, or rather came with its current.[3]

In the summer of 1834, which season was remarkable in England for its swarms and shoals of insects, the air was

---

[1] Lyell's *Princ. of Geol.*, p. 656.
[2] Southey's *Com. Place Bk.*, 1st S. p. 567.
[3] *Mag. of Nat. Hist.*, v. 302.

constantly filled, says a writer in The Mirror, with millions of small delicate Flies, and the sea in many places, particularly on the Norfolk coasts, was perfectly blackened by the amazing shoals. The length of these masses was not determined; but they were, it is asserted, at least a league broad. It is said the oldest fishermen of those seas never remembered having seen or heard of such a phenomenon.[1]

Capt. Dampier calls the natives of New Holland the "poor winking people of New Holland," and concludes his description of them with the following observations: "Their eyelids are always half closed, to keep the Flies out of their eyes, they being so troublesome here that no fanning will keep them from coming to one's face; and without the assistance of both hands to keep them off they will creep into one's nostrils, and mouth, too, if the lips are not shut very close. So that from their infancy, being thus annoyed with these insects, they do never open their eyes as other people, and therefore they cannot see far, unless they hold up their heads, as if they were looking at something over them."[2]

In a house at Zaffraan-craal, Dr. Sparrman suffered so much from the common House-fly, *Musca domestica*, which, in the south of Africa, frequently appears in such prodigious numbers as to cover almost entirely the walls and ceilings, that, as he asserts, it was impossible for him to keep within doors for any length of time. To get rid of these troublesome pests, the natives resort to a very ingenious contrivance. It is thus related by the above-mentioned traveler: "Bunches of herbs are hung up all over the ceiling, on which the Flies settle in great numbers; a person then takes a linen net or bag, of a considerable depth, fixed to a long handle, and, inclosing in it every bunch, shakes it about, so that the Flies fall down to the bottom of the bag: when, after several applications of it in this manner, they are killed by a pint or a quart at a time, by dipping the bag into scalding hot water."[3]

Rhasis, Avicen, and Albertus say: "Bury the tail of a wolf in the house, and the Flies will not come into it."[4]

---

[1] *The Mirror*, xxvii. 68.
[2] Damp. *Voy.* O (vol. i.), 464.
[3] *Travels.* i. 211.
[4] Moufet's *Theat. Ins.*, p. 78.

Berytius says: "Flies will never rest on dumb animals if they are rubbed with the fat of a lion."[1]

Pliny says: "At Rome yee shall not have a Flie or dog that will enter into the chappell of Hercules standing in the beast market."[2]

Plutarch, in the Eighth Book of his Symposiaques, learnedly discourses upon the tamableness of the Fly. His opinion is that it cannot be tamed.[3]

Moufet, in his Theater of Insects, says: "Many ways doth nature also by Flies play with the fancies of men in dreams, if we may credit Apomasaris in his Apotelesms. For the Indians, Persians, and Ægyptians do teach, that if Flies appear to us in our sleep, it doth signifie an herauld at arms, or an approaching disease. If a general of an army, or a chief commander, dream that at such or such a place he should see a great company of Flies, in that very place, wherever it shall be, there he shall be in anguish and grief for his soldiers that are slain, his army routed, and the victory lost. If a mean or ordinary man dream the like, he shall fall into a violent fever, which likely may cost him his life. If a man dream in his sleep that Flies went into his mouth or nostrils, he is to expect with great sorrow and grief imminent destruction from his enemies."[4]

In an English North country chap-book, entitled the Royal Dream-book, we find: "To dream of Flies or other vermin, denotes enemies of all sorts."[5]

"When we see," says Hollingshed, "a great number of Flies in a yeare, we naturallie iudge it like to be a great plague."[6]

Among the deep-sea fishermen of Greenock (Scotland), there is a most comical idea that if a Fly falls into a glass from which any one has been drinking, or is about to drink, it is considered a sure omen of good luck to the drinker, and is always noticed as such by the company.[7] Has this

---

[1] Owen's *Geoponika*, ii. 152.
[2] *Nat. Hist.*, x. 29. Holland, p. 285. D.
[3] Holl. *Trans.*, p. 631.
Vide Pierius' *Hieroglyph.*, p. 268–9. Importunitas ac impudentia; Pertinacia; Res gesta cominus; Indocilitas; Cynici.
[4] *Theatr. Ins.*, p. 70. Topsel's *Hist. of Beasts*, p. 945.
[5] Brand's *Pop. Antiq.*, iii. 134.
[6] *Chron. of Eng.*, iii. 1002.
[7] *N. and Q.*, xii. 488.

any connection with our saying of "taking a glass with a
fly in it ?"

If Flies die in great numbers in a house, it is believed by
the common people to be a sure sign of death to some one
in the family occupying it; if throughout the country, an
omen of general pestilence. It is positively asserted that
Flies always die before the breaking out of the cholera, and
believed that they die of this disease.

Moufet, in his Theater of Insects, says : " When the Flies
bite harder than ordinary, making at the face and eyes of
men, they foretell rain or wet weather, from whence Politian
hath it :

> Thirsty for blood the Fly returns,
> And with his sting the skin he burns.

Perhaps before rain they are most hungry, and therefore,
to asswage their hunger, do more diligently seek after their
food. This also is to be observed, that a little before a
showre or a storme comes, the Flies descend from the upper
region of the air to the lowest, and do fly, as it were, on the
very surface of the earth. Moreover, if you see them very
busie about sweet-meats or unguents, you may know that it
will presently be a showre. But if they be in all places
many and numerous, and shall so continue long (if Alexander Benedict and Johannes Damascenus say true), they
foretell a plague or pestilence, because so many of them
could not be bred of a little putrefaction of the air."[1] Elsewhere Moufet states : "Neither are Flies begotten of dung
only, but of any other filthy matter putrefied by heat in the
summer time, and after the same way spoken of before, as
Grapaldus and Lonicerus have very well noted."[2]

Willsford, in his Nature's Secrets, p. 135, says : "Flies in
the spring or summer season, if they grow busier or blinder
than at other times, or that they are observed to shroud
themselves in warm places, expect then quickly to follow
either hail, cold storms of rain, or very much wet weather ;
and if these little creatures are noted early in autumn to
repair into their winter quarters, it presages frosty mornings, cold storms, with the approach of hoary winter.
Atomes of Flies swarming together, and sporting themselves in the sunbeams, is a good omen of fair weather."[3]

---

[1] *Theatr. Ins.*, p. 70. Topsel's *Hist. of Beasts*, p. 944.
[2] *Ibid.*, p. 55. Topsel, p. 933.
[3] Brand's *Pop. Antiq.*, iii. 191.

In Gayton's Pleasant Notes upon Don Quixot, 1654, p.
99, speaking of Sancho Panza's having converted a cassock
into a wallet, our pleasant annotator observes : " It was serviceable, after this greasie use, for nothing but to preach at
a carnivale or Shrove Tuesday, and to tosse Pancakes in
after the exercise ; or else, if it could have been conveighed
thither, nothing more proper for a man that preaches the
Cook's sermon at Oxford, when that plump society rides
upon their governour's horses to fetch in the Enemie, the
Flie." That there was such a custom at Oxford, let Peshall,
in his history of that city, be a voucher, who, speaking of
St. Bartholomew's Hospital, p. 280, says : " To this Hospital cooks from Oxford flocked, bringing in on Whitsun-
week the Fly." Aubrey saw this ceremony performed in
1642. He adds : " On Michaelmas-day, they rode thither
again to carry the Fly away."[1]

Plutarch, in his disquisition on the Art of Discerning a
Flatterer from a Friend, makes the following curious comparison : " The Gad-Flie (as they say) which useth to
plague bulles and oxen, setteth about their eares, and
so doth the tick deal by dogges : after the same manner,
flatterers take hold of ambitious mens eares, and possesse
them with praises; and being once set fast there, hardly
are they to be removed and chased away."[2]

Plautus twice compares envious and inquisitive persons
to Flies.[3]

In a narrative of unheard-of Popish cruelties toward

---

[1] Brand's *Pop. Antiq.*, i. 84.

[2] Holl. *Trans.*, p. 76. There was one time a law at Athens, which
a good deal nonplussed these sponging gentlemen so appropriately
called Flies. "It was decreed that not more than thirty persons
should meet at a marriage feast: and a wealthy citizen, desirous of
going as far as the law would allow him, had invited the full complement. An honest Fly, however, who respected no law that in-
terfered with his stomach, contrived to introduce himself, and took
his station at the lower end of the table. Presently the magistrate
appointed for the purpose entered, and espying his man at a glance,
began counting the guests, commencing on the other side and end-
ing with the parasite. 'Friend,' said he, 'you must retire. I find
there is one more than the law allows.' 'It is quite a mistake, sir,'
replied the Fly, 'as you will find if you will have the goodness to
count again, beginning *on this side*.' "—St. John's *Man. and Cust. of
Anct. Grec.*, ii. 172.

[3] Vide *Mercator*, A. ii. Sc. 4, and the *Young Carthag.*, A. iii. Sc. 3.

Protestants beyond Seas, printed in 1680, we find the insinuating detectives of the Spanish Inquisition under the name of Flies.[1]

Flies are mentioned somewhere in Lyndwood as the emblem of unclean thoughts.[2]

Flies were driven away when a woman was in labor, for fear she should bring forth a daughter.[3]

Flies are found represented in the pottery of the ancient Egyptians.[4]

Flies (*Cuspi*) were sacrificed to the Sun by the ancient Peruvians.[5]

"To let a Flee (Fly) stick i' the wa'" is, in Scotland, not to speak on some particular topic, to pass it over without remark.[6]

"Certes, a strange thing it is of these Flies," says Pliny, "which are taken to be as senselesse and witlesse creatures, yea, and of as little capacity and understanding as any other whatsoever: and yet at the solemne games and plaies holden every fifth yeare at Olympia, no sooner is the bull sacrificed there to the Idoll or god of the Flies called Myiodes, but a man shall see (a wonderful thing to tell) infinit thousand of flies depart out of that territorie by flights, as it were thick clouds."[7]

This Myiodes or Maagrus, the "Fly-catcher," was the name of a hero, invoked at Aliphera, at the festivals of Athena, as the protector against Flies. It was also a surname of Hercules.

The following rendering of the second verse of the first chapter of the Second Book of Kings, by Josephus, contains an allusion to the worship of Baalzebub under the form of a Fly: "Now it happened that *Ahaziah*, as he was coming down from the top of his house, fell down from it, and in his sickness sent to the *Fly* (Baalzebub), which was the god of *Ekron*, for that was this god's name, to enquire about his recovery."[8]

With reference to this worship, we read in Purchas's

---

1 *Harleian Miscel.*, viii. 423.
2 Fosbr. *Encycl. of Antiq.*, ii. 738.   3 *Ibid.*
4 Wilkinson's *Anct. Egypt.*, 2d S. ii. 126, 260.
5 Hawk's *Peruvian Antiq*, p. 197.
6 Jamieson's *Scottish Dict.*
7 *Nat. Hist.*, xxix. 6.   Holl. *Trans.*, p. 364. K.
8 *Antiq. of the Jews*, B. ix. c. 2.   Whiston's *Trans.*, p. 274.

Pilgrims: "At Accaron was worshipped *Baalzebub*, that is, the Lord of the Flies, either of contempt of his idolatrie, so called; or rather of the multitude of Flies, which attended the multitude of his sacrifices, when from the sacrifices at the Temple of Jerusalem, as some say, they were wholly free : or for that hee was their Larder-god (as the Roman *Hercules*) to drive away flies : or for that from a forme of a Flie, in which he was worshipped. . . . But for Beelzebub, he was their *Æsculapius* or Physicke god, as appeareth by Ahaziah who sent to consult with him in his sickness. And perhaps from this cause the blaspheming Pharisies, rather applyed the name of this then any other Idoll to our blessed Saviour (Math. x. 25) whom they saw indeed to performe miraculous cures, which superstition had conceived of *Baalzebub:* and if any thing were done by that Idoll, it could by no other cause bee effected but by the Devill, as tending (like the popish miracles) to the confirmation of Idolatrie."[1]

This god of the Flies was so called, thinks Whiston, as was Jove among the Greeks, from his supposed power over Flies, in driving them away from the flesh of their sacrifices, which otherwise would have been very troublesome to them.[2]

It has been conjectured that the Fly, under which Baalzebub was represented, was the Tumble-bug, *Scarabæus pilluarius;* in which case, says Dr. Smith, Baalzebub and Beelzebub might be used indifferently.[3]

"Urspergensis saith that the Devil did very frequently appear in the form of a Fly; whence it was that some of the heathens called their familiar spirit *Musca* or Fly : perchance alluding to that of Plautus :

> Hic pol musca est, mi pater,
> Sive profanum, sive publicum, nil clam
>   illum haberi potest :
> Quin adsit ibi illico, et rem omnem tenet.—

This man, O my father, is a Fly, nothing can be concealed from him, be it secret or publick, he is presently there, and knowes all the matter."[4]

---

[1] *Pilg.*, v. 81. Fol. 1626.
[2] Whiston's *Trans. of Josephus*, p. 274, note.
[3] *Dict. of Bible.*
[4] Moufet, *Theatr. Ins.*, p. 79. Topsel's *Transl.*, p. 951.

Loke, the deceiver of the gods, is fabled in the Northern Mythology, to have metamorphosed himself into a Fly: and demons, in the shape of Flies, were kept imprisoned by the Finlanders, to be let loose on men and beasts.[1]

In Scotland, a tutelary Fly, believed immortal, presided over a fountain in the county of Banff: and here also a large blue Fly, resting on the bark of trees, was distinguished as a witch.[2]

Among the games and plays of the ancient Greeks was the Χαλκη Μυια, or Brazen Fly:—a variety of blind-man's-buff, in which a boy having his eyes bound with a fillet, went groping round, calling out, "I am seeking the Brazen Fly." His companions replied, "You may seek, but you will not find it"—at the same time striking him with cords made of the inner bark of the papyros ; and thus they proceeded till one of them was taken.[3]

This is most probably an allusion to some species of Fly of a bronze color which is most difficult to catch, as, for instance, the little fly found in summer beneath arbors, apparently standing motionless in the air.

Petrus Ramus tells us of an iron Fly, made by Regiomontanus, a famous mathematician of Nuremberg, which, at a feast, to which he had invited his familiar friends, flew forth from his hand, and taking a round, returned to his hand again, to the great astonishment of the beholders. Du Bartas thus expresses this :

> Once as this artist, more with mirth than meat,
> Feasted some friends whom he esteemed great,
> Forth from his hand an iron Fly flew out;
> Which having flown a perfect round-about,
> With weary wings return'd unto her master:
> And as judicious on his arm he plac'd her.
> O! wit divine, that in the narrow womb
> Of a small fly, could find sufficient room
> For all those springs, wheels, counterpoise and chains,
> Which stood instead of life, and blood, and veins![4]

We find also in a work bearing the title "Apologie pour les Grands Homines Accusés de Magie," that "Jean de Montroyal presented to the Emperor Charles V. an iron

---

[1] Dalyell's *Darker Superst. of Scotland*, p 562. Edinbgh. 1834.
[2] *Ibid.*
[3] *St. John's Man. and Cust. of Anct. Grec.*, i. 150.
[4] Wanley's *Wonders*, i. 377.

Fly, which made a solemn circuit round its inventor's head, and then reposed from its fatigue on his arm."—Probably the same automaton, since Regiomontanus and Montroyal are the same.

Such a Fly as the above is rather extraordinary, yet I have something better to tell—still about a Fly.

Gervais, Chancellor to the Emperor Otho III., in his book entitled "Otia Imperatoris," informs us that "the sage Virgilius, Bishop of Naples, made a brass Fly, which he placed on one of the city gates, and that this mechanical Fly, trained like a shepherd's dog, prevented any other fly entering Naples; so much so, that during eight years the meat exposed for sale in the market was never once tainted !"[1]

"Varro affirmeth," says Pliny, "that the heads of Flies applied fresh to the bald place, is a convenient medicine for the said infirmity and defect. Some use in this case the bloud of flies: others mingle their ashes with the ashes of paper used in old time, or els of nuts; with this proportion, that there be a third part only of the ashes of flies to the rest, and herewith for ten days together rubb the bare places where the hair is gone. Some there be againe, who temper and incorporat togither the said ashes of Flies with the juice of colewort and brest-milke: others take nothing thereto but honey."[2]

Mucianus, who was thrice consul, carried about him a living Fly, says Pliny, wrapped in a piece of white linen, and strongly asserted that to the use of this expedient he owed his preservation from ophthalmia.[3]

Ferdinand Mendez Pinto says: "In our travels with the ambassador of the King of Bramaa to the Calaminham, we saw in a grot men of a sect of one of their Saints, named Angemacur: these lived in deep holes, made in the mider of the rock, according to the rule of their wretched order, eating nothing but Flies, Ants, Scorpions, and Spiders, with the juice of a certain herb, much like to sorrel."[4]

Says Moufet, in his Theater of Insects : " Plutarch, in his Artaxerxes, relates that it was a law amongst a certain

<hr>

1 *Mem. of Robt. Houdin*, p. 156. Philad. 1859.
2 *Nat. Hist.*, xxix. 6. Holland's *Trans.*, p. 364. I.
3 *Ibid.*, xxviii. 2 (5).
4 *Voy.*, C. 56, p. 222. Wanley's *Wonders*, ii. 373.

people, that whosoever should be so bold as to laugh at and deride their lawes and constitutions of state, was bound for twenty daies together in an open chest naked, all besmeared with honey and milk, and so became a prey to the Flies and Bees, afterward when the days were expired he was put into a woman's habit and thrown headlong down a mountain. . . Of which kinde of punishment also Suidas makes mention in his Epicurus. There was likewise for greater offenders, a punishment of Boats, so called. For that he that was convict of high treason, was clapt between two boats, with his head, hands, and feet hanging out : for his drink he had milk and honey powred down his throat, with which also his head and hands were sprinkled, then being set against the sun, he drew to him abundance of stinging Flies, and within being full of their worms, he putrefied by little and little, and so died. Which kinde of examples of severity as the ancients shewed to the guilty and criminous offenders; so on the other side the Spaniards in the Indies, used to drive numbers of the innocents out of their houses, as the custome is among them, naked, all bedawbed with honey, and expose them in open air to the biting of most cruel Flies."[1]

Mr. Henry Mayhew, in that part of his interesting work on London Labor and London Poor devoted to the London Street-folk, has given us the narratives of several " Catch-'em-Alive" sellers—a set of poor boys who sell prepared papers for the purpose of catching Flies. He discovered, as he relates, a colony of these " Catch-'em-alive" boys residing in Pheasant-court, Gray's-inn-lane. They were playing at "pitch-and-toss" in the middle of the paved yard, and all were very willing to give him their statements; indeed, the only difficulty he had was in making his choice among the youths.

" Please, sir," said one with teeth ribbed like celery, to him, " I've been at it longer than him."

" Please, sir, he ain't been out this year with the papers," said another, who was hiding a handful of buttons behind his back.

" He's been at shoe-blacking, sir; I'm the only reg'lar fly-boy," shouted a third, eating a piece of bread as dirty as London snow.

---

[1] *Theatr. Ins.*, p. 79.   Topsel's *Hist. of Beasts*, p. 951.

A big lad with a dirty face, and hair like hemp, was the first of the "catch-'em-alive" boys who gave him his account of his trade. He was a swarthy featured boy, with a broad nose like a negro's, and on his temple was a big half-healed scar, which he accounted for by saying that "he had been runned over" by a cab, though, judging from the blackness of one eye, it seemed to Mr. Mayhew to have been the result of some street fight. He said:

"I'm an Irish boy, and nearly turned sixteen, and I've been silling fly-papers for between eight and nine year. I must have begun to sill them when they first come out. Another boy first tould me of them, and he'd been silling them about three weeks before me. He used to buy them of a party as lives in a back-room near Drury-lane, what buys paper and makes the catch 'em alive for himself. When they first come out they used to charge sixpence a dozen for 'em, but now they've got 'em to twopence ha'penny. When I first took to silling 'em, there was a tidy lot of boys at the business, but not so many as now, for all the boys seem at it. In our court alone I should think there was about twenty boys silling the things.

"At first, when there was a good time, we used to buy three or four gross together, but now we don't no more than half a gross. As we go along the streets we call out different cries. Some of us says, 'Fly-papers, fly-papers, ketch 'em all alive.' Others make a kind of song of it, singing out, 'Fly-paper, ketch 'em all alive, the nasty flies, tormenting the baby's eyes. Who'd be fly-blow'd, by all the nasty blue-bottles, beetles, and flies?' People likes to buy of a boy as sings out well, 'cos it makes 'em laugh.

"I don't think I sell so many in town as I do in the borders of the country, about Highbury, Croydon, and Brentford. I've got some regular customers in town about the City-prison and the Caledonian-road; and after I've served them and the town custom begins to fall off, then I goes to the country. We goes two of us together, and we takes about three gross. We keep on silling before us all the way, and we comes back the same road. Last year we sould very well in Croydon, and it was the best place for gitting the best price for them; they'd give a penny a piece for 'em there, for they didn't know nothing about them. I went off one day at ten o'clock and didn't come home till two in the morning. I sould eighteen dozen out in that d'rection the other day, and got rid of them before I had got half-way. But flies are very scarce at Croydon this year, and we haven't done so well. There ain't half as many flies this summer as last.

" Some people says the papers draws more flies than they ketches, and that when one gets in, there's twenty others will come to see him. It's according to the weather as the flies is about. If we have a fine day it fetches them out, but a cold day kills more than our papers.

" We sills the most papers to little cook-shops and sweet-meat shops. We don't sill so many at private houses. The public-houses is pretty good customers, 'cos the beer draws the flies. I sould nine dozen at one house—a school—at Highgate, the other day. I sould 'em two for three-ha' pence. That was a good hit, but then t'other days we loses. If we can make a ha'penny each we thinks we does well.

" Those that sills their papers at three a-penny buys them at St. Giles's, and pays only three ha'pence a dozen for them, but they ain't half as big and good as those we pays tup-pence-ha'penny a dozen for.

" Barnet is a good place for fly-papers; there's a good lot of flies down there. There used to be a man at Barnet as made 'em, but I can't say if he do now. There's another at Brentford, so it ain't much good going that way.

" In cold weather the papers keep pretty well, and will last for months with just a little warming at the fire; for they tears on opening when they are dry. You see we always carry them with the stickey sides doubled up together like a sheet of writing-paper. In hot weather, if you keep them folded up, they lasts very well ; but if you opens them, they dry up. It's easy opening them in hot weather, for they comes apart as easy as peeling a horrange. We gener-ally carries the papers in a bundle on our arm, and we ties a paper as is loaded with flies round our cap, just to show the people the way to ketch 'em. We get a loaded paper given to us at a shop.

" When the papers come out first, we use to do very well with fly-papers ; but now it's hard work to make our own money for 'em. Some days we used to make six shillings a day regular. But then we usen't to go out every day, but take a rest at home. If we do well one day, then we might stop idle another day, resting. You see, we had to do our twenty or thirty miles silling them to get that money, and then the next day we was tired.

"The silling of papers is gradual falling off. I could go out and sill twenty dozen wonst where I couldn't sill one now. I think I does a very grand day's work if I yearns a shilling. Perhaps some days I may lose by them. You see, if it's a very hot day, the papers gets dusty; and besides, the stuff gets melted and oozes out; though that don't do much harm, 'cos we gets a bit of whitening and rubs 'em over. Four years ago we might make ten shillings a day at the papers, but now, taking from one end of the fly-season to the other, which is about three months, I think we makes about one shilling a day out of papers, though even that ain't quite certain. I never goes out without getting rid of mine, somehow or another, but then I am obleeged to walk quick and look about me.

"When it's a bad time for silling the papers, such as a wet, could day, then most of the fly-paper boys goes out with brushes, cleaning boots. Most of the boys is now out hopping. They goes reg'lar every year after the season is give over for flies.

"The stuff as they puts on the paper is made out of boiled oil and turpentine and resin. It's seldom as a fly lives more than five minutes after it gets on the paper, and then it's as dead as a house. The blue-bottles is tougher, but they don't last long, though they keeps on fizzing as if they was trying to make a hole in the paper. The stuff is only p'isonous for flies, though I never heard of anybody as ever eat a fly-paper."

A second lad, in conclusion, said: "There's lots of boys going selling 'ketch-'em-alive oh's' from Golden-lane, and White-chapel and the Borough. There's lots, too, comes out of Gray's-inn-lane and St. Giles's. Near every boy who has nothing to do goes out with fly-papers. Perhaps it ain't that the flies is falled off that we don't sill so many papers now, but because there's so many boys at it."

A third, of the lot the most intelligent and gentle in his demeanor, though the smallest in stature, said:

"I've been longer at it than the last boy, though I'm only getting on for thirteen, and he's older than I'm; 'cos I'm little and he's big, getting a man. But I can sell them quite as well as he can, and sometimes better, for I can holler out just as loud, and I've got reg'lar places to go to. I was a very little fellow when I first went out with them, but I could sell them pretty well then, sometimes three or four dozen a day. I've got one place, in a stable, where I can sell a dozen at a time to country people.

"I calls out in the streets, and I goes into the shops, too, and calls out, 'Ketch 'em alive, ketch 'em alive; ketch all the nasty black-beetles, blue-bottles, and flies; ketch 'em from teasing the baby's eyes.' That's what most of us boys cries out. Some boys who is stupid only says, 'Ketch 'em alive,' but people don't buy so well from them.

"Up in St. Giles's there is a lot of fly-boys, but they're a bad set, and will fling mud at gentlemen, and some prigs the gentlemen's pockets. Sometimes, if I sell more than a big boy, he'll get mad and hit me. He'll tell me to give him a halfpenny and he won't touch me, and that if I don't he'll kill me. Some of the boys takes an open fly-paper, and makes me look another way, and then they sticks the ketch-'em-alive on my face. The stuff won't come off without soap and hot water, and it goes black, and looks like mud. One day a boy had a broken fly-paper, and I was taking a drink of water, and he come behind me and slapped it up in my face. A gentleman as saw him give him a crack with a stick and me twopence. It takes your breath away, until a man comes and takes it off. It all sticked to my hair, and I couldn't rack (comb) right for some time . . . .

"I don't like going along with other boys, they take your customers away; for perhaps they'll sell 'em at three a penny to 'em, and spoil the customers for you. I won't go with the big boy you saw, 'cos he's such a blackgeyard; when he's in the country he'll go up to a lady and say, 'Want a fly-paper, marm?' and if she says 'No,' he'll perhaps job his head in her face—butt at her like.

"When there's no flies, and the ketch-'em-alive is out, then I goes tumbling. I can turn a cat'enwheel over on one hand. I'm going to-morrow to the country, harvesting and hopping—for, as we says, 'Go out hopping, come in jumping.' We start at three o'clock to-morrow, and we shall get about twelve o'clock at night at Dead Man's Barn. It was left for poor people to sleep in, and a man was buried there in a corner. The man had got six farms of hops; and if his son hadn't buried him there, he wouldn't have had none of the riches.

"The greatest number of fly-papers I've sold in a day is

about eight dozen. I never sells no more than that; I wish I could. People won't buy 'em now. When I'm at it I makes, taking one day with another, about ten shillings a week. You see, if I sold eight dozen, I'd make four shillings. I sell 'em at a penny each, at two for three-ha'pence, and three for twopence. When they gets stale I sells 'em for three a penny. I always begin by asking a penny each, and perhaps they'll say, 'Give me two for three ha'pence?' I'll say, 'Can't, ma'am,' and then they pulls out a purse full of money and gives a penny.

"The police is very kind to us, and don't interfere with us. If they see another boy hitting us they'll take off their belts and hit 'em. Sometimes I've sold a ketch-'em-alive to a policeman; he'll fold it up and put it into his pocket to take home with him. Perhaps he's got a kid, and the flies teazes its eyes.

"Some ladies like to buy fly-cages better than ketch-'em-alive's, because sometimes when they're putting 'em up they falls in their faces, and then they screams."

The history of the manufacture of Fly-papers was thus given to Mr. Mayhew by a manufacturer, whom he found in a small attic-room near Drury-lane: "The first man as was the inventor of these fly-papers kept a barber's shop in St. Andrew-street, Seven Dials, of the name of Greenwood or Greenfinch, I forget which. I expect he diskivered it by accident, using varnish and stuff, for stale varnish has nearly the same effect as our composition. He made 'em and sold 'em at first at threepence and fourpence a piece. Then it got down to a penny. He sold the receipt to some other parties, and then it got out through their having to employ men to help 'em. I worked for a party as made 'em, and then I set to work making 'em for myself, and afterwards hawking them. They was a greater novelty then than they are now, and sold pretty well. Then men in the streets, who had nothing to do, used to ask me where I bought 'em, and then I used to give 'em my own address, and they'd come and find me."[1]

---

[1] *London Lab. and London Poor*, iii. 28–33.

Oestridæ—Bot-flies.

The larvæ of Bots, *Œstris ovis*, found in the heads of sheep and goats, have been prescribed, and that, from the tripod of Delphos, as a remedy for the epilepsy. We are told so on the authority of Alexander Trallien; but whether Democritus, who consulted the oracle, was cured by this remedy, does not appear; the story shows, however, that the ancients were aware that these maggots made their way even into the brain of living animals.[1] The oracle answered Democritus as follows:

> Take a tame goat that hath the greatest head,
> Or else a wilde goat in the field that's bred,
> And in his forehead a great worm you'l finde,
> This cures all diseases of that kinde. [2]

The common saying that a whimsical person is *maggoty*, or has got *maggots in his head*, perhaps arose from the freaks the sheep have been observed to exhibit when infested by their Bots.[3]

The following "charme for the Bots[4] in a horse" is found in Scots' Discovery of Witchcraft, printed in 1651: "You must both say and do thus upon the diseased horse three days together, before the sun rising: *In nomine pa†tris & fi†lii & Spiritus†sancti, Exorcize te vermen per Deum pa†trem & fi†lium & Spiritum†sanctum*: that is, In the name of God the father, the sonne, and the Holy Ghost, I conjure thee O worm bv God, the Father, the sonne, and the Holy Ghost; that thou neither eate nor drink the flesh, blood, or bones of this horse; and that thou hereby maiest be made as patient as Job, and as good as S. John Baptist, when he baptized Christ in Jordan, *In nomine pa†tris & fi†lii et spiritus†sancti*. And then say three *Pater nosters*, and three *Aves*, in the right eare of the horse, to the glory of the holy trinity. Do†minus fili†us spirit†us Mari†a."[5]

There is a popular error in England respecting the *Œstrus* (*Gasterophilus*) *equi* (*hæmorrhoidalis*), which Shakspeare

---

[1] Kirb. and Sp. *Introd.*, i. 158.
[2] *Theatr. Ins.*, p. 284. Topsel's *Hist. of Beasts*, p 1107, 1122.
[3] Kirby and Spence, *Introd.*, i. 158.
[4] *Gasterophilus equi.*
[5] Reg. Scot's *Disc. of Witchcraft*, p. 179.

has followed, and which has been judiciously explained by Mr. Clark. Shakspeare makes the carrier at Rochester observe: "Peas and oats are as dank here as a dog, and that's the next way to give *poor jades the bots.*"[1]

The larvæ of this insect, says Mr. Clark, are mostly known among the country people by the name of *wormals, wormuls, warbles,* or, more properly, *Bots.* And our ancestors erroneously imagined that poverty or improper food engendered them in horses. The truth, however, seems to be, that when the animal is kept without food the Bots are also, and are then, without doubt, most troublesome; whence it was very naturally supposed that poverty or bad food was the parent of them.[2]

A cow with its hide perforated by Warbles, in England, was said to be elf-shot: the holes being made by the arrows of the little malignant fairies. In the Northern Antiquities, p. 404, we find the following:

"If at such a time you were to look through an elf-bore in wood, where a thorter knot has been taken out, or through the hole made by an elf-arrow (which has probably been made by a Warble) in the skin of a beast that has been elf-shot, you may see the elf-bull naiging (butting) with the strongest bull or ox in the herd; but you will never see with that eye again."

In the Scottish history of the trials of witches, we find the following: Alexander Smaill offended Jonet Cock, who threatened him, "deare sall yow rewe it! and within half ane howre therafter, going to the pleugh,—befoir he had gone one about, their came ane great Wasp or Bee, so that the foir horses did runne away with the pleugh, and wer liklie to have killed themselves, and the said Alexander and the boy that was with him, narrowlie escaped with their lyves."[3] Possibly the incident is not exaggerated, as a single Œstrus will turn the oxen of a whole herd, and render them furious.

Spencer, in his Travels in Circassia, speaks of a poisonous Fly, known in Hungary under the name of the Golubaeser-fly, which is singularly destructive to cattle. The Hungarian peasants, to account for the severity of the bite

---

1 Henry IV., Pt. I. Act ii. Sc. 1.
2 Newell's *Zool. of the Poets,* p. 29.
3 Dalyell's *Superstitions of Scotland,* p. 564.

of this insect, tell us that in the caverns, near the Castle of Golubacs, the renowned champion, St. George, killed the dragon, and that its decomposed remains have continued to generate these insects down to the present day. So firmly did they believe this, that they closed up the mouths of the caverns with stone walls.[1]

---

[1] *Saturday Mag.*, xviii. 153.

ORDER X.

APHANIPTERA.

Pulicidæ—Fleas.

The name *Pulex*, given to the Flea by the Romans, is stated by Isodorus to have been derived from *pulvis*, dust, *quasi pulveris filius*. Our English name *Flea*, and the German *Flock*, are evidently deduced from the quick motions of this insect.

As to the origin of Fleas, Moufet had a similar notion to that contained in the word Pulex, if we adopt the etymology of Isodorus, for he says they are produced from the dust, especially when moistened with urine, the smallest ones springing from putrid matter. Scaliger relates that they are produced from the moistened humors among the hairs of dogs.[1] Conformable to the curious notion of Moufet, Shakspeare says:

2 Car. I think this be the most villainous house in all London road for fleas: I am stung like a tench.

1 Car. Like a tench? by the mass, there is ne'er a king in Christendom could be better bit than I have been since the first cock.

2 Car. Why, they will allow us ne'er a jorden, and then we leak in your chimney; and your *chamber-ley breeds fleas* like a loach.[2]

"Martyr, the author of the Decads of Navigation, writes, that in Perienna, a countrey of the Indies, the drops of sweat that fall from their slaves' bodies will presently turn to Fleas."[3]

Ewlin, in his book of Travels in Turkey, has recorded a singular tradition of the history of the Flea and its confraternity, as preserved among a sect of Kurds, who dwelt in his time at the foot of Mount Sindshar. "When Noah's

---

[1] *Hist. of Ins.* (Murray, 1838), ii. 313.
[2] Henry IV. Pt. I., Act ii. Sc. 1.
[3] Moufet, *Theatr. Ins.*, p. 276. Topsel's *Hist. of Beasts*, p. 1102.

Ark," says the legend, "sprung a leak by striking against a rock in the vicinity of Mount Sindshar, and Noah despaired altogether of safety, the serpent promised to help him out of his mishap if he would engage to feed him upon human flesh after the deluge had subsided. Noah pledged himself to do so; and the serpent coiling himself up, drove his body into the fracture and stopped the leak. When the pluvious element was appeased, and all were making their way out of the ark, the serpent insisted upon the fulfillment of the pledge he had received; but Noah, by Gabriel's advice, committed the pledge to the flames, and scattering its ashes in the air, there arose out of them Fleas, Flies, Lice, Bugs, and all such sort of vermin as prey upon human blood, and after this fashion was Noah's pledge redeemed."[1]

The Sandwich Islanders have the following tradition in regard to the introduction of Fleas into their country: Many years ago a woman from Waimea went out to a ship to see her lover, and as she was about to return, he gave her a bottle, saying that there was very little valuable property (*waiwai*) contained in it, but that she must not open it, on any account, until she reached the shore. As soon as she gained the beach, she eagerly uncorked the bottle to examine her treasure, but nothing was to be discovered,—the Fleas hopped out, and "they have gone on hopping and biting ever since."[2]

Our pigmy tormentor, *Pulex irritans*, in the opinion of some, seems to have been regarded as an agreeable rather than a repulsive object. "Dear Miss," said a lively old lady to a friend of Kirby and Spence (who had the misfortune to be confined to her bed by a broken limb, and was complaining that the Fleas tormented her), "don't you like *Fleas?* Well, I think they are the prettiest little merry things in the world.—I never saw a dull Flea in all my life."[3] Dr. Townson, as mentioned by the above writers, from the encomium which he bestows upon these vigilant little vaulters, as supplying the place of an alarum and driving us from the bed of sloth, should seem to have regarded them with the same happy feelings.[4]

When Ray and Willughby were traveling, they found "at

---

[1] *Hist. of Ins* (Murray, 1838), ii. 312.
[2] Jenkin's *Voy. of the U. S. Explor. Exped.*, p. 385.
[3] *Introd.*, i. 100.   [4] *Ibid.*

Venice and Augsburg Fleas for sale, and at a small price too, decorated with steel or silver collars around their necks, of which Willughby purchased one. When they are kept in a box amongst wool or cloth, in a warm place, and fed once a day, they will live a long time. When they begin to suck they erect themselves almost perpendicularly, thrusting their sucker, which originates in the middle of the forehead, into the skin. The itching is not felt immediately, but a little afterwards. As soon as they are full of blood, they begin to void a portion of it, and thus, if permitted, they will continue for many hours sucking and voiding. After the first itching no uneasiness is subsequently felt. Willughby's Flea lived for three months by sucking in this manner the blood of his hand; it was at length killed by the cold of winter."[1]

We read in Purchas's Pilgrims that a city of the Miantines is said to have been dispeopled by Fleas;[2] and Messrs. Lewis and Clarke, who found these insects more tormenting than all the other plagues of the Missouri country, say they sometimes here compel even the natives to shift their quarters.[3]

Dr. Clarke was informed by an Arab Sheikh that "the king of the Fleas held his court at Tiberias."[4]

To prevent Fleas from breeding, Pliny gives the following curious recipe: "Since I have made mention of the cuckow," says this writer, "there comes into my mind a strange and miraculous matter that the said magicians report of this bird; namely, that if a man, the first time that he heareth her to sing, presently stay his right foot in the very place where it was when he heard her, and withal mark out the point and just proportion of the said foot upon the ground as it stood, and then digg up the earth under it within the said compasse, look what chamber or roume of the house is strewed with the said mould, there will no Fleas bread there."[5]

Thomas Hill, in his Naturall and Artificiall Conclusions,

---

[1] Ray, *Hist. of Ins.*, p. 8.

[2] *Pilgr.*, iii. 997.

Myas, a principal city of Ionia, was abandoned on account of Fleas.—*Wanley's Wonders*, ii. 507.

[3] K. and S. *Introd.*, i. 100.

[4] *Travels*, vol. ii.

[5] *Nat. Hist.*, xxx. 10. Holl. *Trans.*, p. 387.

printed 1650, quotes this passage from Pliny, calling it "A very easic and merry conceit to keep off fleas from your beds or chambers."[1]

The Hungarian shepherds grease their linen with hogs' lard, and thus render themselves so disgusting even to the Fleas and Lice, as to put them effectually to flight.[2]

There is still shown in the Arsenal at Stockholm a diminutive piece of ordnance, four or five inches in length, with which, report says, on the authority of Linnæus, the celebrated Queen Christiana used to cannonade Fleas.[3]

But, seriously, if you wish for an effectual remedy, that prescribed by old Tusser, in his Points of Goode Husbandry, in the following lines, will answer your purpose:

> While wormwood hath seed, get a handfull or twaine,
> To save against March, to make flea to refraine:
> Where chamber is sweeped and wormwood is strown,
> No flea for his life dare abide to be known.

The inhabitants of Dalecarlia place the skins of hares in their apartments, in which the Fleas willingly take refuge, so that they are easily destroyed by the immersion of the skin in scalding water.[4]

Pamphilius among others gives the following remedies against Fleas: If a person, he says, sets a dish in the middle of the house, and draws a line around it with an iron sword (it will be better if the sword has done execution), and if he sprinkles the rest of the house, excepting the place circumscribed, with an irrigation of staphisagria, or of powdered leaves of the bay-tree, they having been boiled in brine or in sea-water, he will bring all the Fleas together into the dish. A jar also being set in the ground with its edge even with the pavement, and smeared with bulls' fat, will attract all the Fleas, even those that are in the wardrobe. If you enter a place where there are Fleas, express the usual exclamation of distress, and they will not touch you. Make a small trench under a bed, and pour goats' blood into it, and it will bring all the Fleas together, and it will allure those from your clothing. Fleas may be removed

---

[1] Brand's *Pop. Antiq.*, ii. 198.
[2] K. and S. *Introd.*, i. 101.
[3] *Lach. Lapp.*, ii. 32, note.
[4] *Hist. of Ins.*, iii. 319, Murray, 1838.

also, concludes this writer, from the most villous and from the thickest pieces of tapestry, whither they betake themselves when full, if goats' blood is set in a vessel or in a cork.[1]

Moufet says: "A Gloeworm, set in the middle of the house, drives away Fleas."[2]

On the subject of destroying Fleas, the following pleasant piece of satire, by Poor Humphrey, will be read with a smile: "A notable projector became notable by one project only, which was a certain specific for the killing of fleas, and it was in form of a powder, and sold in papers, with plain directions for use, as followeth: The flea was to be held conveniently between the thumb and finger of the left hand; and to the end of the trunk or proboscis, which protrudeth in the flea, somewhat as the elephant's doth, a very small quantity of the powder was to be put from between the thumb and finger of the right hand. And the deviser undertook, if any flea to whom his powder was so administered should prove to have afterwards bitten a purchaser who used it, then that purchaser should have another paper of the said powder gratis. And it chanced that the first paper thereof was bought idly, as it were, by an old woman, and she, without meaning to injure the inventor, or his remedy, but, of her mere harmlessness, did innocently ask him, whether, when she had caught the flea, and after she had got it, as before described, if she should kill it with her nail it would not be as well. Whereupon the ingenious inventor was so astonished by the question, that, not knowing what to answer on the sudden occasion, he said with truth to this effect, that without doubt her way would do, too. And according to the belief of Poor Humphrey, there is not as yet any device more certain or better for destroying a flea, when thou hast captured him, than the ancient manner of the old woman's, or instead thereof, the drowning of him in fair water, if thou hast it by thee at the time."[3]

The old English hunters report that foxes are full of Fleas, and they tell the following queer story how they get rid of them: "The fox," say they, as recorded by Mouffet, "gathers

---

[1] Owen's *Geoponika*, ii. 155-6.
[2] *Theatr. Ins.*, p. 277. Topsel's *Hist. of Beasts*, p. 1103.
[3] *Hist. of Ins.*, ii. 318. Murray, 1838.

some handfuls of wool from thorns and briars, and wrapping it up, he holds it fast in his mouth, then goes by degrees into a cold river, and dipping himself close by little and little, when he finds that all the Fleas are crept so high as his head for fear of drowning, and so for shelter crept into the wool, he barks and spits out the wool, full of Fleas, and so very froliquely being delivered from their molestation, he swims to land."[1]

Ramsay thus alludes to this story:

> Then sure the lasses, and ilk gaping coof,
> Wad rin about him, and had out their loof.
> *M.* As fast as fleas skip to the tale of woo,
> Whilk slee Tod Lowrie (the fox) hads without his mow,
> When he to drown them, and his hips to cool,
> In summer days slides backward in a pool.[2]

Preceding this story, Mouffet makes the following observations: "The lesser, leaner, and younger they are, the sharper they bite, the fat ones being more inclined to tickle and play; and then are not the least plague, especially when in greater numbers, since they molest men that are sleeping, and trouble wearied and sick persons; from whom they escape by skipping; for as soon as they find they are arraigned to die, and feel the finger coming, on a sudden they are gone, and leap here and there, and so escape the danger; but so soon as day breaks, they forsake the bed. They then creep into the rough blankets, or hide themselves in rushes and dust, lying in ambush for pigeons, hens, and other birds, also for men and dogs, moles and mice, and vex such as passe by."[3]

It is frequently affirmed that asses are never troubled with Fleas or other vermin; and, among the superstitious, it is said that it is all owing to the riding of Christ upon one of these animals.[4]

Willsford, in his Nature's Secrets, printed 1658, p. 130, says: "The little sable beast (called a *Flea*), if much thirsting after blood, it argues rain."[5]

It is related that the Devil, teasing St. Domingo in the shape of a Flea, skipped upon his book, when the saint

---

[1] *Theatr. Ins.*, p. 102.
[2] Ramsay's *Poems*, ii. 143.
[3] *Theatre of Insects*, p. 102.
[4] Brookes' *Nat. Hist. of Ins.*, p. 284.
[5] Brand's *Pop. Antiq.*, iii. 204.

fixed him as a mark where he left off, and continued to use him so through the volume.[1]

Fleas infesting beds were attributed to the envy of the Devil.[2]

Giles Fletcher says that Iwan Vasilowich sent to the City of Moscow to provide for him a measure full of Fleas for a medicine. They answered that it was impossible, and if they could get them, yet they could not measure them because of their leaping out. Upon which he set a mulct upon the city of seven thousand rubles.[3]

We read in Purchas's Pilgrims that the Jews were not permitted to burn Fleas in the flame of their lamps on Sabbath evenings.[4]

The muscular power of the Flea is so great that it can leap to the distance of two hundred times its own length, which will appear the more surprising when we consider that a man, were he endowed with equal strength and agility, would be able to leap between three and four hundred yards. Aristophanes, in his usual licentious way, ridicules the great Socrates for his pretended experiments on this great muscular power :

Disciple.   That were not lawful to reveal to strangers.
Strepsiades. Speak boldly then as to a fellow-student ;
             For therefore am I come.
Disc.   Then I will speak ;
             But set it down among our mysteries.
             It is a question put to Chærophon
             By our great master Socrates to answer,
             How many of his own lengths at a spring
             A Flea can hop; for one by chance had skipp'd
             Straight from the brow of Chærophon to th' head
             Of Socrates.
Streps.   And how did then the sage
             Contrive to measure this?
Disc.   Most dext'rously.
             He dipp'd the insect's feet in melted wax,
             Which hard'ning into slippers as it cool'd,
             By these computed he the question'd space.
Streps.   O Jupiter, what subtilty of thought ![5]

---

[1] Southey's *Com. Place Bk.*, 2d S. p. 406.
[2] Fosbr. *Encycl. of Antiq.*, ii. 539.
[3] Southey's *Com. Place Bk.*, 4th S. p. 470.
[4] *Pilgr* , v. 192.
[5] Aristoph. *Clouds*, A. i. Sc. 2.

The witty Butler has also commemorated the same circumstance in his justly celebrated poem of Hudibras:

How many scores a Flea will jump
Of his own length, from head to rump;
Which Socrates and Chærophon
In vain assay'd so long agon.

As illustrative of the strength of the Flea, the following facts may also be given: We read in a note to Purchas's Pilgrims that "one Marke Scaliot, in London, made a lock and key and chain of forty-three links, all which a Flea did draw, and weighed but a grain and a half."[1] Mouffet, who also records this fact, says he had heard of another Flea that was harnessed to a golden chariot, which it drew with the greatest ease.[2] Bingley tells us that Mr. Boverick, an ingenious watchmaker in the Strand, exhibited some years ago a little ivory chaise with four wheels, and all its proper apparatus, and the figure of a man sitting on the box, all of which were drawn by a single Flea. The same mechanic afterward constructed a minute landau, which opened and shut by springs, with the figures of six horses harnessed to it, and of a coachman on the box, a dog between his legs, four persons inside, two footmen behind it, and a postillion riding on one of the fore horses, which were all easily dragged along by a single Flea. He likewise had a chain of brass, about two inches long, containing two hundred links, with a hook at one end and a padlock and key at the other, which a Flea drew nimbly along.[3] At a fair of Charlton, in Kent, 1830, a man exhibited three Fleas harnessed to a carriage in form of an omnibus, at least fifty times their own bulk, which they pulled along with great ease; another pair drew a chariot, and a single Flea a brass cannon. The exhibitor showed the whole first through a magnifying glass, and then to the naked eye; so that all

[1] *Pilg.*, ii. 840, note.
[2] *Ins. Theatr.*, p. 275.
[3] *Anim. Biog.*, iii. 462.
The hand-bill, published by Mr. Boverick, in the Strand, in the year 1745, and another nearly of the same date, ran thus: " To be seen at MR. BOVERICK's, Watchmaker, at the DIAL, facing Old Round Court, near the New Exchange, in the Strand, at One Shilling each person." Then follows a descriptive list of the articles to be seen, among which are mentioned the above. — Kirby's *Wonderful Museum*, i. 101.

were satisfied there was no deception.[1]   Latrielle also mentions a Flea of a moderate size, which dragged a silver cannon, mounted on wheels, that was twenty-four times its own
weight, and which being charged with gunpowder was fired
off without the Flea appearing in the least alarmed.[2]

It is recorded in Purchas's Pilgrims that an Egyptian
artisan received a garment of cloth of gold for binding a
Flea in a chain.[3]

The Flea is twice mentioned in the Bible, and in both
cases David, in speaking to Saul, applies it to himself as a
term of humility.[4]

A Prussian poet, quoted by Jaeger,[5] gives us the song of
a young Flea who had emigrated to this country from Prussia, and thus expresses his dissatisfaction to his sweetheart:

> Kennst de nunmehr das Land, wo Dorngestripp und Disteln
>   blüh'n,
> Im frost'gen Wald nur eckelhafte Tannenzapfen glüh'n,
> Der Schierling tief, und hoch der Sumach steht,
> Ein rauher Wind vom schwarzen Himmel weht;
> Kennst du es wohl?   O lass uns eilig zieh'n,
> Und schnell zurück in unsre Hiemath flieh'n!

An English prose translation of which is: "Know'st
thou now this country, where only briars and thistles bloom;
where ugly fur-nuts only glow in the icy forest; where down
in the vale the fetid hemlock grows, and on the hills the
poisonous sumach; where heavy winds blow from black
clouds over desolate lands?   Dost thou not know of this
country?   Oh, then, let us fly in haste and return to our
own fatherland!"

"To send one away with a Flea in his ear," is a very old
English phrase, meaning to dismiss one with a rebuke.[6]
"Flea-luggit" is the Scottish—to be unsettled or confused.[7]

There.is a collection of poems called "La Puce des
grands jours de Poitiers"—the Flea of the carnival of Poitiers.   The poems were begun by the learned Pasquier, who

---

[1] *Ins. Misc.*, p. 188.
[2] *Nouv. Dict. d'Hist. Nat.*, xxviii. 249.
[3] *Pilg.*, ii. 840.
[4] 1 Saml. xxiv. 14; xxvi. 20.
[5] *Hist. of Ins.*, p. 310.
[6] Wright's *Provincial Dict.*
[7] Jamieson's *Scottish Dict.*

edited the collection, upon a Flea which was found one morning in the bosom of the famous Catherine des Roches.[1]

During the winter of 1762, at Norwich, England, after a chilling storm of snow and wind that had destroyed many lives, myriads of Fleas were found skipping about on the snow.[2]

To the Pulicidæ belongs also a native of the West Indies and South America, the *Pulex penetrans*, variously named in the countries where it is found, Chigoe, Jigger, Nigua, Tungua, and Pique. According to Stedman, this "is a kind of small sand-flea, which gets in between the skin and the flesh without being felt, and generally under the nails of the toes, where, while it feeds, it keeps growing till it becomes of the size of a pea, causing no further pain than a disagreeable itching. In process of time, its operation appears in the form of a small bladder, in which are deposited thousands of eggs, or nits, and which, if it breaks, produce so many young Chigoes, which, in course of time, create running ulcers, often of very dangerous consequence to the patient ; so much so, indeed, that I knew a soldier the soles of whose feet were obliged to be cut away before he could recover; and some men have lost their limbs by amputation—nay, even their lives—by having neglected in time to root out these abominable vermin. The moment, therefore, that a redness and itching more than usual are perceived, it is time to extract the Chigoe that occasions them. This is done with a sharp-pointed needle, taking care not to occasion unnecessary pain, and to prevent the Chigoe from breaking in the wound. Tobacco ashes are put into the orifice, by which in a little time the sore is perfectly healed."[3] The female slaves are generally employed to extract these pests, which they do with uncommon dexterity. Old Ligon tells us he had ten Chigoes taken out of his feet in a morning "by the most unfortunate Yarico,"[4] whose tragical story is now so celebrated in prose and verse. Mr. Southey says that many of the first settlers of Brazil, before they knew the remedy to extract the Chigoes, lost their feet in the most dreadful manner.[5]

---

[1] D'Israeli, *Curios. of Lit.*, i. 339.
[2] *Gent. Mag*, xxxii. 208.
[3] Stedman's *Surinam.*
[4] *Hist. of Barbados*, p. 65.
[5] *Hist. of Brazil*, i. 326.

Walton, in his Present State of the Spanish Colonies, tells us of a Capuchin friar, who carried away with him a colony of Chigoes in his foot as a present to the Scientific Colleges in Europe; but, unfortunately for himself and for science, the length of the voyage produced mortification in his leg, that it became necessary to cut it off to save the zealous missionary's life, and the leg, with all its inhabitants, were tumbled together into the sea.[1]

Humboldt observes "that the whites born in the torrid zone walk barefoot with impunity in the same apartment where a European, recently landed, is exposed to the attack of this animal. The *Nigua*, therefore, distinguishes what the most delicate chemical analysis could not distinguish, the cellular membrane and blood of an European from those of a Creole white."[2]

---

[1] Vol. i. p. 128.
[2] *Pers. Narrative*, E. T. v. 101.

ANOPLEURA.

Pediculidæ—Lice.

At Hurdenburg, in Sweden, Mr. Hurst tells us the mode of choosing a burgomaster is this: The persons eligible sit around, with their beards upon a table; a Louse is then put in the middle of the table, and the one, in whose beard this insect first takes cover, is the magistrate for the ensuing year.[1]

Respecting the revenue of Montecusuma, which consisted of the natural products of the country, and what was produced by the industry of his subjects, we find the following story in Torquemada: "During the abode of Montecusuma among the Spaniards, in the palace of his father, Alonzo de Ojeda one day espied in a certain apartment of the building a number of small bags tied up. He imagined at first that they were filled with gold dust, but on opening one of them, what was his astonishment to find it quite full of Lice? Ojeda, greatly surprised at the discovery he had made, immediately communicated what he had seen to Cortes, who then asked Marina and Anguilar for some explanation. They informed him that the Mexicans had such a sense of their duty to pay tribute to their monarch, that the poorest and meanest of the inhabitants, if they possessed nothing better to present to their king, daily cleaned their persons, and saved all the Lice they caught, and that when they had a good store of these, they laid them in bags at the feet of their monarch." Torquemada further remarks, that his reader might think these bags were filled with small worms (gasanillos), and not with Lice; but appeals to Alonzo de

---

[1] Bayle, iii. 484. Southey's *Com. Place Bk.*, 4th S. p. 439.

(316)

Ojeda, and another of Cortes' soldiers, named Alonzo de Mata, who were eye-witnesses of the fact.[1]

Oviedo pretends to have observed that Lice, at the elevation of the tropics, abandon the Spanish sailors that are going to the Indies, and attack them again at the same point on their return. The same is reported in Purchas's Pilgrims.[2] One of the supplementary writers to Cuvier's History of Insects says: "This is an observation that has need of being corroborated by more certain testimonies than we are yet in possession of. But, if true, there would be nothing in the fact very surprising. A degree of considerable heat, and a more abundant perspiration, might prove unfavorable to the propagation of the *Pediculi corporis*. As their skin is more tender, the influence of the air might prove detrimental to them in those burning climates."[3]

We read in Purchas's Pilgrims, that "if Lice doe much annoy the natives of Cambaia and Malabar, they call to them certain Religious and holy men, after their account: and these Observants y will take upon them all those Lice which the other can find, and put them on their head, there to nourish them. But yet for all this lousie scruple, they stick not to coozenage by falese weights, measures, and coyne, nor at usury and lies."[4]

In a side-note to this curious passage, we find: "The like lousie trick is reported in the Legend of S. *Francis*, and in the life of Ignatius, of one of the Jesuitical pillars, by .Moffæus."

Steedman says of the Caffres, that "except an occasional plunge in a river, they never wash themselves, and consequently their bodies are covered with vermin. On a fine day their karosses are spread out in the sun, and as their tormentors creep forth they are doomed to destruction. It often happens that one Caffir performs for another the kind office of collecting these insects, in which case he preserves the entomological specimens, carefully delivering them to the person

---

[1] Bernal Diaz' *Conquest of Mexico*, i. 394, note 54. This story, no doubt, is founded on something like truth, and most probably these bags were filled with the *Coccus cacti*, the Cochineal insect, then unknown to the Spaniards, who might have easily mistaken them in a dried state for Lice.

[2] *Pilg.*, iii. 975.

[3] Cuv. *An. King.*—*Ins.*, i. 163.

[4] *Pilg.*, v. 542.

to whom they originally appertained, supposing, according to their theory, that as they derived their support from the blood of the man from whom they were taken, should they be killed by another, the blood of his neighbor would be in his possession, thus placing in his hands the power of some superhuman influence."[1]

Kolben says the Hottentots eat the largest of the Lice with which they swarm; and that if asked how they can devour such detestable vermin, they plead the law of retaliation, and urge that it is no shame to eat those who would eat them—"They suck our blood, and we devour 'em in revenge."[2]

We are assured in Purchas's Pilgrims, that Lice and "long wormes" were sold for food in Mexico.[3] From this ancient collection of Travels, we learn that when the Indians of the Province of Cuena are infected with Lice, "they dresse and cleanse one another; and they that exercise this, are for the most part women, who eate all that they take, and have herein (eating?) such dexterity by reason of their exercise, that our own men cannot lightly attaine thereunto."[4]

The Budini, a people of Scythia, commonly feed upon Lice and other vermin bred upon their bodies.[5]

Mr. Wafer, in his description of the Isthmus of America, says: "The natives have Lice in their Heads, which they feel out with their Fingers, and eat as they catch them."[6] Dobrizhoffer also mentions that Lice are eaten by the Indian women of South America.[7]

The disgusting practice of eating these vermin is not confined to the Hottentots, the Negroes of Western Africa, the Simiæ, and the American Indians, for it has been observed to prevail among the beggars of Spain and Portugal.[8]

Schroder, in his History of Animals that are useful in Physic, says: "Lice are swallowed by country people

---

[1] *Wand. and Adv. in S. Africa*, i. 266.

[2] Kolb. *Trav.*, ii. 179.   Astley's *Col. of Voy. and Trav.*, iii. 352.

[3] *Pilg.*, iii. 1133.

[4] *Ibid.*, iii. 975.

[5] Wanley's *Wonders*, ii. 373.

[6] Dampier's *Voy.*, iii. 331.   Lond. 1729.

[7] Dobriz., ii: 396.   Southey's *Com. Place Bk.*, 2d S. p. 527.

[8] Cuvier, *An. Kingd.—Ins.*, i. 163.

against the jaundice."[1] As a specific against this disease, Beaumont and Fletcher thus allude to them :

Die of the jaundice, yet have the cure about you; lice, large lice, begot of your own dust and the heat of the brick kilns.[2]

Lice were also made use of in cases of Atrophy, and Dioscorides says they were employed in suppressions of urine, being introduced into the canal of the urethra.[3]

In the Gentleman's Magazine for 1746, there is a curious letter on "a certain *creature*, of rare and extraordinary qualities"—a Louse, containing many humorous observations on this "*lover* of the human race," and concluding with some queries as to its origin and pedigree. "Was it," the writer asks, "created within the six days assigned by *Moses* for the formation of all things? If so, where was its habitation? We can hardly suppose that it was quartered on *Adam* or his lady, the neatest, nicest pair (if we believe *John Milton*) that ever joyned hands. And yet, as it disdained to graze the fields, or lick the dust for sustenance, where else could it have had its subsistence?"[4]

In a modern account of Scotland, written by an English gentleman, and printed in the year 1670, we find the following: "In that interval between Adam and Moses, when the Scottish Chronicle commences, the country was then baptized (and most think with the sign of the cross) by the venerable name of Scotland, from Scota, the daughter of Pharaoh, King of Egypt. Hence came the rise and name of these present inhabitants, as their Chronicle informs us, and is not to be doubted of, from divers considerable circumstances; the plagues of Egypt being entailed upon them, that of Lice (being a judgment unrepealed) is an ample testimony, these loving animals accompanied them from Egypt, and remain with them to this day, never forsaking them (but as rats leave a house) till they tumble into their graves."[5]

Linnæus, seemingly very anxious to become an apologist for the Lice, gravely observes that they probably preserve

---

[1] Southey's *Com. Place Bk.*, 4th S. p. 430.
[2] *Thierry and Theod.*, A. v. Sc. 1.
[3] James's *Med. Dict.*
[4] *Gent. Mag.*, xvi. 534.
[5] *Harleian Miscel.*, vii. 435.

children who are troubled with them, from a variety of complaints to which they would be liable![1]

As an attempt toward discovering the intention of Providence in permitting the frequency of these tormenting animals, the following lines of Serenus may be given:

> See nature, kindly provident ordain
> Her gentle stimulants to harmless pain;
> Lest Man, the slave of rest, should waste away
> In torpid slumber life's important day!

Of the horrible disease, Phthiriasis, occasioned by myriads of Lice, *Pediculi*, and sometimes by Mites, *Acari*, and *Larvæ* in general, I shall but mention that the inhuman Pheretrina, Antiochus Epiphanes, the Dictator Sylla, the two Herods, the Emperor Maximin, and Philip the Second were among the number carried off by it.

Quintus Serenus speaks thus of the death of Sylla:

> Great Sylla too the fatal scourge hath known;
> Slain by a host far mightier than his own.

According to Pliny, Nits are destroyed by using dog's fat, eating serpents cooked like eels, or else taking their sloughs in drink.[2]

In Leyden's Notes to Complaynt of Scotland are recorded the following few rhymes of the Gyre-carlin—the bug-bear of King James V.

> The Mouse, the Louse, and Little Rede,
> Were a' to mak' a gruel in a lead.

The two first associates desire Little Rede to go to the door, to "see what he could see." He declares that he saw the gyre-carlin coming,

> With spade, and shool, and trowel,
> To lick up a' the gruel.

Upon which the party disperse:

> The Louse to the claith,
>   And the Mouse to the wa',
> Little Rede behind the door,
>   And licket up a'.[3]

---

[1] Shaw, *Zool.*, vi. 454.
[2] *Nat. Hist.*, xxix. 6 (75).
[3] Chambers' *Pop. Rhymes of Scotl.*, p. 282-3. Edit. of 1841, p. 243.

ARACHNIDA.[1]

Acaridæ—Mites.

THE white spot on the back of a certain species of Wood-tic (*Acarus*) is said to be the spot where the pin went through the body when Noah pinned it in the Ark to keep it from troubling him.

Phalangidæ—Daddy-Long-legs.

A superstition obtains among our cow-boys that if a cow be lost, its whereabouts may be learned by inquiring of the Daddy-Long-legs (Phalangium), which points out the direction of the lost animal with one of its fore legs.

In England, the Phalangium has been christened the Harvest-man, from a superstitious belief that if it be killed there will be a bad harvest.[2]

Pedipalpi—Scorpions.

Concerning the generation of the Scorpion, Topsel, in his History of Four-footed Beasts and Serpents, printed in 1658, treats as follows:

"Now, then, it followeth that we inquire about the manner of their (Scorpions') breed or generation, which I find to be double, as divers authors have observed, one way is

---

[1] Properly the second *Class* of the sub-kingdom *Articulata*.
[2] Chambers' *Book of Days*, i. 687.

by putrefaction, and the other by laying of egges, and both these ways are consonant to nature, for Lacinius writeth that some creatures are generated only by propagation of seed — such are men, vipers, whales, and the palm-tree; some again only by putrefaction, as mice, Scorpions, Emmets, Spiders, purslain, which, first of all, were produced by putrefaction, and since their generation are conserved by the seed and egges of their own kinde. Now, therefore, we will first of all speak of the generation of Scorpions by putrefaction, and afterward by propagation.

" Pliny saith[1] that when Sea-crabs dye, and their bodies are dried upon the earth, when the sun entereth into Cancer and Scorpius, out of the putrefaction thereof ariseth a Scorpion; and so out of the putrefied body of the crefish burned arise Scorpions, which caused Ovid thus to write:

> Concava littoreo si demas brachia cancro,
> Cætera supponas terræ, de parte sepulta
> Scorpius exibit, caudaque minabitur unca.

And again:

> Obrutus exemptis cancer tellure lacertis,
> Scorpius exiguo tempore factus erit.

In English thus:

> If that the arms you take from Sea-crab-fish,
> And put the rest in earth till all consumed be,
> Out of the buried part a Scorpion will arise,
> With hooked tayl doth threaten for to hurt thee.

"And therefore it is reported by Ælianus that about Estamenus, in India, there are abundance of Scorpions generated only by corrupt rain-water standing in that place. Also out of the Basalisk beaten into pieces and so putrefied are Scorpions engendered. And when as one had planted the herb basilica on a wall, in the room or place thereof he found two Scorpions. And some say that if a man chaw in his mouth fasting this herb basill before he wash, and afterward lay the same abroad uncovered where no sun cometh at it for the space of seven nights, taking it in all the daytime, he shall at length finde it transmuted into a Scorpion, with a tayl of seven knots.[2]

---

[1] *Nat. Hist.*, xx. 12.
[2] Cf. Pliny. x. 12; and Moufet's *Theatr. Ins.*, p. 205.

" Hollerius,[1] to take away all scruple of this thing, writeth that in Italy in his dayes there was a man that had a Scorpion bred in his brain by continuall smelling to this herb basill ; and Gesner, by relation of an apothecary in France, writeth likewise a story of a young maid who, by smelling to basill, fell into an exceeding headache, whereof she died without cure, and after death, being opened, there were found little Scorpions in her brain.

"Aristotle remembreth an herb which he calleth sissimbria, out of which putrefied Scorpions are engendered, as he writeth. And we have shewed already, in the history of the Crocodile, that out of the Crocodile's egges do many times come Scorpions, which at their first egression do kill their dam that hatched them, which caused Archelaus, which wrote epigrams of wonders unto Ptolemæus, to sing of Scorpions in this manner :

> In vos dissolvit morte, et redigit crocodilum
> Natura extinctum, Scorpii omnipotens.

Which may be Englished thus :

> To you by Scorpions death the omnipotent
> Ruines the crocodil in nature's life extinct." [2]

The remarks referred to by Topsel in the last paragraph in his history of the Crocodile are as follows :

" It is said by Philes that, after the egge is laid by the crocodile, many times there is a cruel Stinging Scorpion which cometh out thereof, and woundeth the crocodile that laid it.[3]

" The Scorpion also and the crocodile are enemies one to

---

[1] B. i. ch. 1.

[2] *Hist. of Four-footed Beasts and Serpents*, p. 753.—Scorpions are bred "from the carkass of the crocodile, as Antigonus affirms, *lib. de mirab. hist. cong.* 24. For in Archelaus there is an epigram of a certain Egyptian in these words:

> In vos dissolvit morte, et redigit crocodilum,
> Natura extinctum (Scorpioli) omniparens.

In English :

> The carkass of dead crocodiles is made the feed,
> By common nature, whence Scorpions breed."

Moufet's *Theatr. Ins.*, p. 208. Topsel's *Trans.*, p. 1052.

[3] *Qua supra*, p. 685.

the other, and therefore when the Egyptians will describe
the combat of two notable enemies, they paint a crocodile
and a Scorpion fighting together, for ever one of them killeth another; but if they will decipher a speedy overthrow
to one's enemy, then they picture a crocodile; if a slow and
slack victory, they picture a Scorpion."[1]

"Some maintain," says Moufet, "that they (Scorpions)
are not bred by copulation, but by exceeding heat of the
sun.   Ælian, *lib.* 6, *de Anim. cap.* 22, among whom Galen
must first be blamed, who in his Book *de fœt. form.* will
not have nature, but chance to be the parent of Scorpions,
Flies, Spiders, Worms of all sorts, and he ascribes their beginning to the uncertain constitutions of the heavens, place,
matter, heat, etc."[2]

Topsel further says: "The principall of all other subjects of their (the Scorpions') hatred are virgins and women,
whom they do not only desire to harm, but also when they
have harmed are never perfectly recovered. (Albertus) . . .

"The lion is by the Scorpion put to flight wheresoever he
seeith it, for he feareth it as the enemy of his life, and therefore writeth S. Ambrose, *Exiguo Scorpionis aculeo exagi-
tatur leo*, the lion is much moved at the small sting of a
Scorpion."[3]

Naude tells us that there is a species of Scorpions in
Italy, which are so domesticated as to be put between sheets
to cool the beds during the heat of summer.[4]   Pliny mentions that the Scorpions of Italy are harmless.[5]

Among the curious things recorded by Pliny concerning
the Scorpion, the following have been selected: Some
writers, he says, are of opinion that the Scorpion devours
its offspring, and that the one among the young which is the
most adroit avails itself of its sole mode of escape by placing
itself on the back of the mother, and thus finding a place
where it is in safety from the tail and the sting.   The one
that thus escapes, they say, becomes the avenger of the rest,

---

[1] *Qua supra*, p. 689.
[2] *Ibid.*, p. 207.   Topsel's *Trans.*, p. 1051.
[3] *Ibid.*, p. 754.
[4] Andrew's *Anecdotes*, p. 427.
[5] *Nat. Hist.*, xi. 25.   Pliny here probably alludes to the Panorpis,
or Scorpion-fly, the abdomen of which terminates in a forceps,
which resembles the tail of the Scorpion.

and at last, taking advantage of its elevated position, puts its parent to death.[1]

According to Pliny, those who carry the plant "tricoccum," or, as it is also called, "scorpiuron,"[2] about their person are never stung by a Scorpion, and it is said, he continues, that if a circle is traced on the ground around a Scorpion with a sprig of this plant, the animal will never move out of it, and that if a Scorpion is covered with it, or even sprinkled with the water in which it has been steeped, it will die that instant.[3]

Attalus assures us, says Pliny, that if a person, the moment he sees a Scorpion, says "Duo,"[4] the reptile will stop short and forbear to sting.[5]

Concerning Scorpions, Diophanes, contemporary with Cæsar and Cicero, has collected the following several opinions of the more ancient writers: If you take a Scorpion, he says, and burn it, the others will betake themselves to flight: and if a person carefully rubs his hands with the juice of radish, he may without fear and danger take hold of Scorpions, and of other reptiles: and radishes laid on Scorpions instantly destroy them. You will also cure the bite of a Scorpion, by applying a silver ring to the place. A suffumigation of sandarach[6] with galbanum, or goat's fat, will drive away Scorpions and every other reptile. If a person will also boil a Scorpion in oil, and will rub the place bit by a Scorpion, he will stop the pain.[7] But Apuleius says, that if a person bit by a Scorpion sits on an ass, turned toward its tail, that the ass suffers the pain, and that it is destroyed.[8] Democritus says that a person bit by a Scorpion, who instantly says to his ass, "A Scorpion has bit me," will suffer no pain, but it passes to the ass.[9] The newt

---

[1] *Nat. Hist.*, xi. 25.

[2] "Scorpion's tail." Dioscorides gives this name to the Helioscopium, or great Heliotropium.

[3] *Nat. Hist.*, xxii. 29.

[4] "Two."

[5] *Nat. Hist.*, xxviii. 5.

[6] The red arsenic of the Greeks was called by this name.—*Matthiol,* vi. 81.

[7] This prescription is given at the present day in Italy and the Levant.

[8] Zoroaster also mentions this. Vide Owen's *Geoponika*, ii. 194.

[9] Pliny relates the same story, *Nat. Hist.*, xxviii. 10 (42); also Zoroaster, *qua supra.*

has an antipathy to the Scorpion: if a person, therefore, melts a newt in oil, and applies the oil to the person that is bitten, he frees him from pain. The same author also says that the root of a rose-tree being applied, cures persons bit by Scorpions. Plutarch recommends to fasten small nuts to the feet of the bed, that Scorpions may not approach it. Zoroaster says that lettuce-seed, being drunk with wine, cures persons bit by Scorpions. Florentinus says, if one applies the juice of the fig to the wound of a person just bitten, that the poison will proceed no farther; or, if the person bit eat squill, he will not be hurt, but he will say that the squill is pleasant to his palate. Tarentinus also says that a person holding the herb sideritis may take hold of Scorpions, and not be hurt by them.[1] Dioscorides, among many other remedies for the sting of the Scorpion, prescribes "a fish called *Lacerta*, salted and cut in pieces; the barbel fish cut in two; the flesh of a fish called *Smaris*; house-mice cut asunder; horse or ass dung; the shell of an Indian small nut; ram's flesh burnt; mummie, four grains, with butter and cow's milk; a broiled Scorpion eaten; river-crabs raw and bruised, and drank with asses' milk: locusts broiled and eaten," etc. Rabby Moyses prescribes pigeon's dung dried; Constantinus, hens' dung, or the heart applied outwardly; Anatolius, crows' dung; Averrhois, the bezoar-stone; Monus, silver; Silvaticus, from Serapis, pewter; and Orpheus, coral.

"Quintus Serenus writes thus, and adviseth:

> These are small things, but yet their wounds are great,
> And in pure bodies lurking do most harm,
> For when our senses inward do retreat,
> And men are fast asleep, they need some charm,
> The Spider and the cruel Scorpion
> Are wont to sting, witnesse great Orion,
> Slayn by a Scorpion, for poysons small
> Have mighty force, and therefore presently
> Lay on a Scorpion bruised. to recall
> The venome, or sea-water to apply
> Is held full good, such virtue is in brine,
> And 'tis approved to drink your fill of wine.

"And Macer writes of houseleek thus:

---

[1] Owen's *Geoponika*, ii 146-8.

> Men say that houseleek hath so soveraign a might,
> Who carries but that, no Scorpion can him bite."[1]

The natives of South Africa, when bitten by a Scorpion, apply, as a remedy, a living frog to the wound, into which animal it is supposed the poison is tranferred from the wound, and it dies; then they apply another, which dies also: the third perhaps only becomes sickly, and the fourth no way affected. When this is observed, the poison is considered to be extracted, and the patient cured. Another method is to apply a kidney, scarlet, or other bean, which swells; then apply another and another, till the bean ceases to be affected, when they consider the poison extracted.[2]

There is a vast desert tract, says Pliny, on this side of the Ethiopian Cynamolgi—the "dog-milkers"—the inhabitants of which were exterminated by Scorpions and venomous ants.[3]

Navarette tells us, in the account of his voyage to the Philippine Islands, that there was there in practice a good and easy remedy against the Scorpions which abound in that country. This was, when they went to bed, to make a commemoration of St. George. He himself, he says, for many years continued this devotion, and, "God be praised," he adds, "the Saint always delivered me both there and in other countries from those and such like insects." He confesses, however, they used another remedy besides, which was to rub all about the beds with garlic.[4]

Navarette[5] and Barbot[6] both tell us that a certain remedy against the sting of a Scorpion, is to rub the wound with a child's private member. This, the latter adds, immediately takes away the pain, and then the venom exhales. The moisture that comes from a hen's mouth, Barbot says, is also good for the same.

The Persians believe that Scorpions may be deprived of the power of stinging, by means of a certain prayer which they make use of for that purpose. The person who has the power of "binding the Scorpion," as it is called, turns his

---

[1] Moufet's *Theatr. Ins.*, 210–215. Topsel's *Hist. of Beasts and Serpents*, p. 1053-7.

[2] Campbell's *Travels in S. Africa*, p. 325.

[3] *Nat. Hist.*, viii. 29 (43).

[4] Churchill's *Col. of Voy. and Trav.*, i. 212.

[5] *Ibid.*   [6] *Ibid.*, v. 221.

face toward the sign Scorpio, in the heavens, and repeats this prayer; while every person present, at the conclusion of a sentence, claps his hands. After this is done they think that they are perfectly safe; nor, if they should chance to see any Scorpions during that night, do they scruple to take hold of them, trusting to the efficacy of this fancied all-powerful charm. "I have frequently seen," says Francklin, "the man in whose family I lived, repeat the above-mentioned prayer, on being desired by his children to bind the Scorpions; after which the whole family has gone quietly and contentedly to bed, fully persuaded that they could receive no hurt by them."[1]

Bell says the Persians "have such a dread of these creatures, that, when provoked by any person, they wish a Kashan Scorpion may sting him."[2]

An old story is, that a Scorpion surrounded with live coals, finding no method of escaping, grows desperate from its situation, and stings itself to death. This, though considered a mere fable of antiquity, may still have some truth, if we believe the following from the pen of Ulloa: "We more than once," says this traveler, "entertained ourselves with an experiment of putting a Scorpion into a glass vessel, and injecting a little smoke of tobacco, and immediately by stopping it found that its aversion to this smell is such, that it falls into the most furious agitations, till, giving itself several stings on the head, it finds relief by destroying itself."[3] There is also told a story in the East Indies, that "the Scorpion is sometimes so pestered with the pismires, that he stings himself to death in the head with his tail, and so becomes a prey to the pismires."[4]

The Scorpion was an emblem of the Egyptian goddess Selk; and she is usually found represented with this animal bound upon her head.[5]

Ælian mentions Scorpions of Coptos, which, though inflicting a deadly sting, and dreaded by the people, so far respected the Egyptian goddess Isis, who was particularly worshiped in that city, that women, in going to express

---

[1] Pinkerton's *Col. of Voy. and Trav.*, ix. 261.
[2] *Ibid.*, vii. 298.
[3] *Ibid.*, xiv. 348.
[4] Churchill's *Coll. of Voy. and Trav.*, ii. 316.
[5] Wilkinson's *Anct. Egypt.*, v. 52, 254.

their grief before her, walked with bare feet, or lay upon the ground, without receiving any injury from them.[1]

The Ethiopians that dwell near the River Hydaspis commonly eat Scorpions and serpents without the slightest harm, "which certainly proceeds from no other thing than a secret and wonderful constitution of the body!" says Mercurialis.[2]

Lutfullah, the learned Mohammedan gentleman, in his Autobiography, relates the following:

"On the morning of the 11th (April, 1839), I ordered my servant boy to shake my bedding and put it in the sun for an hour or so, that the moisture imbibed by the quilt might be dried. As soon as the quilt was removed from its place, what did I behold but an immense Scorpion, tapering towards its tail of nine vertebræ, armed with a sting at the end, crawling with impunity at the edge of the carpet. I had never seen such a large monster before. It was black in the body, with small bristles all over, dark green in the tail, and red at the sting. This hideous sight rendered me and the servant horror-struck. In the mean time, an Afghan friend of mine, by name Ata Mohamed Khan Kakar, a respectable resident of the town, honoured me with a visit, and, seeing the reptile, observed, 'Lutfullah, you are a lucky man, having made a narrow escape this morning. This cursed worm is called Jerrara, and its fatal sting puts a period to animal life in a moment; return, therefore, your thanks to the Lord, all merciful, who gave you a new life in having saved you from the mortal sting of this evil bed-companion of yours.' 'I have no fear of the worm,' replied I, 'for it dare not sting me unless it is written in the book of fate to be stung by it.' Saying this, I made the animal crawl into a small earthen vessel, and stopped the mouth of it with clay; and then making a large fire, I put the vessel therein for an hour or so, to turn the reptile into ashes, which, administered in doses of half a grain to adults, are a specific remedy for violent colicky pains."[3]

The ashes of burnt Scorpions, besides being good for colicky pains, as Lutfullah says, were often prescribed by the ancient physicians for stone in the bladder;[4] and Topsel,

---

[1] Ælian, xvi. 41, and xii. 38. Wilkinson's *Anct. Egypt.*, v. 254.
[2] Wanley's *Wonders*, ii. 459.
[3] *Autobiog.*, Lond. 1858, p. 304–5.
[4] Prescribed by Galen, Pliny, Lanfrankus, etc.

quoting Kiranides, has the following: "If a man take a
vulgar Scorpion and drown the same in a porringer of oyl
in the wane of the moon, and therewithall afterward anoynt
the back from the shoulders to the hips, and also the head
and forehead, with the tips of the fingers and toes of one
that is a demoniack or a lunatick person, it is reported, that
he shall ease and cure him in short time.   And the like is
reported of the Scorpion's sting joyned with the top of basil
wherein is seed, and with the heart of a swallow, all included in a piece of harts skin."[1]   The oil of Scorpions,
Brassavolus says, "drives out worms miraculously;" and
oil of Scorpions' and vipers' "tongues is a most excellent
remedy against the plague, as Crinitus testifies, i. 7."[2]
Galen prescribes Scorpions for jaundice, and Kiranides the
same for the several kinds of ague.   "Plinius Secundus saith,
that a quartan ague, as the magicians report, will be cured
in three daies by a Scorpion's four last joynts of his tail, together with the gristle of his ear, so wrapped up in a black
cloth, that the sick patient may neither perceive the Scorpion that is applied, nor him that bound it on . . . . . Sa-
monicus commends Scorpions against pains in the eyes, in
these verses:

> If that some grievous pain perplex thy sight,
> Wool wet in oyl is good bound on all night.
> Carry about thee a live Scorpion's eye,
> Ashes of coleworts if thou do apply,
> With bruised fankincense, goat's milk, and wine,
> One night will prove this remedy divine."[3]

The following Asiatic fable of the Scorpion and the Tortoise is from the Beharistan of Jamy: A Scorpion, armed
with pernicious sting and filthy poison, undertook a journey.
Coming to the bank of a wide river, he stopped in great
perplexity, wanting height of leg to cross over, yet very un-
willing to return.   A Tortoise, seeing his situation, and
moved with compassion, took him on his back, sprang into
the river, and was swimming toward the opposite shore,
when he heard a noise on his shell as of something striking
him; he called out to know what it was; the ungrateful
Scorpion answered, "It is the motion of my sting only, I

---

[1] *Hist. of Beasts and Serpents*, p. 757.
[2] So also Manardus.—Moufet, p. 210.   Topsel's *Trans.*, p. 1053.
[3] *Ibid.*

know it cannot affect you, but it is a habit which I cannot relinquish." "Indeed," replied the Tortoise, "then I cannot do better than free so evil-minded a creature from his bad disposition, and secure the good from his malevolence." Saying which he dived under the water, and the waves soon carried the Scorpion beyond the bourn of existence.

> When, in this banquet house of vice and strife,
> A knave oft strikes the various stings of fraud,
> 'Tis best the sea of death ingulf him soon,
> That he be freed from man, and man from him.[1]

Topsel, in his History of Four-footed Beasts and Serpents, has the following in his chapter on the Scorpion :

"There is a common adage, *Cornix Scorpium*, a Raven to a Scorpion, and it is used against them that perish by their own inventions: when they set upon others, they meet with their matches, as a raven did when it preyed upon a Scorpion, thus described by Alciatus, under his title *Justa ultio*, just revenge, saying as followeth :

> Raptabat volucer captum pede corvus in auras
> Scorpion, audaci præmia parta gulæ.
> Ast ille infuso sensim per membra venemo,
> Raptorem in stygias compulit ultor aquas.
> O risu res digna ! aliis qui fata parabat,
> Ipse periit, propriis succubuitque dolis.

Which may be Englished thus :

> The ravening crow for prey a Scorpion took
> Within her foot, and therewithal aloft did flie.
> But he empoysoned her by force and stinging stroke,
> So ravener in the Stygian Lake did die.
> O sportfull game ! that he which other for bellyes sake did kill,
> By his own deceit should fall into death's will.

"There be some learned writers, who have compared a Scorpion to an epigram, or rather an epigram to a Scorpion, because as the sting of the Scorpion lyeth in the tayl, so the force and vertue of an epigram is in the conclusion, for *vel acriter salse mordeat, vel jucunde atque dulciter delectet*, that is, either let it bite sharply at the end, or else delight pleasingly."[2]

---

[1] *Asiatic Miscellany*, ii. 451.
[2] Topsel's *Hist. of Beasts and Serpents*, p. 755-6.

Araneidæ—True Spiders.

A little head and body small,
With slender feet and very tall,
Belly great, and from thence come all
The webs it spins.—MOUFET.[1]

"Domitian sometime," says Hollingshed incidentally in his Chronicles of England, "and an other prince yet living, delited so much to see the iollie combats betwixt a stout Flie and an old Spider. . . . . Some parasites also in the time of the aforesaid emperour (when they were disposed to laugh at his follie, and yet would seem in appearance to gratifie his fantasticall head with some shew of dutiful demenour) could devise to set their lorde on worke, by letting a fresh flie privilie into his chamber, which he foorthwith would egerlie have hunted (all other businesse set apart) and never ceased till he had caught him in his fingers: whereupon arose the proverbe 'ne musca quidem,' altered first by Vitius Priscus, who being asked whether anie bodie was with Domitian, answered 'ne musca quidem,' whereby he noted his follie. There are some cockes combs here and there in England, learning it abroad as men transregionate, which make account also of this pastime, as of a notable matter, telling what a fight is seene betweene them, if either of them be lustie and couragious in his kind. One also hath mad a booke of the Spider and the Flie, wherein he dealeth so profoundlie, and beyond all measure of skill, that neither he himself that made it, neither anie one that readeth it can reach unto the meaning thereof."[2]

Chapelain, the author of Pucelle, was called by the academicians the Knight of the Order of the Spider, because he was so avaricious, that though he had an income of 13,000 livres, and more than 240,000 in ready money, he wore an old coat so patched, pieced, and threadbare, that the stitches exhibited no bad resemblance to the fibers produced by that insect. Being one day present at a large party given by the great Condé, a Spider of uncommon size fell from the ceiling upon the floor. The company thought it could not

---

[1] Topsel's *Trans.—Hist. of Beasts and Serpents*, p. 1058.
[2] *Chronicles*, i. 885.

have come from the roof, and all the ladies at once agreed that it must have proceeded from Chapelain's wig ;—the wig so celebrated by the well-known parody.[1]

The often-told anecdote of the Scottish monarch, Robert Bruce, and the cottage Spider, is thus related in Chambers' Miscellany : While wandering on the wild hills of Carrick, in order to escape the emissaries of Edward, Robert the Bruce on one occasion passed the night under the shelter of a poor deserted cottage. Throwing himself down on a heap of straw, he lay upon his back, with his hands placed under his head, unable to sleep, but gazing vacantly upward at the rafters of the hut, disfigured with cobwebs. From thoughts long and dreary about the hopelessness of the enterprise in which he was engaged, and the misfortunes he had already encountered, he was roused to take interest in the efforts of a poor industrious Spider, which had begun to ply its vocation with the first gray light of morning. The object of the animal was to swing itself, by its thread, from one rafter to another; but in the attempt it repeatedly failed, each time vibrating back to the point whence it had made the effort. Twelve times did the little creature try to reach the desired spot, and as many times was it unsuccessful. Not disheartened with its failure, it made the attempt once more, and, lo! the rafter was gained. "The thirteenth time," said Bruce, springing to his feet; "I accept it as a lesson not to despond under difficulties, and shall once more venture my life in the struggle for the independence of my beloved country." The result is well known.[2]

It is related in the life of Mohammed, that when he and Abubeker were fleeing for their lives before the Coreishites, they hid themselves for three days in a cave, over the mouth of which a Spider spread its web, and a pigeon laid two eggs there, the sight of which made the pursuers not go in to search for them.[3]

A similar story is told in the Lives of the Saints, of St. Felix of Nola: "But the Saint," says Butler, "in the mean time had slept a little out of the way, and crept through a

---

[1] Keddie's *Cyclop. of Anecd.*, p. 288.
[2] *Chamb. Misc.*, vol. xi. No. 100. Compare this story with that of Timour and the Ant.
[3] Ockley's *Hist. of the Saracens*, i. 86.

hole in a ruinous old wall, which was instantly closed up by Spiders' webs. His enemies, never imagining anything could have lately passed where they saw so close a Spider's web, after a fruitless search elsewhere, returned in the evening without their prey. Felix finding among the ruins, between two houses, an old well half dry, hid himself in it for six months; and received during that time wherewithal to subsist by means of a devout Christian woman."[1]

It is said of Heliogabalus, that, for the purpose of estimating the magnitude of the City of Rome, he commanded a collection of Spiders to be made.[2]

Illustrative of the singularly pleasurable effect of music upon Spiders, in the Historie de la Musique, et de ses Effets, we find the following relation:

"Monsieur de ———, captain of the Regiment of Navarre, was confined six months in prison for having spoken too freely of M. de Louvois, when he begged leave of the governor to grant him permission to send for his lute to soften his confinement. He was greatly astonished after four days to see at the time of his playing the mice come out of their holes, and the Spiders descend from their webs, who came and formed in a circle round him to hear him with attention. This at first so much surprised him, that he stood still without motion, when having ceased to play, all those Spiders retired quietly into their lodgings; such an assembly made the officer fall into reflections upon what the ancients had told of Orpheus, Arion, and Amphion. He assured me he remained six days without again playing, having with difficulty recovered from his astonishment, not to mention a natural aversion he had for this sort of insects, nevertheless he began afresh to give a concert to these animals, who seemed to come every day in greater numbers, as if they had invited others, so that in process of time he found a hundred of them about him. In order to rid himself of them he desired one of the jailors to give him a cat, which he sometimes shut up in a cage when he wished to have this company and let her loose when he had a mind to dismiss them, making it thus a kind of comedy that alleviated his imprisonment. I long doubted the truth of this story, but it was confirmed to me six months ago by M. P———, intendant

<hr>

[1] *Lives of the Saints*, i 177–8. Cf. Wanley's *Wonders*, ii. 402.
[2] Bucke *on Nature*, ii. 103.

of the duchy of V——————, a man of merit and probity, who played upon several instruments to the utmost excellence. He told me that being at ————, he went into his chamber to refresh himself after a walk, and took up a violin to amuse himself till supper time, setting a light upon the table before him ; he had not played a quarter of an hour before he saw several spiders descend from the ceiling, who came and ranged themselves round about the table to hear him play, at which he was greatly surprised, but this did not interrupt him, being willing to see the end of so singular an occurrence. They remained on the table very attentively till somebody came to tell him that supper was ready, when having ceased to play, he told me these insects remounted to their webs, to which he would suffer no injury to be done. It was a diversion with which he often entertained himself out of curiosity.'[1]

The Abbé Olivet has described an amusement of Pelisson during his confinement in the Bastile for refusing to betray to the government certain secrets intrusted to him by a friend who was a leading politician at the court of Louis XIV., which consisted in feeding a Spider, which he discovered forming its web across the only air-hole of his cell. For some time he placed his flies at the edge of the window, while a stupid Basque, his sole companion, played on a bagpipe. Little by little the Spider used itself to distinguish the sound of the instrument, and issued from its hole to run and catch its prey. Thus calling it always by the same sound, and placing the flies at a still greater distance, he succeeded, after several months, to drill the Spider by regular exercise, so that at length it never failed appearing at the first sound to seize on the fly provided for it, at the extremity of the cell, and even on the knees of the prisoner.[2]

---

[1] *Hist. de la Mus.*, i. 321. Hawkins' *Hist. of Music*, iii. 117, note.

[2] *Biogr. Univers.*, tome xxxiii. See also Arvine's *Anecdotes*, p. 402.

To this account, in the *Hist. of Insects* printed by John Murray, 1830, i. 269, is added: "The governor of the Bastile hearing that this unfortunate prisoner had found a solace in the society of a Spider, paid Pelisson a visit, desiring to see the manœuvres of the insect. The Basque struck up his notes, the Spider instantly came to be fed by his friend ; but the moment it appeared on the floor of the cell, the governor placed his foot on its body, and crushed it to death.''

At a ladies' school at Kensington, England, an immense species of Spider is said to be uncomfortably common; and that when the young ladies sing their accustomed hymn or psalm before morning and evening prayers, these Spiders make their appearance on the floor, or suspended overhead from their webs in the ceiling, obviously attracted by the "concord of sweet sounds."

The following lines "to a Spider which inhabited a cell," [1] are from the Anthologia Borealis et Australis:

In this wild, groping, dark, and drearie cove,
  Of wife, of children, and of health bereft,
I hailed thee, friendly Spider, who hadst wove
  Thy mazy net on yonder mouldering raft:
Would that the cleanlie housemaid's foot had left
  Thee tarrying here, nor took thy life away;
For thou, from out this scare old ceiling's cleft,
  Came down each morn to hede my plaintive lay;
Joying like me to heare sweete musick play,
Wherewith I'd fein beguile the dull, dark, lingering day. [2]

"When the great and brilliant Lauzun was held in captivity, his only joy and comfort was a friendly Spider: she came at his call; she took her food from his finger, and well understood his word of command. In vain did jailors and soldiers try to deceive his tiny companion; she would not obey their voices, and refused the tempting bait from their hand. Here, then, was not only an ear, but a keen power of distinction. The despised little animal listened with sweet affection, and knew how to discriminate between not unsimilar tones." [3]

Quatremer Disjonval, a Frenchman by birth, was an adjutant-general in Holland, and took an active part on the side of the Dutch patriots when they revolted against the Stadtholder. On the arrival of the Prussian army under the Duke of Brunswick, he was immediately taken, tried, and, having been condemned to twenty-five years' imprisonment, was incarcerated in a dungeon at Utrecht, where he remained eight years. During this long confinement, by many curious observations upon his sole companions, Spiders, he discovered that they were in the highest degree

---

[1] *The Mirror*, xxvii. 69.
[2] Hone's *Ev. Day Book*, i 334.
[3] *Stray Leaves from the Book of Nature.*

sensitive of approaching changes in the atmosphere, and that their retirement and reappearance, their weaving and general habits, were intimately connected with the changes of the weather. In the reading of these living barometers he became wonderfully accurate, so much so, that he could prognosticate the approach of severe weather from ten to fourteen days before it set in, which is proven by the following remarkable fact, which led to his release: "When the troops of the French republic overran Holland in the winter of 1794, and kept pushing forward over the ice, a sudden and unexpected thaw, in the early part of December, threatened the destruction of the whole army unless it was instantly withdrawn. The French generals were thinking seriously of accepting a sum offered by the Dutch, and withdrawing their troops, when Disjonval, who hoped that the success of the republican army might lead to his release, used every exertion, and at length succeeded in getting a letter conveyed to the French general in 1795, in which he pledged himself, from the peculiar actions of the Spiders, of whose movements he was enabled to judge with perfect accuracy, that within fourteen days there would commence a most severe frost, which would make the French masters of all the rivers, and afford them sufficient time to complete and make sure of the conquest they had commenced, before it should be followed by a thaw. The commander of the French forces believed his prognostication, and persevered. The cold weather, which Disjonval had predicted, made its appearance in twelve days, and with such intensity, that the ice over the rivers and canals became capable of bearing the heaviest artillery. On the 28th of January, 1795, the French army entered Utrecht in triumph; and Quatremer Disjonval, who had watched the habits of his Spiders with so much intelligence and success, was, as a reward for his ingenuity, released from prison."[1]

In Bartholomæus, De Proprietatibus Rerum (printed by Th. Berthelet, 27th Henry VIII.), lib. xviii. fol. 314, speaking of Pliny, we read: "Also he saythe, spynners (Spiders) ben tokens of divynation and of knowing what wether shal fal, for oft by weders that shal fal, some spin and weve

---

[1] *Quart. Rev.* for Jan. 1814.

higher or lower.   Also he saythe, that multytude of spynners is token of moche reyne."[1]

Willsford, in his Nature's Secrets, p. 131, tells us: "Spiders creep out of their holes and narrow receptacles against wind or rain; Minerva having made them sensible of an approaching storm."[2]

Hone, in his Every Day Book, also mentions that from Spiders prognostications as to the weather may be drawn; and gives the following instructions to read this animal-barometer: "If the weather is likely to become rainy, windy, or in other respects disagreeable, they fix the terminating filaments, on which the whole web is suspended, unusually short; and in this state they await the influence of a temperature which is remarkably variable. On the contrary, if the terminating filaments are uncommonly long, we may, in proportion to their length, conclude that the weather will be serene, and continue so at least for ten or twelve days. But if the Spiders be totally indolent, rain generally succeeds; though, on the other hand, their activity during rain is the most certain proof that it will be only of short duration, and followed with fair and constant weather. According to further observations, the Spiders regularly make some alterations in their webs or nets every twenty-four hours; if these changes take place between the hours of six and seven in the evening, they indicate a clear and pleasant night."[3]

Pausanias tells us that after the slaughter at Chæronea, the Thebans were obliged to place a guard within the walls of their city; but which, however, after the death of Philip, and during the reign of Alexander, they drove out. For this action, this historian continues, it was that Divinity gave them tokens in the webs of Spiders of the destruction that awaited them. For, during the battle at Leuctra, the Spiders in the temple of Ceres Thesmophoros wove white

---

[1] This passage from Pliny is thus translated by Bostock and Riley: "Presages are also drawn from the Spider, for when a river is about to swell, it will suspend its web higher than usual. In calm weather these insects do not spin, but when it is cloudy they do, and hence it is, that a great number of cobwebs is a sure sign of showery weather."—*Nat. Hist.*, xi. 24 (28). *Trans.*, iii. 28.

[2] Brande's *Pop. Antiq.*, iii 223.

[3] *Ev. Day Bk.*, i. 981. Quot. also in Chamb. *Journ.*, 1st Ser., vi. 95.

webs about the doors; but when Alexander and the Macedonians attacked their dominions, their webs were found to be black.[1]

It was thought by the Classical Ancients and the old English unlucky to kill Spiders; and prognostications were made from their manner of weaving their webs.[2] It is still thought unlucky to injure these animals.

Park has the following note in his copy of Bourne and Brande's Popular Antiquities, p. 93: "Small Spiders, termed *money-spinners*, are held by many to prognosticate good luck, if they are not destroyed or injured, or removed from the person on whom they are first observed."

In Teviotdale, Scotland, "when Spiders creep on one's clothes, it is viewed as betokening good luck; and to destroy them is equivalent to throwing stones at one's own head."[3]

In Maryland, this superstition is thus expressed: If you kill a Spider upon your clothing, you destroy the presents they are then weaving for you.

In the Secret Memoirs of Mr. Duncan Campbell, p. 60, in the chapter of omens, we read that "others have thought themselves secure of receiving money, if by chance a little Spider fell upon their clothes."[4]

"When a Spider is found upon your clothes, or about your person," says a writer in the Notes and Queries,[5] "it signifies that you will shortly receive some money. Old Fuller, who was a native of Northamptonshire, thus quaintly moralizes this superstition: 'When a Spider is found upon your clothes, we used to say some money is coming toward us. The moral is this: such who imitate the industry of that contemptible creature may, by God's blessing, weave themselves into wealth and procure a plentiful estate.' "[6]

A South Northamptonshire superstition of the present day is, that, in order to propitiate money-spinners, they must be thrown over the left shoulder.[7]

It is most probable that Euclio, in Plautus' Aulularia,

---

[1] Paus. *Hist. of Greece*, B. 9, c. 6.
[2] Fosbr. *Encycl. of Antiq.*
[3] Jamieson's *Scottish Dict.*
[4] Brande's *Pop. Antiq.*, iii. 223.
[5] N. and Q., iii. 3.
[6] *Worthies*, p. 58. Pt. II. Ed. 1662.
[7] N. and Q., ii. 165.

would not suffer the Spiders to be molested because they were considered to bring good luck.

Staphyla. Here in our house there's nothing else for thieves to gain, so filled is it with emptiness and cobwebs.

Euclio. You hag of hags, I choose those cobwebs to be watched for me [1]

A superstition prevails among us that if a Spider approaches, either by crawling toward or descending from the ceiling to a person, it forebodes good to such person; and, on the contrary, if the Spider runs hurriedly away, it is an omen of bad luck. But if the Spider be a poisonous one, or a Fly catcher, and it approaches you, some evil is about to befall you, which to avert you must cross your heart thrice.

If you kill a Spider crossing your path, you will have bad luck.

A Spider should not be killed in your house, but out of doors; if in the house, our country people say you are " pulling down your house."

If a Spider drops down from its web or from a tree directly in front of a person, such person will see before night a dear friend.

A variety of this superstition is, that, if the Spider be white, it foretells the acquaintance of a friend; and if black, an enemy.

In the Netherlands, a Spider seen in the morning forebodes good luck; in the afternoon, bad luck.[2]

There is a common saying at Winchester, England, that no Spider will hang its web on the roof of Irish oak in the chapel or cloisters;[3] and the cicerone, who shows the cathedral church at St. David's, points out to the visitor that the choir is roofed with Irish oak, which does not harbor Spiders, though cobwebs are plentifully seen in other parts of the cathedral.[4] This superstition (for it certainly is nothing more)[5] probably originated with the old story of St. Patrick's having exorcised and banished all kinds of vermin from Ireland.

The same virtue of repelling Spiders is attributed also to

---

[1] *Aulul.*, A. i. Sc. 3.
[2] Thorpe's *North. Antiq.*, iii. 329.
[3] N. and Q., 2d ed. iv. 298.
[4] *Ibid.*, iv. 377.
[5] *Gent. Mag.*, June, 1771, xli. 251.

chestnut and cedar wood;[1] and the old roof at Turner's Court, in Gloucestershire, four miles from Bath, which is of chestnut, is said to be perfectly free from cobwebs;[2] hence also are the cloisters of New College, and of Christ's Church, in England, roofed with chestnut.[3]

A small Spider of a red color, called a Tainct in England, is accounted, by the country people, a deadly poison to cows and horses; so when any of their cattle die suddenly and swell up, to account for their deaths, they say they have "licked a Tainct." Browne thinks this is, most probably, but a vulgar error.[4]

It is a very ancient and curious belief that there exists a remarkable enmity between the Spider and serpents,[5] and more especially between the Spider and the toad; and many curious stories are told of the combats between these animals. The following, related by Erasmus, which he asserts he had directly from one of the spectators, is probably the most remarkable, and we insert it in the words of Dr. James: "A person (a monk)[6] lying along upon the floor of his chamber in the summer-time to sleep in a supine posture, when a toad, creeping out of some green rushes, brought just before in to adorn the chimney, gets upon his face and with his feet sits across his lips. To force off the toad, says the historian, would have been accounted death to the sleeper; and to leave her there, very cruel and dangerous; so that upon consultation, it was concluded to find out a Spider, which, together with her web and the window she was fastened to, was brought carefully, and so contrived as to be held perpendicularly to the man's face; which was no sooner done but the Spider, discovering his enemy, let himself down and struck in his dart, afterward betaking himself up again to his web: the toad swelled, but as yet kept his station.

---

[1] N. and Q., 2d ed. iv. 523.   [2] *Ibid.*, iv. 421.   [3] *Ibid.*, iv. 298.

[4] *Vulg. Err.*, B. iii. c. 277.   *Works*, ii. 527.

[5] Pliny says the Spider, poised in its web, will throw itself upon the head of a serpent as it lies stretched beneath the shade of the tree where it has built, and with its bite pierce its brain; such is the shock, he continues, that the creature will hiss from time to time, and then, seized with vertigo, coil round and round, while it finds itself unable to take to flight, or so much as to break the web of the Spider, as it hangs suspended above; this scene, he concludes, only ends with its death.—*Nat. Hist.*, x. 95.

[6] Browne's *Works*, ii. 524, note.

The second wound is given quickly after by the Spider, upon which he swells yet more, but remained alive still. The Spider, coming down again by his thread, gives the third blow, and the toad, taking off his feet from over the man's mouth, fell off dead."[1]

The following cosmogony is found in the sacred writings of the Pundits of India: A certain immense Spider was the origin, the first cause of all things; which, drawing the matter from its own bowels, wove the web of this universe, and disposed it with wonderful art; she, in the mean time, sitting in the center of her work, feels and directs the motion of every part, till at length, when she has pleased herself sufficiently in ordering and contemplating this web, she draws all the threads she had spun out again into herself; and, having absorbed them, the universal nature of all creatures vanishes into nothing.[2]

Among the Chululahs of our western coast, Capt. Stuart informs me there is a vague superstition that the Spider is connected with the origin of the world. To what extent this curious notion prevails, or anything more concerning it, I have been unable to learn.

The natives of Guinea, says Bosman, believe that the first men were created by the large black Spider, which is so common in their country, and called in their jargon "Anause;" nor is there any reasoning, continues this traveler, a great number of them out of it.[3] Barbot also remarks that, in the belief of the Guinea negroes, the black Anause created the first man.[4]

That the Spider should be connected with the origin of the world and man in the several beliefs of the Hindoos, Chululahs, and negroes, races so widely different and separated from one another, is a coincidence most remarkable.

A large and hideous species of Spider, said to be only found in the palace of Hampton Court, England, is known by the name of the "Cardinals." This name has been given them from a superstitious belief that the spirits of Car-

---

[1] *Med. Dict.*, sub *Araneus*.

[2] *Univers. Hist.*, i. 48, also *Gent. Mag.*, xli. 400.

[3] *Trav.*, p. 322, and Astley's *Col. of Voy. and Trav.*, ii. 726. Bosman says this "was the greatest piece of ignorance and stupidity he observed in the negroes."

[4] Churchill's *Col. of V. and T.*, v. 222.

dinal Wolsey and his retinue still haunt the palace in their shape.[1]

In running across the carpet in an evening, with the shade cast from their large bodies by the light of the lamp or candle, these "Cardinals" have been mistaken for mice, and have occasioned no little alarm to some of the more nervous inhabitants of the palace.[2]

The story of the gigantic Spider found in the Church of St. Eustace, at Paris, in Chambers' Miscellany, is related as follows: It is told that the sexton of this church was surprised at very often discovering a certain lamp extinguished in the morning, notwithstanding it had been duly replenished with oil the preceding evening. Curious to learn the cause of this mysterious circumstance, he kept watch several evenings, and was at last gratified by the discovery. During the night he observed a Spider, of enormous dimensions, come down the chain by which the lamp was suspended, drink up the oil, and, when gorged to satiety, slowly retrace its steps to a recess in the fretwork above. A similar Spider is said to have been found, in 1751, in the cathedral church of Milan. It was observed to feed also on oil. When killed, it weighed four pounds! and was afterward sent to the imperial museum at Vienna.[3]

The following remarkable anecdote is translated from the French : "M. F—— de Saint Omer laid on the chimney-piece of his chamber, one evening on going to bed, a small shirt-pin of gold, the head of which represented a fly. Next day, M. F—— would have taken his pin from the place where he had put it, but the trinket had disappeared. A servant-maid, who had only been in M. F——'s service a few days, was solely suspected of having carried off the pin, and sent away. But, at length, M. F——'s sister, putting up some curtains, was very much surprised to find the lost pin suspended from the ceiling in a Spider's web! And thus was the disappearance of the *bijou* explained : A Spider, deceived by the figure of the fly which the pin presented, had drawn it into his web."[4]

In the Treasvrie of Avncient and Moderne Times, it is

---

[1] *N. and Q.*, vii. 431.
[2] Chamb *Misc.*, vol. xi. No. 100.   [3] *Ibid.*
[4] *The Mirror*, xxvii. 69.

stated that "Spiders do shun all such wals as run to ruine, or are like to be ouerthrowne."[1]

A Spider hanging from a tree is said to have made both Turenne and Gustavus Adolphus shudder![2]

M. Zimmerman relates the following instance of antipathy to Spiders: "Being one day in an English company," says he, "consisting of persons of distinction, the conversation happened to fall on antipathies. The greater part of the company denied the reality of them, and treated them as old women's tales; but I told them that antipathy was a real disease. Mr. William Matthew, son of the Governor of Barbados, was of my opinion, and, as he added that he himself had an extreme antipathy to Spiders, he was laughed at by the whole company. I showed them, however, that this was a real impression of his mind, resulting from a mechanical effect. Mr. John Murray, afterward Duke of Athol, took it into his head to make, in Mr. Matthew's presence, a Spider of black wax, to try whether this antipathy would appear merely on the sight of the insect. He went out of the room, therefore, and returned with a bit of black wax in his hand, which he kept shut. Mr. Matthew, who in other respects was a sedate and amiable man, imagining that his friend really held a Spider, immediately drew his sword in a great fury, retired with precipitation to the wall, leaned against it, as if to run him through, and sent forth horrible cries. All the muscles of his face were swelled, his eye-balls rolled in their sockets, and his whole body was as stiff as a post. We immediately ran to him in great alarm, and took his sword from him, assuring him at the same time that Mr. Murray had nothing in his hand but a bit of wax, and that he himself might see it on the table where it was placed. He remained some time in this spasmodic state, and I was really afraid of the consequences. He, however, gradually recovered, and deplored the dreadful passion into which he had been thrown, and from which he still suffered. His pulse was exceedingly quick and full, and his whole body was covered with a cold sweat. After taking a sedative, he was restored to his former tranquillity, and his agitation was attended with no other bad consequences."[3]

---

[1] B. 7, c. xv. p. 665. Printed 1613.
[2] Eliz. Cook's *Journ.*, vii. 378.
[3] Wanley's *Wonders*, i. 20.

In Batavia, New York, on the evening of the 13th of September, 1834, Hon. David E. Evans, agent of the Holland Land Company, discovered in his wine-cellar a live striped snake, about nine inches in length, suspended between two shelves, by the tail, by Spiders' web. From the shelves being two feet apart, and the position of the web, the witnesses were of opinion the snake could not have fallen by accident into it, and thus have become inextricably entangled, but that it had been actually captured, and drawn up so that its head could not reach the shelf below by about an inch, by Spiders, and of a species much smaller than the common fly, three of which at night were seen feeding upon it, while it was yet alive.

Hon. S. Cummings, first Judge of the Court of Common Pleas in his county, and also Postmaster of Batavia, and Mr. D. Lyman Beecher have described this phenomenon, and given the names of quite a number of gentlemen who witnessed it, and will testify to the accuracy of their accounts. Says Mr. Cummings: "Upon a critical examination through a magnifying glass, the following curious facts appeared. The mouth of the snake was fast tied up, by a great number of threads, wound around it so tight that he could not run out his tongue. His tail was tied in a knot, so as to leave a small loop, or ring, through which the cord was fastened; and the end of the tail, above this loop, to the length of something over half an inch, was lashed fast to the cord, to keep it from slipping. As the snake hung, the length of the cord, from his tail to the focus to which it was fastened, was about six inches; and a little above the tail, there was observed a round ball, about the size of a pea. Upon inspection, this appeared to be a green fly, around which the cord had been wound as a windlass, with which the snake had been hauled up; and a great number of threads were fastened to the cord above, and to the rolling side of this ball to keep it from unwinding, and letting the snake down. The cord, therefore, must have been extended from the focus of this web to the shelf below where the snake was lying when first captured; and being made fast to the loop in his tail, the fly was carried and fastened about midway to the side of the cord. And then by rolling this fly over and over, it wound the cord around it, both from above and below, until the snake was raised to the proper height, and then was fastened, as before mentioned.

"In this situation the suffering snake hung, alive, and furnished a continued feast for several large Spiders, until Saturday forenoon, the 16th, when some persons, by playing with him, broke the web above the focus, so as to let part of his body rest upon the shelf below. In this situation he lingered, the Spiders taking no notice of him, until Thursday, eight days after he was discovered, when some large ants were found devouring his body."[1]

At a recent meeting of the Academy of Natural Sciences, Philadelphia, Mr. Lesley read the following extract from a letter written by Mr. E. A. Spring, of Eagleswood, N. J.:

"I was over on the South Amboy shore with a friend, walking in a swampy wood, where a dyke was made, some three feet wide, when we discovered in the middle of this ditch a large black Spider making very queer motions for a Spider, and, on examination, it proved that he had *caught a fish*.

"He was biting the fish, just on the forward side of the dorsal fin, with a deadly gripe, and the poor fish was swimming round and round slowly, or twisting its body as if in pain. The head of its black enemy was sometimes almost pulled under water, but never entirely, for the fish did not seem to have had enough strength, but moved its fins as if exhausted, and often rested. At last it swam under a floating leaf at the shore, and appeared to be trying, by going under that, to scrape off the Spider, but without effect. They then got close to the bank, when suddenly the long black legs of the Spider came up out of the water, where they had possibly been embracing a fish (I have seen Spiders seize flies with all their legs at once), reached out behind, and fastened upon the irregularities of the side of the ditch. The Spider then commenced tugging to get his prize up the bank. My friend stayed to watch them, while I went to the nearest house for a wide-mouthed bottle. During the six or eight minutes that I was away, the Spider had drawn the fish entirely out of the water, when they had both fallen in again, the bank being nearly perpendicular. There had been a great struggle; and now, on my return, the fish was already hoisted head first more than half his length out on the land. The fish was very much exhausted, hardly making any move-

---

ment, and the Spider had evidently gained the victory, and was slowly and steadily tugging him up. He had not once quitted his hold during the quarter to half an hour that we had watched them. He held, with his head toward the fish's tail, and pulled him up at an angle of forty-five degrees by stepping backward. . . . . The Spider was three-fourths of an inch long, and weighed fourteen grains; the fish was three and one-fourth inches long, and weighed sixty-six grains."[1]

The following interesting account of the rarely-witnessed phenomenon of 'a shower of webs of the Gossamer-spider, *Aranea obtextrix*, is given us by Mr. White : "On the 21st of September, 1741, being intent on field diversions, I rose," says this gentleman, "before daybreak; when I came into the enclosures, I found the stubbles and clover grounds matted all over with a thick coat of cobweb, in the meshes of which a copious and heavy dew hung so plentifully, that the whole face of the country seemed, as it were, covered with two or three setting-nets, drawn one over another. When the dogs attempted to hunt, their eyes were so blinded and hood-winked that they could not proceed, but were obliged to lie down and scrape the incumbrances from their faces with their fore-feet. . . . . As the morning advanced, the sun became bright and warm, and the day turned out one of the most lovely ones which no season but the autumn produces; cloudless, calm, serene, and worthy of the south of France itself.

"About nine an appearance very unusual began to demand our attention, a shower of cobwebs falling from very elevated regions, and continuing, without any interruption, till the close of the day. These webs were not single filmy threads, floating in the air in all directions, but perfect flakes of rags; some near an inch broad, and five or six long. On every side, as the observer turned his eyes, might he behold a continual succession of fresh flakes falling into his sight, and twinkling like stars."[2]

The Times of October 9th, 1826, records another shower of gossamer as follows: "On Sunday, Oct. 1st, 1826, a phenomenon of rare occurrence in the neighborhood of Liverpool was observed in that vicinage, and for many miles dis-

---

[1] *Annual of Sci. Disc.*, 1862, p. 335.
[2] *Nat. Hist. of Selborne*, p. 285.

tant, especially at Wigan. The fields and roads were covered with a light filmy substance, which, by many persons, was mistaken for cotton; although they might have been convinced of their error, as staple cotton does not exceed a few inches in length, while the filaments seen in such incredible quantities extended as many yards. In walking in the fields the shoes were completely covered with it, and its floating fibres came in contact with the face in all directions. Every tree, lamp-post, or other projecting body had arrested a portion of it. It profusely descended at Wigan like a sleet, and in such quantities as to affect the appearance of the atmosphere. On examination it was found to contain small flies, some of which were so diminutive as to require a magnifying glass to render them perceptible. The substance so abundant in quantity, was the gossamer of the garden, or field Spider, often met with in fine weather in the country, and of which, according to Buffon, it would take 663,552 Spiders to produce a single pound."[1]

"In the yeare that L. Paulus and C. Marcellus were Consuls," says Pliny, "it rained wool about the castle Carissa, neare to which a yeare after, T. Annius Milo was slaine."[2] This rain of wool was doubtless a shower of gossamer.

It was an old and strange notion that the gossamer webs were composed of dew burned by the sun. Says Spenser:

> More subtle web Arachne cannot spin;
> Nor *the fine nets*, which oft we woven see,
> Of *scorched dew*, do not in th' ayre more lightly flee.[3]

Thomson also:

> How still the breeze! save what *the filmy threads*
> Of *dew evaporate* brushes from the plain.[4]

And Quarles:

> And now *autumnal dews* were seen
> To *cobweb* every green.[5]

Likewise Blackmore:

---

[1] Hone's *Ev. Day Bk.*, p. 1332.
[2] *Nat. Hist.*, ii. 54.   Holl. *Trans.*, p. 27. F.
[3] *Fuerie Queene*, B. 2, c. xii. s. 77.
[4] *Seasons: Summer*, l. 1209.
[5] *Emblems*, p. 375.

How part is spun in *silken threads*, and clings,
Entangled in the grass, in *gluey strings*.[1]

Henry More also mentions this old belief; but suspected, however, the true origin and use of the filmy threads:

As light and thin as *cobwebs* that do fly
In the blue air caused by th' *autumnal sun*,
That *boils the dew*, that on the earth doth lie;
May seem this whitish rag then is the scum;
Unless that wiser men mak't the *field-spider's loom*.[2]

Jamieson, in his Scottish Dictionary, gives *sun-dew webs* as a name given in the South of Scotland to the gossamer.

The Swedes call a cobweb *dwaergsnaet*, from *dwaerg*, a species of malevolent fairy or demon; very ingenious, and supposed often to assume the appearance of a Spider, and to form these nets. The peasants of that country say, *Jorden naetjar sig*, "the earth covers itself with a net," when the whole surface of the ground is covered with gossamer, which, it is commonly believed, indicates the seed-time.[3]

Voss, in a note on his Luise (iii. 17), says that the popular belief in Germany is, that the gossamers are woven by the Dwarfs. Keightley thinks the word gossamer is a corruption of *gorse*, or *goss samyt*, *i.e.* the *samyt*, or finely-woven silken web that lies on the *gorse* or furze.[4]

A learned man and good natural philosopher, and one of the first Fellows of the Royal Society, Robert Hooke, the author of *Micrographia*, gravely remarked in his scientific disquisition on the gossamer, that it "was not unlikely, but those great white clouds, that appear all the summer time, may be of the same substance!!"[5]

The following well-authenticated incident is told by Turner as having occurred when he was a young practitioner: A certain young woman was accustomed, when she went

---

[1] Blackmore, *Prince Arthur*.

[2] Quot. in the *Athenæum*, v. 126.

[3] Jamieson's *Scot. Dict.*, iv. 133.

[4] Keightley's *Fairy Mythol.*, p. 514.

[5] *Microgr.*, p. 202. It has been objected, say Kirby and Spence, to the excellent primitive writer, Clemens Romanus, that he believed the absurd fable of the phœnix. But surely this may be allowed for in him, who was no naturalist, when a scientific natural philosopher could believe that the clouds are made of Spiders' web!—*Introd.*, ii. 331, note.

into the vault after night, to go Spider-hunting, as she called it, setting fire to the webs of Spiders, and burning the insects with the flame of the candle. It happened at length, however, after this whimsey had been indulged a long time, one of the persecuted Spiders sold its life much dearer than those hundreds she had destroyed, and most effectually cured her of her idle cruel practice; for, in the words of Dr. James, "lighting upon the melted tallow of her candle, near the flame, and his legs becoming entangled therein, so that he could not extricate himself, the flame or heat coming on, he was made a sacrifice to his cruel persecutor, who, delighting her eyes with the spectacle, still waiting for the flame to take hold of him, he presently burst with a great crack, and threw his liquor, some into her eyes, but mostly upon her lips; by means of which, flinging away her candle, she cried out for help, as fancying herself killed already with the poison." In the night the woman's lips swelled excessively, and one of her eyes was much inflamed. Her gums and tongue were also affected, and a continual vomiting attended. For several days she suffered the greatest pain, but was finally cured by an old woman with a preparation of plantain leaves and cobwebs applied to the eyes, and taken inwardly two or three times a day.

Before this accident happened to her, this woman asserted that the smell of the Spiders burning oftentimes so affected her head, that objects about her seemed to turn round; she grew faint also with cold sweats, and sometimes a light vomiting followed, yet so great was her delight in tormenting these creatures, and driving them from their webs, that she could not forbear, till she met with the above narrated accident.[1]

A similar story is related by Nic. Nicholas of a man he saw at his hotel in Florence, who, burning a large black Spider in the flame of a candle, and staying for some time in the same room, from the fumes arising, grew feeble, and fell into a fainting fit, suffering all night great palpitation at the heart, and afterward a pulse so very low as to be scarcely felt.[2]

Several monks, in a monastery in Florence, are said to

---

[1] James's *Med. Dict.*

[2] *Ibid.*

have died from the effects of drinking wine from a vessel in which there was afterward found a drowned Spider.[1]

There are two animals to which the Italians give the name Tarantula: the one is a species of Lizard, whose bite is reputed mortal, found about Fondi, Cajeta, and Capua; the other is a large Spider, found in the fields in several parts of Italy, and especially at Tarentum—hence the name. "Such as are stung by this creature (the *Aranea Tarantula*)," says Misson, "make a thousand different gestures in a moment; for they weep, dance, tremble, laugh, grow pale, cry, swoon away, and, after a few days of torment, expire, if they be not assisted in time. They find some relief by sweating and antidotes, but *music* is the great and specific remedy. A learned gentleman of unquestionable credit told me at Rome, that he had been twice a witness both of the disease and of the cure. They are both attended with circumstances that seem very strange; but the matter of fact is well attested, and undeniable."[2] Such is the story generally told, believed, and unquestioned, that has found its way into the works of many learned travelers and naturalists, but which is without the slightest shadow of truth.

"I think I could produce," continues the deluded Misson, "natural and easy reasons to explain this effect of music; but without engaging myself in a dissertation that would carry me too far, I shall content myself with relating some other instances of the same kind: Every one knows the efficacy of David's harp to restore Saul to the use of his reason. I remember Lewis Guyon, in his Lessons, has a story of a lady of his acquaintance, who lived one hundred and six years without ever using any other remedy than music; for which purpose she allowed a salary to a certain musician, whom she called her physician; and I might add that I was particularly acquainted with a gentleman, very much subject to the gout, who infallibly received ease, and sometimes was wholly freed from his pains by a loud noise. He used to make all his servants come into his chamber, and beat with all their force upon the table and floor; and the noise they made, in conjunction with the sound of the violin, was his sovereign remedy."[3]

In the Treasvrie of Avncient and Moderne Times, printed

---

[1] James's *Med. Dict.*
[2] Harris's *Coll. of Voy. and Trav.*, ii. 580-7.   [3] *Ibid.*

in London, the year 1619. we find the following: "*Alexander Alexandrinus* proceedeth farther, affirming that he beheld one wounded by this Spider, to dance and leape about incessantly, and the Musitians (finding themselves wearied) gave over playing: whereupon, the poore offended dancer, hauing vtterly lost all his forces, fell downe on the ground, as if he had bene dead. The Musitians no sooner began to playe againe, but hee returned to himselfe, and mounting vp vpon his feet, danced againe as lustily as formerly hee had done, and so continued dancing still, til hee found the harme asswaged, and himselfe entirely recovered. Heereunto he addeth, that when it hath happened, that a man hath not beene thorowly cured by Musique in this manner; within some short while after, hearing the sound of Instruments, hee hath recouered footing againe, and bene enforced to hold on dancing, and never to ceasse, till his perfect and absolute healing, which (questionlesse) is admirable in nature."[1]

Robert Boyle, in his Usefulness of Natural Philosophy, among other stories of the power of music upon those bitten by Tarantulas, mentions the following: "*Epiphanius Ferdinandus* himself not only tells us of a man of 94 years of age, and weak, that he could not go, unless supported by his staff, who did, upon the hearing of musick after he was bitten, immediately fall a dancing and capering like a kid; and affirms that Tarantulas themselves may be brought to leap and dance at the sound of lutes, small drums, bagpipes, fiddles, etc.; but challenges those, that believe them not, to come and try, promising them an occular conviction: and adds what is very memorable and pleasant, that not only men, in whom much may be ascribed to fancy, but other animals being bitten, may likewise, by musick, be reduced to leap or dance: for he saith, he saw a Wasp, which being bitten by a Tarantula, whilst a lutanist chanced to be by; the musician, playing upon his instrument gave them the sport of seeing both the Wasp and Spider begin to dance: Annexing, that a bitten Cock did the like."[2]

In an Italian nobleman's palace, Skippon saw a fellow who was bitten by a Tarantula; "he danced," says this traveler, "very antickly, with naked swords, to a tune played

<hr>

[1] *Treasvrie of Anct. and Mod. Times*, p. 393.
[2] Boyle's *Works*, ii. 181–2.

on an instrument." The Italians say that if the Spider be immediately killed, no such effects will appear; but as long as it lives, the person bitten is subject to these paroxysms, and when it dies he is free. Skippon says that usually they are the poorer sort of people who say they are bitten, and they beg money while they are in these dancing fits.[1]

Bell was informed at Buzabbatt (in Persia) that the celebrated Kashan Tarantula "neither stings nor bites, but drops its venom upon the skin, which is of such a nature that it immediately penetrates into the body, and causes dreadful symptoms; such as giddiness of the head, a violent pain in the stomach, and a lethargic stupefaction. The remedy is the application of the same animal when bruised to the part affected, by which the poison is extracted. They also make the patient," continues this traveler, "drink abundance of sweet milk, after which he is put in a kind of tray, suspended by ropes fixed in the four corners; it is turned round till the ropes are twisted hard together, and, when let go at once, the untwining causes the basket to run round with a quick motion, which forces the patient to vomit."[2]

Skippon was shown by Corvino, in his Museum at Rome, "a *Tarantula Apula*, which he kept some time alive; and the poison of it, he said, broke two glasses."[3]

In the Treasvrie of Avncient and Moderne Times, it is stated of "Harts, that when they are bitten or stung by a venomous kinde of Spiders, called *phalanges;* they heale themselves by eating *Creuisses*, though others do hold, that it is by an Hearb growing in the water."[4]

Diodorus Siculus tells us that there border upon the country of the Acridophagi a large tract of land, rich in fair pastures, but desert and uninhabited; not that there were never any people there, but that formerly, when it was inhabited, an immoderate rain fell, which bred a vast host of Spiders and Scorpions: that these implacable enemies of the country increased so, that though at first the whole nation attempted to destroy them (for he who was bitten or stung by them, immediately fell dead), so that, not knowing where to remain, or how to get food, they were forced to fly

---

1 Astley's *Col of Voy. and Trav.*, vi. 607.
2 Pinkerton's *Col. of Voy. and Trav.*, vii. 290.
3 Astley's *Col. of Voy. and Trav.*, vi. 656.
4 B. 7, c. 15, p. 664. Printed 1613.

to some other place for relief.[1] Strabo has inserted also this miraculous story in his Geography.[2]

Mr. Nichols mentions Spiders as having been embroidered on the white gowns of ladies in the time of Queen Elizabeth.[3]

Sloane tells us the housekeepers of Jamaica keep large Spiders in their houses to kill cockroaches.[4]

Captain Dampier, after minutely describing in his quaint way the "teeth" of a "sort of Spider, some near as big as a Man's Fist," which are found in the West Indies, says: "These Teeth we often preserve. Some wear them in their Tobacco-pouches to pick their Pipes. Others preserve them for tooth-pickers, especially such as are troubled with the toothache; for by report they will expell that Pain."[5] These teeth, which are of a finely polished substance, extremely hard, and of a bright shining black, are often, in the Bermudas, for these qualities set in silver or gold and used also for tooth-picks.[6]

Dr. Sparrman says that Spiders form an article of the Bushman's dainties;[7] and Labillardiere tells us that the inhabitants of New Caledonia seek for and eat with avidity large quantities of a Spider nearly an inch long (which he calls *Aranea edulis*) and which they roast over the fire.[8] Spiders are also eaten by the American Indians and Australians.[9] Molien says: "The people of Maniana, south of Gambia and Senegal, are cannibals. They eat Spiders, Beetles, and old men."[10] In Siam, also, we learn from Turpin, the egg-bags of Spiders are considered a delicate food. The bags of certain poisonous species which make holes in the ground in the woods are preferred.[11]

And Peter Martyr, in his History of the West Indies, makes the following statement: "The Chiribichenses (Carib-

---

[1] Diod., B. 3, c. 2.

[2] Strabo, B. 16, c. 6, § 13.

[3] Fosbr. *Encyc. of Antiq*, ii. 738.

[4] Sloane's *Hist. of Jamaica*, ii. 195.

[5] Damp. *Voy. Camp.*, p. 64.

[6] Harris's *Col. of Voy. and Trav.*, ii. 242. Cf. Smith's *Nature and Art*, x. 257.

[7] *Travels*, i. 201.

[8] *Voyage à la recherche de la Perouse*, ii. 240. K. & S. *Introd.*, i. 311.

[9] *New Amer Cyclop*.

[10] *Trav. in Africa.* Bucke *on Nature*, ii. 297.

[11] Pinkerton's *Col. of Voy. and Trav.*, ix. 612.

beans) eate Spiders, Frogges, and whatsoever woormes, and lice also without loathing, although in other thinges they are so queasie stomaked, that if they see anything that doth not like them, they presently cast upp whatsoever is in their stomacke."[1]

Reaumur tells us of a young lady who when she walked in her grounds never saw a Spider that she did not take and eat upon the spot.[2] Another female, the celebrated Anna Maria Schurman, used to crack them between her teeth like nuts, which she affirmed they much resembled in taste, excusing her propensity by saying that she was born under the sign Scorpio.[3] "When Alexander reigned, it is reported that there was a very beautiful strumpet in Alexandria, that fed alwayes from her childhood on Spiders, and for that reason the king was admonished that he should be very carefull not to embrace her, lest he should be poysoned by venome that might evaporate from her by sweat. Albertus Magnus also makes mention of a certain noble mayd of Collen, that was fed with Spiders from her childhood. And we in England have a great lady yet living, who will not leave off eating of them. And Phaerus, a physician, did often eat them without any hurt at all."[4]

La Lande, the celebrated French astronomer, we are told by Disjonval, ate as delicacies Spiders and Caterpillars. He boasted of this as a philosophic trait of character, that he could raise himself above dislikes and prejudices; and, to cure Madame Lepaute of a very annoying fear of, and antipathy to Spiders, it is said he gradually habituated her to look upon them, to touch, and finally to swallow them as readily as he himself.[5]

A German, immortalized by Rösel, used to eat Spiders by handfuls, and spread them upon his bread like butter, observing that he found them very useful, "*um sich auszulaxiren.*"[6]

---

[1] *Hist. of West Indies*, p. 301.

[2] Reaum., ii. 342. K & S. *Introd.*, i. 311.

[3] *Phil. Trans.* Southey's *Com. Place Bk.*, 3d S. p. 731. Shaw, *Nat. Misc.*

[4] Moufet, *Theatr. Ins.*, p. 220. Topsel's *Hist. of Beasts and Serpents*, p. 789, 1067. Wanley's *Wonders*, ii. 459.

[5] *Biogr. Univers.*, tome xxiii. p. 230, note.

[6] Rösel, iv. 257. K. & S. *Introd.*, i. 311.

The satirist, Peter Pindar, records the same of Sir Joshua
Banks :

> How early Genius shows itself at times,
> Thus Pope, the prince of poets, lisped in rhymes,
> And our Sir Joshua Banks, most strange to utter,
>   To whom each cockroach-eater is a fool,
>   Did, when a very little boy at school,
> Eat Spiders, spread upon his bread and butter.

Conradus, bishop of Constance, at the sacrament of the
Lord's Supper, drank off a Spider that had fallen into his
cup of wine, while he was busied in the consecration of the
elements; "yet did he not receive the least hurt or damage
thereby."[1]

We learn from Poggio, the Florentine, that Zisca, the
great and victorious reformer of Bohemia, was such an epi-
cure, that he only asked for, as his share of the plunder, what
he was pleased to call "the cobwebs, which hung from the
roofs of the farmers' houses." It is said, however, that this
was but one of his witty circumlocutions to express the hams,
sausages, and pig-cheeks, for which Bohemia has always
been celebrated.[2]

For the bite of all Spiders, according to Pliny, the best
remedies are "a cock's brains, taken in oxycrate with a little pepper; five ants, swallowed in drink; sheep's dung applied in vinegar; and Spiders of any kind, left to putrify
in oil."[3] Another proper remedy, says this writer, is, "to
present before the eyes of a person stung another Spider of
the same description, a purpose for which they are preserved
when found dead. Their husks also," he continues, "found
in a dry state, are beaten up and taken in drink for a similar
purpose. The young of the weasel, too, are possessed of a
similar property."[4]

Among the remedies given by Pliny for diseases of eyes,
is mentioned "the cobweb of the common fly-Spider, that
which lines its hole more particularly. This," he continues,
"applied to the forehead across the temples, in a compress
of some kind or other, is said to be marvellously useful for
the cure of defluxions of the eyes; the web must be taken,
however, and applied by the hands of a boy who has not

---

[1] Wanley's *Wonders*, ii. 459.
[2] Andrew's *Anecd.*, p. 37. App.
[3] *Nat. Hist.*, xxix. 27. Bost. & Riley.   [4] *Ibid.*

arrived at the years of puberty; the boy, too, must not show himself to the patient for three days, and during those three days neither of them must touch the ground with his feet uncovered. The white Spider with very elongated, thin legs, beaten up in old oil, forms an ointment which is used for the cure of albugo. The Spider, too, whose web, of remarkable thickness, is generally found adhering to the rafters of houses, applied in a piece of cloth, is said to be curative of defluxions of the eyes."[1]

As a remedy for the ears, Pliny says: "The thick pulp of a Spider's body, mixed with oil of roses, is used for the ears; or else the pulp applied by itself with saffron or in wool."[2]

For fractures of the cranium, Pliny says, cobwebs are applied, with oil and vinegar; the application never coming away till a cure has been effected. Cobwebs are good, too, he continues, for stopping the bleeding of wounds made in shaving.[3] They are still used for this purpose, as also the fur from articles made of beaver.

In Ben Jonson's Stable of News, Almanac says of old Penny boy (as a skit upon his penuriousness), that he

> Sweeps down no cobwebs here,
> But sells 'em for cut fingers; and the Spiders,
> As creatures rear'd of dust, and cost him nothing,
> To fat old ladies' monkies.[4]

And Shakspeare, in his Midsummer-Night's Dream, makes Bottom say to the fairy Cobweb:

"I shall desire you of more acquaintance, good master Cobweb. If I cut my finger, I shall make bold with you."[5]

Pills formed of Spiders' webs are still considered an infallible cure for the ague.[6] Dr. Graham, in his Domestic Medicine, prescribes it for ague and intermittent fever. And Spiders themselves, with their legs pinched off, and then

---

[1] *Nat. Hist.*, xxix. 38.
[2] *Ibid.*, xxix. 39.   [3] *Ibid.*, xxix. 36.
[4] *Staple of News*, A. ii. Sc. 1, vol. v. p. 219. Lond. 1816. "A Spider is usually given to monkeys, and is esteemed a sovereign remedy for the disorders those animals are principally subject to." —*James's Med. Dict.* Spiders are also fed to mocking-birds, not only as food, but also as an aperient.
[5] *Mid. Night's Dream*, Act iii. Sc. 1.
[6] Vide *Eventful Life of a Soldier.* Edinbg. 1852.

powdered with flour, so as to resemble a pill, are also sometimes given for ague.[1] Dr. Chapman, of Philadelphia, states
that in doses of five grains of Spiders' web, repeated every
fourth or fifth hour, he has cured some obstinate intermit-
tents, suspended the paroxysms of hectic, overcome morbid
vigilance from excessive nervous mobility, and quieted irrita-
tion of the system from various causes, and not less as connected with protracted coughs and other chronic pectoral
affections.[2]

Mrs. Delany, in a letter dated March 1st, 1743–4, gives
two infallible recipes for ague.

1st. Pounded ginger, made into paste with brandy, spread
on sheep's leather, and a plaister of it laid over the navel.

2d. A Spider put into a goose-quill, well sealed and secured, and hung about the child's neck as low as the pit of
its stomach.

Upon this Lady Llanover notes: "Although the prescription of the Spider in the quill will probably create
amusement, considered as an old charm, yet there is no
doubt of the medicinal virtues of Spiders and their webs,
which have been long known to the Celtic inhabitants of
Great Britain and Ireland."[3]

The above mentioned Dr. Graham states that he has
known of a Spider having been sewed up in a rag and worn
as a periapt round the neck to charm away the ague.[4]

In the Netherlands, it is thought good for an ague, to inclose a Spider between the two halves of a nut-shell, and
wear it about the neck.[5]

"In the diary of Elias Ashmole, 11th April, 1681, is preserved the following curious incident: 'I took early in the
morning a good dose of elixir, and hung three Spiders about
my neck, and they drove my ague away. Deo gratias!'
Ashmole was a judicial astrologer, and the patron of the
renowned Mr. Lilly. Par nobile fratrum."[6]

"Among the approved remedies of Sir Matthew Lister, I
find," says Dr. James, "that the distilled water of black

---

[1] *N. and Q.*, 2d ed. x. 138.
[2] *Elements of Mat. Med. and Therap.*, Philad. 1825.
[3] Chamb. *Bk. of Days*, i. 732.
[4] Grah. *Domest. Med.*
[5] Thorpe's *North. Mythol.*, iii. 329.
[6] Brand's *Pop. Antiq.*, iii. 287.

Spiders is an excellent cure for wounds, and that this was one of the choice secrets of Sir Walter Raleigh. . . .

"The Spider is said to avert the paroxisms of fevers, if it be applied to the pulse of the wrist, or the temples; but it is peculiarly recommended against a quartan, being enclosed in the shell of a hazlenut. . . .

"The Spider, which some call the catcher, or wolf, being beaten into a plaister, then sewed up in linen, and applied to the forehead and temples, prevents the return of the tertian. . . . . . There is another kind of Spider, which spins a white, fine, and thick web. One of this sort, wrapped in leather, and hung about the arm, will, it is said, avert the fit of a quartan. Boiled in oil of roses, and distilled into the ears, it eases (says Dioscorides, ii. 68) pains in those parts. . . .

"The country people have a tradition, that a small quantity of Spiders' web, given about an hour before the fit of an ague, and repeated immediately before it, is effectual in curing that troublesome, and sometimes obstinate distemper. . . . . . The Indians about North Carolina have great dependence on this remedy for ague, to which they are much subject."[1]

"Of the cod or bags of Spiders, M. Bon caused a sort of drops to be made, in imitation of those of Goddard, because they contain a great quantity of volatile salt."[2]

Moufet, in Theatrum Insectorum, has the following: "Also that knotty whip of God, and mock of all physicians, the Gowt, which learned men say can be cured by no remedy, findes help and cure by a Spider layed on, if it be taken at that time when neither sun nor moon shine, and the hinder legs pulled off, and put into a deer's skin and bound to the pained foot, and be left on it for some time. Also for the most part we finde those people to be free from the gowt of hands or feet (which few medicaments can doe), in whose houses the Spiders breed much, and doth beautifie them with her tapestry and hangings. . . . . Our chirurgeons cure warts thus: They wrap a Spider's ordinary web into the fashion of a ball, and laying it on the wart, they set it on fire, and so let it burn to ashes; by this means the wart is rooted out by the roots, and will never grow again. . . . . . I cannot but repeat a history that I formerly heard from

---

[1] James's *Med. Dict.*

[2] Geoffroy's *Substances used in Med.*, p. 383.

our dear friend worthy to be believed, Bruerus.   A lustfull nephew of his, having spent his estate in rioting and brothel-houses, being ready to undertake anything for money, to the hazzard of his life ; when he heard of a rich matron of London, that was troubled with a timpany, and was forsaken of all physicians as past cure, he counterfeited himself to be a physician in practice, giving forth that he would cure her and all diseases.   But as the custom is, he must have half in hand, and the other half under her hand, to be payed when she was cured.   Then he gave her a Spider to drink, as supposing her past cure, promising to make her well in three dayes, and so in a coach with four horses he presently hastes out of town, lest there being a rumor of the death of her (which he supposed to be very neer) he should be apprehended for killing her.   But the woman shortly after by the force of the venome was cured, and the ignorant physician, who was the author of so great a work, was not known. After some moneths this good man returns, not knowing what had happened, and secretly enquiring concerning the state of that woman, he heard she was recovered.   Then he began to boast openly, and to ask her how she had observed her diet, and he excused his long absence, by reason of the sickenesse of a principal friend, and that he was certain that no harm could proceed from so healthful physick ; also he asked confidently for the rest of his reward, and to be given him freely."[1]

"A third kind of Spiders," says Pliny, "also known as the 'phalangium,' is a Spider with a hairy body, and a head of enormous size.   When opened, there are found in it two small worms, they say : these, attached in a piece of deer's skin, before sunrise, to a woman's body, will prevent conception, according to what Cæcilius, in his Commentaries, says.   This property lasts, however, for a year only ; and, indeed, it is the only one of all the anti-conceptives that I feel myself at liberty to mention, in favour of some women whose fecundity, quite teeming with children (plena liberis), stands in need of some such respite."[2]

Mr. John Aubrey, in the chapter of his Miscellanies devoted to Magick, gives the following : "To cure a Beast that is sprung, (that is) poisoned (It mostly lights upon Sheep):

<hr>

[1] Moufet, *Theatr. Insect.*, p. 237.   Topsel's *Hist. of Beasts and Serpents*, p. 1073.
[2] *Nat. Hist.*, xxix. 27.

Take the little red Spider, called a tentbob (not so big as a great pin's-head), the first you light upon in the spring of the year, and rub it in the palm of your hand all to pieces: and having so done, make water on it, and rub it in, and let it dry; then come to the beast and make water in your hand, and throw it in his mouth. It cures in a matter of an hour's time. This rubbing serves for a whole year, and it is no danger to the hand. The chiefest skill is to know whether the beast be poisoned or no."[1] Mr. Aubrey had this receipt from Mr. Pacy.

In the year 1709, M. Bon, of Montpellier, communicated to the Royal Academy of that city a discovery which he had made of a new kind of silk, from the very fine threads with which several species of Spiders (probably the *Aranea diadema* and others closely allied to it) inclose their eggs; which threads were found to be much stronger than those composing the Spider's web. They were easily separated, carded, and spun, and then afforded a much finer thread than that of the silk-worm, but, according to Reaumur, inferior to this both in luster and strength. They were also found capable of receiving all the different dyes with equal facility. M. Bon carried his experiments so far as to obtain two or three pairs of stockings and gloves of this silk, which were of an elegant gray color, and were presented, as samples, to the Academy. As the Spiders also were much more prolific, and much more hardy than silk-worms, great expectations were formed of benefit of the discovery. Reaumur accordingly took up and prosecuted the inquiry with zeal. He computed that 663,522 Spiders would scarcely furnish a single pound of silk; and conceived that it would be impossible to provide the necessarily immense numbers with flies, their natural food. This obstacle, however, was soon removed, by his finding that they would subsist very well upon earth-worms chopped, and upon the soft ends or roots of feathers. But a new obstacle arose from their unsocial propensities, which proved insurmountable; for though at first they seemed to feed quietly, and even work together, several of them at the same web, yet they soon began to quarrel, and the strongest devoured the weakest, so that of several hundred, placed together in a box, but three or four remained alive after a few days; and nobody could propose to keep and feed each separately. The silk was found to be

[1] *Miscellanies.* p. 138.

naturally of different colors; particularly white, yellow, gray, sky-blue, and coffee-colored brown.[1]

A Spider raiser in France, more recently, is said to have tamed eight hundred Spiders, which he kept in a single apartment for their silk.[2]

De Azara states that in Paraguay a Spider forms a spherical cocoon for its eggs, an inch in diameter, of a yellow silk, which the inhabitants spin on account of the permanency of the color.[3]

The ladies of Bermuda make use of the silk of the Silk-Spider, *Epeira clavipes*, for sewing purposes.[4]

The Spider-web fabric has been carried so nearly to transparency (in Hindostan) that the Emperor Aurengzebe is said to have reproved his daughter for the indelicacy of her costume, while she wore as many as seven thicknesses of it.[5]

Astronomers employ the strongest thread of Spiders, the one, namely, that supports the web, for the divisions of the micrometer. By its ductility this thread acquires about a fifth of its ordinary length.[6]

Topsel, in his History of Four-footed Beasts and Serpents, has the following, which he calls an "old and common verse:

> Nos aper auditu præcellit, Aranea tactu,
> Vultur odoratu, lynx visu, simia gustu.

Which may be Englished thus:

> To hear, the boar, to touch, the Spider us excells,
> The lynx to see, the ape to taste, the vulture for the smells."[7]

"It is manifest," says Moufet, "that Spiders are bred of some aereall seeds putrefied, from filth and corruption, because that the newest houses the first day they are whited will have both Spiders and cobwebs in them."[8] This theory of generation from putrefaction was a favorite one among the ancient writers; see the history of the Scorpion.

---

[1] Vide *Hist. and Mem. de l'Acad. Royale des Sciences*, ann. 1710; Dissert. by M. Bon, *Sur l'utilité de la soye des Arraignées*, 8vo. Also, Bancroft *on Permanent Colors*, i. 101; and Shaw's *Nat. Hist.*, vi. 481.

[2] *New Amer. Cyclop.*

[3] *Voy. dans l'Amer. Merid.*, i. 212. K. and S. *Introd.*, i. 337.

[4] *Naturalist in Bermuda*, p. 126.

[5] *Atlantic Monthly*, June, 1858, p. 92.

[6] *Nouv. Dict. d'Hist. Nat.*, ii. 280. K. and S. *Introd.*, i. 337, note.

[7] *Hist. of Beasts and Serpents*, p. 778.

[8] *Theatr. Ins.*, p. 235. Topsel's *Trans.*, p. 1072.

MISCELLANEOUS.

IT may be new to many of our readers, who are familiar with the Elegy in a Country Church-yard, to be told that its author was at the pains to turn the characteristics of the Linnæan orders of insects into Latin hexameters, the manuscript of which is still preserved in his interleaved copy of the "Systema Naturæ."[1]

It is related by Boerhaave, in his Life of Swammerdam, that when the Grand Duke of Tuscany was visiting with Mr. Thevenot the curiosities of Holland, in 1668, he found nothing more worthy of his admiration than the great naturalist's account of the structure of caterpillars,—for Swammerdam, by the skillful management of instruments of wonderful delicacy and fineness, showed the duke in what manner the future butterfly, with all its parts, lies neatly folded up in the caterpillar, like a rose in the unexpanded bud. He was, indeed, so struck with this and other wonders of the insect world, disclosed to him by the great naturalist, that he made him the offer of twelve thousand florins to induce him to reside at his court; but Swammerdam, from feelings of independence, modestly declined to accept it, preferring to continue his delightful studies at home.[2]

There is an epitaph in the church of St. Hilary at Poictiers, beginning "Vermibus hic ponor," which the people interpreted to mean that a Saint was buried there who undertook to cure children of the worms. Women, accordingly used to scrape the tomb and administer the powder; but the clergy, to prevent this absurdity (for Luther had arisen), erected a barrier to keep them off. They soon began, however, to carry away for the same purpose pieces of the wooden bars.[3]

---

[1] *Ins. Archit.*, p. 7.

[2] Swammerdam, *Hist. of Ins.*, p. 5.

[3] Garasse, *Recherches des Recherches de M. Estiene Pasquier*, p. 357. Southey's *Com. Place Bk.*, 3d S. p. 282.

A diseased woman at Patton, drinking of the water in which the bones of St. Milburge were washed, there came from her stomach "a filthie worme, ugly and horrible to behold, having six feete, two hornes on his head, and two on his tayle." Brother Porter, in his Flowers of the Saints, tells this, and adds that the "worme was shutt up in a hollow piece of wood, and reserved afterward in the monasterie as a trophy and monument of S. Milburg, untill, by the lascivious furie of him that destroyed all goodness in England, that with other religious houses and monasteries, went to ruin." Hence the "filthic worme" was lost, and we have nothing now instead but the Reformation.[1]

Capt. Clarke, in his passage from Dublin to Chester, on the 2d of September, 1733, met with a cloud "of flying insects of various sorts," which stuck about the rigging of the vessel in a surprising manner.[2]

De Geer, chamberlain to the King of Sweden, writes (iv. 63) that in January, 1749, at Leufsta, in Sweden, and in three or four neighboring parishes, the snow was covered with living worms and insects of various kinds. The people assured him they fell with the snow, and he was shown several that had dropped on people's hats. He caused the snow to be removed from places where these worms had been seen, and found several which seemed to be on the surface of the snow which had fallen before, and were covered by the succeeding. It was impossible that they could have come there from under the ground, which was then frozen more than three feet deep, and absolutely impervious to such insects. In 1750, he again discovered vast quantities of insects on the snow, which covered a large frozen lake some leagues from Stockholm. Preceding and accompanying both these falls of insects were violent storms that had torn up trees by the roots, and carried away to a great distance the surrounding earth, and at the same time the insects which had taken up their winter quarters in it.[3] These insects were chiefly *Brachyptera* L., *Aphodii*, Spiders, caterpillars, and particularly the larvæ of the *Telephorus fuscus*.[4] Another shower of insects is recorded to have fallen

---

[1] Hone's *Ev. Day Bk.*, i. 204.
[2] *Gent. Mag.*, iii. 492.   [3] *Ibid.*, xxiv. 293.
[4] K. and S. *Introd.*, ii. 415.

in Hungary, November 20, 1672;[1] another, also, in the newspapers of July 2d, 1810, to have fallen in France the January preceding, accompanied by a shower of red snow.[2]

In the Muses Threnodie, p. 213, we read that "many are the instances, even to this day, of charms practised among the vulgar, especially among the Highlands, attended with forms of prayer. In the Miscellaneous MS., written by Baillie Dundee, among several medicinal receipts I find an exorcism against all kinds of worms in the body, in the name of the Father, Son, and Holy Ghost, to be repeated three mornings, as a certain remedy."[3]

The Guahibo, Humboldt says, that "eats everything that exists above, and everything under ground," eats insects, and particularly scolopendras and worms.[4] The same traveler also says he has seen the Indian children drag out of the earth centipedes eighteen inches long, and more than half an inch broad, and devour them [5]

"The seventeene of March, 1586," says John Stow in his Annales of England, "a strange thing happened, the like whereof before hath not beene heard of in our time. Master Dorington, of Spaldwicke, in the countie of Huntington, esquier, one of his maiesties gentlemen Pensioners, had a horse which died sodainly, and, being ripped to see the cause of his death, there was found in the hole of the hart of the same horse a strange worme, which lay on a round heape in a kall or skin of the likeness of a toade, which, being taken out and spread abroad, was in forme and fashion not easie to be described, the length of which worme divided into many greines to the number of fiftie (spread from the bodie like the branches of a tree), was from the snowte to the ende of the longest greine, seventeene inches, having four issues in the greines, from the which dropped foorth a red water; the body in bignes round about was three inches and a halfe, the colour whereof was very like a makerel. This monstrous worme, found in manner aforesaid, crauling to have got away, was stabbed in with a dagger and died, which, after being dried, was shewed to many honorable persons of the realme."[6]

---

[1] *Ephem. Nat. Curios.*, 1673. 80.
[2] K. and S. *Introd.*, ii. 415, note.
[3] Brand's *Pop. Antiq.*, iii. 273.
[4] *Pers. Nar.*, iv. 571.
[5] *Ibid.*, ii. 205.
[6] *Ann. of Eng.*, p. 1210.

Dr. Sparrman, in his journey to Paarl, an inland town at the Cape of Good Hope, having filled his insect-box with fine specimens, was obliged to put a "whole regiment of flies and other insects" round the brim of his hat. Having entered the house of a rich old widow troubled with the gout, for food, he was warned by his servant that if she should happen to see the insects he would certainly be turned out of doors for a conjuror (hexmeester). Accordingly he was very careful to keep his hat always turned away from her, but all would not do—the old lady discovered the "little beasts," and to her greater astonishment that they were run through their bodies with pins. An immediate explanation was demanded; and had the doctor not been just then lamenting with the widow for her deceased husband, and giving dissertations on the dropsy and cough that carried off the poor man, the explanation he gave would hardly have been sufficient to quell the rage of this superstitious boor at the thought of there being a sorcerer in her house.[1]

In several parts of Europe quite a trade is carried on in the way of buying and selling rare insects, chiefly the rare Alpine butterflies and moths. The instant the entomologist steps from his carriage, in the celebrated valley of Chamouni, with net in hand, whence he is known to be a papillionist, he is surrounded by half a dozen Savoyard boys, from the age of fifteen down to eight, each with a large collecting-box full of insects in his hands for sale, and with the scientist bargains for the insects that are found only on the mountains, and which these hardy chaps alone can obtain. There are again insect dealers on a larger scale, who live there, and have many of these boys in their employ; one of which wholesale merchants, Michel Bossonney, at Martigni in the Vallais, in the year 1829, sold 7000 insects, mostly of rare and beautiful species. Another dealer, on a perhaps still larger scale, is M. Provost Duval, of Geneva, a highly respectable entomologist. In 1830, he could supply upwards of 600 species of Lepidoptera, and as many Coleoptera, of the Swiss Alps, the south of France, and Germany, at prices varying from one to fifteen francs each, according to their rarity.

The advantage of this new traffic, both to the individuals engaged in it and to science, is great. Now the *Sphinx*

---

[1] *Voy. to C. of Good Hope*, i. 45.

(*Deilephila*) *hippophaes*, formerly sold at sixty francs each, and of which one of the first discovered specimens was sold for two hundred francs, is so plentiful, in consequence of the numbers collected and reared through their several stages, by the peasants all along the course of the Arve, where the plant, *Hippophae rhamöides*, on which the larvæ feed, and the imago takes its specific name, grows in profusion, that a specimen costs but three francs. A general taste also for the science, and an appreciation for beauty, is spread by the more striking Alpine species, such as *Parnassius apollo* and *Calichroma alpina*, not only among the travelers who buy them for their beauty, who before would hardly deign to look upon an insect, but among the more ignorant Alpine collectors themselves.[1]

Navarette, under the head of "Insects and Vermin," speaks of an animal which the Chinese call Jen Ting, or Wall-dragon, because it runs up and down walls. It is also, says this traveler, called the Guard of the Palace, and this for the following reason: The emperors were accustomed to make an ointment of this insect, and some other ingredients, with which they anointed their concubines' wrists, as the mark of it continued as long as they had not to do with man; but as soon as they did so, it immediately vanished, by which their honesty or falsehood was discovered. Hence it came that this insect was called the Guard of the Court, or, of the court ladies. Navarette laments that all men have not a knowledge of this wonderful ointment.[2]

Navarette tells us he once caught (in China?) a small insect that was injurious to poultry—"a very deformed insect, and of a strange shape"—when, as soon as it was known, several women ran to him to beg its *tail*. He gave it to them, and they told him it was of excellent use when dried, and made into powder, "being a prodigious help to women in labor, to forward their delivery, if they drank it in a little wine."[3]

The Irish have a large beetle of which strange tales are believed; they term it the Coffin-cutter, and deem it in some way connected with the grave and purgatory.[4]

---

[1] *Mag. of Nat. Hist.*, iv. 148–9.

[2] *Hist. of China*, B. I. c. 18, and Churchill's *Col. of Voy. and Trav.*, i. 39.

[3] Churchill's *Col. of Voy. and Trav.*, i. 212.

[4] *The Mirror*, xix. 180.

Turpin, in his History of Siam, says: "There is a very singular animal in Siam . . . . . bred in the dung of elephants. It is entirely black, its wings are strong, and its head extremely curious: it is furnished on the top with several points, in the form of a trunk, and a small horn in the middle: it has four large feet, which raise it more than an inch from the ground: its back seems to be one very hard entire shell. It flies to the very top of the cocoa-trees, of which it eats the heart, and often kills them, if a remedy is not applied. Children play with them, and make them fight."[1]

General Count Déjeau, Aid-de-camp to Napoleon Bonaparte, was so anxious, says Jaeger, in his Life of North American Insects, to increase the number of specimens in his entomological cabinet, that he even availed himself of his military campaigns for this purpose, and was continually occupied in collecting insects and fastening them with pins on the outside of his hat, which was always covered with them. The Emperor, as well as the whole army, were accustomed to see General Déjeau's head thus singularly ornamented, even when in battle. But the departed spirits of those murdered insects once had their revenge on him; for, in the battle of Wagram, in 1809, and while he was at the side of Napoleon, a shot from the enemy struck Déjeau's head, and precipitated him senseless from his horse. Soon, however, recovering from the shock, and being asked by the Emperor if he was still alive, he answered, "I am not dead; but, alas! my insects are all gone!" for his hat was literally torn to pieces.[2]

Professor Jaeger tells also the following anecdote of another passionate naturalist: The celebrated Prince Paul of Würtemberg, whom Mr. Jaeger met in 1829 at Port-au-Prince, being one day at the latter's house, shed tears of envy when he showed him the gigantic beetle Actæon, which, only a short time before, had been presented to him by the Haytien Admiral Banajotti, he having found it at the foot of a cocoa-nut tree on his plantation.[3]

While traveling in Poland, Professor Jaeger visited the highly accomplished Countess Ragowska, at her country residence, when she exhibited her fine, scientifically-arranged

---

[1] Pinkerton's *Col. of Voy. and Trav.*, ix. 632.
[2] *Hist. of Ins.*, p. 53-4.   [3] *Ibid.*

collection of butterflies and other insects, and told him that she had personally instructed her children in botany, history, and geography by means of her entomological cabinet—botany, from the plants on which the various larvæ feed;[*] history, from the names, as Menelaus, Berenice, etc., given as specific names to the perfect insects; and geography, from the native countries of the several specimens.[1] From the scientific names of insects, and the technical terms employed in their study, quite a knowledge of Latin and Greek, and philology in general, might also be gained.

In R. Brookes' "Natural History of Insects, with their properties and uses in medicine," we find the following statement: "There have been the solid shells of a sort of Beetle brought to England, that were found on the eastern coast of Africa, over against part of the Island of Madagascar, which the natives hang to their necks, and make use of them as whistles to call their cattle together."[2] What this "sort of Beetle" is I have not been able yet to determine.

Mr. Fitch W. Taylor, chaplain to the squadron commanded by Commodore Geo. C. Read, gives a translation of several Siamese books, and among others the Siamese Dream-book. It was translated by Mrs. Davenport, and the subject is thus introduced:

"In former times a great prophet and magician, who had much wisdom and could foretell all future events, gave the following interpretation of signs and dreams. Whosoever sees signs and visions, if he wishes to know whether they forebode good or evil, whether happiness or misery, if he dream of any animals, insects, birds, or fishes, and wishes to know the interpretation, let him examine this book."

Of these signs and dreams I make extract of those which refer to insects, as follows:

"If a person be alone, and an insect or reptile fall before the face, but the individual see it only without touching it, it denotes that some heavenly being will bestow great blessings on him. If it fall to the right side, it denotes that all his friends, wherever scattered abroad, shall again meet him in peace. If it fall behind the person, it denotes that he shall be slandered and maliciously talked of by his friends

---

[1] *Hist. of Ins.*, p. 197.
[2] *Nat. Hist. of Ins.*, p. 35.

and acquaintances. If in falling it strike the face, it denotes that the individual will soon be married. If it strike the right arm, it denotes that the individual's wishes, whatever they are, shall be accomplished. If it strike the left hand, it denotes that the individual will lose his friends by death. If it strike the foot, it denotes that whatever trouble the individual may have had, all shall vanish, and he shall reach the summit of happiness. If, after touching the foot, it should crawl upward toward the head, it denotes that the individual shall be raised to high office by the rulers of his country. If it crawl to the right side, it denotes that the person shall hear bad tidings of some absent friend. If the insect or reptile fall without touching the body, and immediately flee toward the northeast, it denotes deep but not lasting trouble; if toward the northwest, it denotes that the person shall receive numerous and valuable presents; if toward the southeast, it denotes that he shall receive great riches, and afterwards go to a distant land; or that he shall go to a distant land, and there amass great wealth.

"If an animal, insect, bird, or reptile, cross the path of any one as he walks along, the animal coming from the right, let him not proceed—some calamity will surely happen to him in the way. If the animal come from the left, let him proceed—good fortune shall surely happen to him. If the animal proceed before him in the same road in which he intends to travel, it denotes good fortune. . . . . . .

"I now beg to interpret the signs of the night. If at midnight an individual hears the noises of animals in the house where he resides, I will show him whether they indicate good or evil. If any insect cry 'click, click, click,' he will possess real treasures while he abides there. If it cry 'kek, kek,' it is an evil omen both to that and the neighboring houses. If it cry 'chit, chit,' it denotes that he shall always feed upon the most sumptuous provisions. If it cry 'keat, keat,' in a loud, shrill voice, it denotes that his residence there shall be attended with evil.

"I now beg to interpret with regard to the Spider. If a Spider on the ceiling utter a low, tremulous moan, it denotes that the individual who hears the noise shall either change his residence or that his goods shall be stolen. If it utter the same voice on the outside of the house, and afterward

the Spider crawl to the head of the bed, it denotes troublesome visitors and quarrels to the residents."[1]

Thevenot, in his Travels into the Levant, relates the following: "But I cannot tell what to say of a Moorish Woman who lives in a corner close by the quarter of France,
and pulls worms out of Children's Ears. When a Child does
nothing but cry, and that they know it is ill, they carry it
to that Woman, who, laying the Child on its side upon her
knee, scratches the ear of it, and then Worms, like those
which breed in musty weevily Flower, seem to fall out of
the Child's Ear; then, turning it on the other side, she
scratches the other Ear, out of which the like Worms drop
also; and in all there may come out ten or twelve, which
she raps up in a Linen-Rag, and gives them to those that
brought the Child to her, who keep them in that Rag at
home in their House; and when she has done so she gives
them back the Child, which in reality cries no more. She
once told me that she performed this by means of some
words that she spake. There was a French Physician and
a Naturalist there, who attentively beheld this, and told me
that he could not conceive how it could be done; but that
he knew very well that if a child had any of these Worms
in its head it would quickly die. In so much that the
Moors and other inhabitants of *Caire* look upon this as a
great Vertue, and give her every time a great many *maidins*
(pieces of money). They say that it is a secret which hath
been long in the Family. There are children every day carried to her, roaring and crying, and as many would see the
thing done, need only to follow them, provided they be not
Musulman Women who carry them, for then it would cost
an *Avanie;* but when they are Christian or Jewish Women,
one may easily enter and give a few *maidins* to that Worm-
drawer."[2]

This is most probably but a sleight-of-hand performance,
since "worms, like those which breed in musty weevily
flower," could easily be obtained and concealed in her hand
or sleeve; imagination would then effect the cure, as probably it had done the disease.

Dr. Livingstone and his party, in traveling in South Af-

[1] *Voy. round the World*, ii. 35–7.
[2] Thevenot's *Travels*, Pt. I. p. 249.

rica, sometimes suffered considerably from scarcity of meat, though not from absolute want of food. And the natives, says this traveler, to show their sympathy, gave the children, who suffered most, a large kind of caterpillar, which they seemed to relish. He concluded these insects could not be unwholesome, for the natives devoured them in large quantities themselves.[1]

---

[1] *Trav. and Res. in S. Africa*, p. 48.

INDEX.

33

ERRATA.

Page 43, line 19 from the top, between the words "is it" and "plain" insert the word "not."

Page 71, line 29, for "*Carabus chrysocephaluo*" read "*Carabus chrysocephalus.*"

Page 131, line 12, for "Mrs. A. L. Ruyter Dufour" read "Mrs. A. L. Ruter Dufour."

www.ingramcontent.com/pod-product-compliance
Lightning Source LLC
Chambersburg PA
CBHW032135110726
47902CB00003B/595